Student Solutions Manual for Hirsch/Goodman's

Understanding Elementary Algebra with Geometry

A Course for College Students

Fifth Edition

Cheryl Cantwell
Seminole Community College

BROOKS/COLE
THOMSON LEARNING

Australia • Canada • Mexico • Singapore • Spain • United Kingdom • United States

Assistant Editor: *Julie Foster*
Marketing Manager: *Leah Thomson*
Marketing Communications: *Samantha Cabaluna*
Marketing Assistant: *Maria Salinas*
Production Coordinator: *Dorothy Bell*

Cover Design: *Andrew Ogus Book Design*
Cover Illustration: *Judith Harkness*
Print Buyer: *Christopher Burnham*
Printing and Binding: *Patterson Printing*

COPYRIGHT © 2002 Wadsworth Group. Brooks/Cole is an imprint of the Wadsworth Group, a division of Thomson Learning, Inc. Thomson Learning™ is a trademark used herein under license.

For more information about this or any other Brooks/Cole product, contact:
BROOKS/COLE
511 Forest Lodge Road
Pacific Grove, CA 93950 USA
www.brookscole.com
1-800-423-0563 (Thomson Learning Academic Resource Center)

ALL RIGHTS RESERVED. No part of this work covered by the copyright hereon may be reproduced or used in any form or by any means—graphic, electronic, or mechanical, including photocopying, recording, taping, Web distribution, or information storage and retrieval systems—without the written permission of the publisher.

For permission to use material from this work, contact us by
Web: www.thomsonrights.com
fax: 1-800-730-2215
phone: 1-800-730-2214

Printed in the United States of America

10 9 8 7 6 5 4 3 2 1

ISBN 0-534-38361-0

TABLE OF CONTENTS

Chapter 1 1
Chapter 2 12
Chapter 3 24
Cumulative Review, Chapters 1-3 54
Cumulative Practice Test, Chapters 1-3 56
Chapter 4 57
Chapter 5 87
Chapter 6 112
Cumulative Review, Chapters 4-6 133
Cumulative Practice Test, Chapters 4-6 137
Chapter 7 139
Chapter 8 159
Chapter 9 172
Cumulative Review, Chapters 7-9 199
Cumulative Practice Test, Chapters 7-9 202
Chapter 10 205
Chapter 11 224
Cumulative Review, Chapters 10-11 261
Cumulative Practice Test, Chapters 10-11 264
Supplementary Chapter on Geometry 266
Appendix A 284
Appendix B 291
Appendix D 292

Chapter 1
The Integers

Exercises 1.1

1. True
3. True
5. True
7. True
9. False
11. $\{1, 2, 3, 4, 5, 6, 7\}$
13. $\{0, 2, 4, 6, 8, 10, 12, 14, 16, 18\}$
15. $\{0, 1, 2, 3, 4, 5, 6\}$
17. $\{0, 1, 2, 3, 4, 5\}$
19. $\{6, 7, 8, 9, \ldots\}$
21. $\{7, 8, 9, 10, \ldots\}$
23. $\{3, 4, 5\}$
25. $\{1, 2, 3, 4, 6, 8, 12, 24\}$
27. $\{1, 2, 3, 4, 5, 6, 12, 15\}$
29. $\{0, 4, 8, 12, \ldots\}$
31. $\{0, 12, 24, 36, \ldots\}$
33. $\emptyset$
35. $A = \{x | x$ is a natural number less than 20 and not divisible by 2$\}$
    $A = \{1, 3, 5, 7, 9, 11, 13, 15, 17, 19\}$
37. $C = \{t | t$ is a natural number less than 50 that is divisible by 5, but not by 10$\}$
    $C = \{5, 15, 25, 35, 45\}$
39. $S = \{t | t > 20$ and that is divisible by 100$\}$ $S = \{100, 200, 300, 400, 500, \ldots\}$
41. $>$
43. $=$
45. $>$
47. $=$
49. $=$
51. $<, \leq, \neq$
53. $\geq, \leq, =$
55. $>, \geq, \neq$
57. $<, \leq, \neq$
59. $>, \geq, \neq$
61. $2 \cdot 7$
63. $3 \cdot 11$
65. $2 \cdot 3 \cdot 5$
67. Prime
69. $2 \cdot 2 \cdot 2 \cdot 2 \cdot 2 \cdot 2$
71. $2 \cdot 2 \cdot 2 \cdot 2 \cdot 2 \cdot 3$
73. $3 \cdot 29$
75. $2 \cdot 2 \cdot 2 \cdot 3 \cdot 3 \cdot 5$
77. $2 \cdot 3 \cdot 3 \cdot 7$
79. $2 \cdot 3 \cdot 11 \cdot 13$
81. $2 \cdot 2 \cdot 2 \cdot 2 \cdot 3 \cdot 19$
83. $2 \cdot 2 \cdot 2 \cdot 2 \cdot 7 \cdot 17$
85. [number line 0–9]
87. [number line 1–11]
89. $\mathbb{W}$ contains the number 0; $\mathbb{N}$ does not.
90. In a sum, the numbers to be added are called terms; in a product, the numbers to be multiplied are called factors. In $2 + 5$, 2 is a term; in $2 \cdot 3$, 2 is a factor.
91. A factor of n is a number that divides exactly into n; a multiple of n is a number that is exactly divisible by n. 3 is a factor of 12; 12 is a multiple of 3.
92. C is the set containing the first three multiples of 6.
93. C is the set containing the first three multiples of 12.
94. (a) Sum of three terms: x, y, and z.
    (b) Product of three factors: x, y, and z.
    (c) Sum of two terms: xy and z.
    (d) Product of two factors: x and $(y + z)$.

Exercises 1.2

1. True (commutative property of addition)
3. True (commutative property of addition)
5. False
7. False
9. True (commutative property of multiplication)
11. True (commutative property of multiplication)
13. False
15. True (associative property of addition)
17. True (associative property of addition)
19. True (associative property of multiplication)
21. True (associative property of multiplication)
23. False
25. False
27. Associative property of addition; addition
29. Associative property of multiplication; multiplication
31. Commutative property of addition; associative property of addition; addition
33. Commutative property of multiplication; associative property of multiplication; multiplication
35. $-(+4) = -4$
37. $-(-4) = +4 = 4$
39. $|4| = 4$
41. $|-4| = 4$
43. $-|4| = -4$
45. $-|-4| = -4$
47. $|10 - 5| = |5| = 5$
49. $|10| - |-5| = 10 - 5 = 5$
51. $|8 - 4 - 2| = |2| = 2$
53. $|15 - 7 + 3| = |11| = 11$
55. $|2| - |-2| + |-5| = 2 - 2 + 5 = 5$
57. $7 + 3 \cdot 4 = 7 + 12 = 19$
59. $4 \cdot 3 + 6 = 12 + 6 = 18$
61. $11 - 4 \cdot 2 = 11 - 8 = 3$
63. $15 - 5 \cdot 3 = 15 - 15 = 0$
65. $6 + 3 \cdot 5 - 2 = 6 + 15 - 2 = 19$
67. $12 - 4 \cdot 3 + 6 = 12 - 12 + 6 = 6$
69. $6 + 4(3 + 2) = 6 + 4 \cdot 5 = 6 + 20 = 26$
71. $6 + (4 \cdot 3 + 2) = 6 + (12 + 2)$
$= 6 + 14 = 20$
73. $6 + (4 \cdot 3) + 2 = 6 + 12 + 2 = 20$
75. $(6 + 4)(3 + 2) = 10 \cdot 5 = 50$
77. $30 \div 5(2) = 6 \cdot 2 = 12$
79. $30 \div (5 \cdot 2) = 30 \div 10 = 3$
81. $15 + 12 \div 6 - 3 = 15 + 2 - 3 = 17 - 3 = 14$
83. $\dfrac{15 + 2 \cdot 3}{3 + 2 \cdot 2} = \dfrac{15 + 6}{3 + 4} = \dfrac{21}{7} = 3$
85. $\dfrac{18 - 4 \cdot 2}{13 - 4 \cdot 2} = \dfrac{18 - 8}{13 - 8} = \dfrac{10}{5} = 2$
87. $\dfrac{24 - 4 \cdot 5 + 8}{2 + 3 \cdot 6 - 4 \cdot 2} = \dfrac{24 - 20 + 8}{2 + 18 - 8} = \dfrac{4 + 8}{20 - 8} = \dfrac{12}{12} = 1$
89. $\dfrac{5 \cdot 2 + 3 \cdot 4 + 6 \cdot 3}{4 \cdot 5 + 2 \cdot 6 + 4 \cdot 2} = \dfrac{10 + 12 + 18}{20 + 12 + 8} = \dfrac{40}{40} = 1$
91. $3 + 2[3 + 2(3 + 2)] = 3 + 2[3 + 2 \cdot 5]$
$= 3 + 2[3 + 10]$
$= 3 + 2 \cdot 13$
$= 3 + 26 = 29$

93. $9 - 4[6 - 2(3 - 1)] = 9 - 4[6 - 2 \cdot 2]$
$= 9 - 4[6 - 4]$
$= 9 - 4 \cdot 2$
$= 9 - 8 = 1$

95. $4 + \{3 + 5[2 + 2(3 + 1)]\} = 4 + \{3 + 5[2 + 2 \cdot 4]\}$
$= 4 + \{3 + 5[2 + 8]\}$
$= 4 + \{3 + 5 \cdot 10\}$
$= 4 + \{3 + 50\} = 4 + 53 = 57$

97. $\dfrac{10 + 2(5 + 3)}{2 \cdot 5 + 3} = \dfrac{10 + 2 \cdot 8}{10 + 3} = \dfrac{10 + 16}{13} = \dfrac{26}{13} = 2$

99. Jane has a checking account with overdraft privileges. If she has $283 in her account and she writes a check for $300, by how much is Jane's account overdrawn?
($283 - $300 = -$17)

100.

Exercises 1.3

1. $+6 + (-8) = -(8 - 6) = -2$
3. $-7 + (-5) = -(7 + 5) = -12$
5. $-5 + (+12) = +(12 - 5) = +7$
7. $+9 + (-4) = +(9 - 4) = +5$
9. $+7 + (-3) = +(7 - 3) = +4$
11. $-7 + (-3) = -(7 + 3) = -10$
13. $-7 + (+3) = -(7 - 3) = -4$
15. $+7 + (+3) = 7 + 3 = 10$
17. $-4 + (-11) = -(4 + 11) = -15$
19. $12 + (-16) = -(16 - 12) = -4$
21. $9 + (-9) = 0$
23. $8 + (-15) = -(15 - 8) = -7$
25. $-20 + (-24) = -(20 + 24) = -44$
27. $-4 + 10 = +(10 - 4) = +6$
29. $-7 + (-3) + (-5) = -(7 + 3) + (-5)$
$= -10 + (-5)$
$= -(10 + 5) = -15$
31. $7 + (-3) + (-5) = +(7 - 3) + (-5)$
$= +4 + (-5)$
$= -(5 - 4) = -1$
33. $-7 + 3 + (-5) = -(7 - 3) + (-5)$
$= -4 + (-5)$
$= -(4 + 5) = -9$
35. $-7 + (-3) + 5 = -(7 + 3) + 5$
$= -10 + 5$
$= -(10 - 5) = -5$
37. $15 + (-12) + (-8) = +(15 - 12) + (-8)$
$= +3 + (-8)$
$= -(8 - 3) = -5$
39. $-6 + 13 + (-6) = +(13 - 6) + (-6)$
$= +7 + (-6)$
$= +(7 - 6) = 1$
41. $-1 + (-9) + (-5) = -(1 + 9) + (-5)$
$= -10 + (-5)$
$= -(10 + 5) = -15$
43. $18 + (-10) + (-2) = +(18 - 10) + (-2)$
$= +8 + (-2)$
$= +(8 - 2) = 6$

45. $-25 + (-5) + 26 = -(25 + 5) + 26$
$= -30 + 26$
$= -(30 - 26) = -4$

47. $22 + (-3) + (-14) + 1 = +(22 - 3) + (-14) + 1$
$= +19 + (-14) + 1 = +(19 - 14) + 1$
$= +5 + 1 = 6$

49. $-9 + 12 + (-14) + 5 = +(12 - 9) + (-14) + 5$
$= +3 + (-14) + 5 = -(14 - 3) + 5$
$= -11 + 5 = -(11 - 5) = -6$

51. $6 + (-3) + (-10) + (-15) + 27 = +(6 - 3) + (-10) + (-15) + 27$
$= +3 + (-10) + (-15) + 27$
$= -(10 - 3) + (-15) + 27 = -7 + (-15) + 27$
$= -(7 + 15) + 27 = -22 + 27 = +(27 - 22) = 5$

53. $32 + (-61) = -(61 - 32) = -29$ 55. $-48 + (-19) = -(48 + 19) = -67$

57. $-29 + (-31) + 51 = -(29 + 31) + 51$
$= -60 + 51$
$= -(60 - 51) = -9$

59. $124 + (-237) + (-102) = -(237 - 124) + (-102)$
$= -113 + (-102)$
$= -(113 + 102) = -215$

61. $-86 + 112 + (-78) + 201 = +(112 - 86) + (-78) + 201$
$= +26 + (-78) + 201$
$= -(78 - 26) + 201$
$= -52 + 201 = 149$

63. $-187 + 455 = +(455 - 187) = 268$

65. $-84 + 127 + (-111) = +(127 - 84) + (-111)$
$= +43 + (-111)$
$= -(111 - 43) = -68$

67. $-6 + [9 + (-4)] = -6 + [+(9 - 4)]$
$= -6 + 5$
$= -(6 - 5) = -1$

69. $-7 + 7 \cdot 4 + (-20) = -7 + 28 + (-20)$
$= +(28 - 7) + (-20)$
$= +21 + (-20)$
$= +(21 - 20) = 1$

71. $|1 + (-6)| + |1| + |-6| = |-(6 - 1)| + |1| + |-6|$
$= |-5| + |1| + |-6|$
$= 5 + 1 + 6 = 12$

73. $|-8| + (-8) = 8 + (-8) = 0$

75. $+548 + [(-56) + (-16) + (-71)] + (+145) + [(-180) + (-67) + (-205)] + (+75)$
    $+ [(-115) + (-45)] = +548 + (-143) + (+145) + (-452) + (+75) + (-160) = +13$
    So the final balance in Niki's checking account is $13.

77. $-47 + 80 = +(80 - 47) = 33$
    The high temperature was 33°F.

79. $+48 + [(-18) + (-22) + (-15)] + (+50) + [(-28) + (-12)] + (+20) + (-17) + (+27)$
    $= +48 + (-55) + (+50) + (-40) + (+20) + (-17) + (+27) = +33$
    So the final balance in Carla's checking account is $33.

81. $+25 + (+8) + (-14) + (-5) = +(25 + 8) + (-14) + (-5)$
    $= +33 + (-14) + (-5) = +(33 - 14) + (-5)$
    $= +19 + (-5) = +(19 - 5) = +14$
    So on fourth down, the team is located on its own 14 yard line.

83. $+\$10,432 + (-\$1,678) + (-\$2,046) + \$7,488 = \$8,754 + (-\$2,046) + \$7,488$
    $= +\$6,708 + \$7,488 = \$14,196 \text{ (profit)}$

85. $10,000 + 380 + 540 + (-275) + (-600) + (-72) = 10,920 + (-947)$
    $= 9,973 \text{ (feet)}$

87. When we add an integer to its opposite, we get 0 as our result. This occurs because an integer and its opposite have the same absolute value.

88. If the sum of two integers is zero, the integers must be opposites. In all other cases, the sum of the integers would either be positive or negative.

89. $4 - 7 = -3$, because $-3 + 7 = 4$ 90. $4 - (-7) = 11$, because $11 + (-7) = 4$

Exercises 1.4

1. $6 - (+10) = 6 + (-10) = -4$ 3. $-7 - (+4) = -7 + (-4) = -11$
5. $3 - (-6) = 3 + (+6) = 9$ 7. $-8 - (-2) = -8 + (+2) = -6$
9. $-5 + (+8) = 3$ 11. $5 - (+8) = 5 + (-8) = -3$
13. $5 - (-8) = 5 + (+8) = 13$ 15. $-5 - (+8) = -5 + (-8) = -13$
17. $-5 - (-8) = -5 + (+8) = 3$ 19. $5 + (-8) = -3$
21. $-5 + (-8) = -13$ 23. $6 - (-7) = 6 + (+7) = 13$
25. $-6 - (-7) = -6 + (+7) = 1$ 27. $2 + (-6) - (+7) = 2 + (-6) + (-7)$
                                               $= -4 + (-7) = -11$
29. $2 - 6 - 7 = 2 + (-6) + (-7)$ 31. $2 - (6 - 7) = 2 - (-1)$
    $= -4 + (-7) = -11$ $= 2 + (+1) = 3$
33. $7 - 9 - 3 + 2 = 7 + (-9) + (-3) + 2$ 35. $11 - 5 + 4 - 7 = 11 + (-5) + 4 + (-7)$
    $= -2 + (-3) + 2$ $= 6 + 4 + (-7)$
    $= -5 + 2 = -3$ $= 10 + (-7) = 3$

37. $2 - 3 - 6 - 2 = 2 + (-3) + (-6) + (-2)$
$= -1 + (-6) + (-2)$
$= -7 + (-2) = -9$

39. $-10 + 4 - 9 - (-3) = -10 + 4 + (-9) + (+3)$
$= -6 + (-9) + (+3)$
$= -15 + (+3) = -12$

41. $3 - 6 + 1 - (-4) = 3 + (-6) + 1 + (+4)$
$= -3 + 1 + (+4)$
$= -2 + (+4) = 2$

43. $-1 - 4 - 2 - (-5) = -1 + (-4) + (-2) + (+5)$
$= -5 + (-2) + (+5)$
$= -7 + (+5) = -2$

45. $4 - 8 - 6 + 3 = 4 + (-8) + (-6) + 3$
$= -4 + (-6) + 3$
$= -10 + 3 = -7$

47. $4 - 8 - (6 + 3) = 4 - 8 - 9$
$= 4 + (-8) + (-9)$
$= -4 + (-9) = -13$

49. $4 - (8 - 6) + 3 = 4 - 2 + 3$
$= 4 + (-2) + 3$
$= 2 + 3 = 5$

51. $4 - (8 - 6 + 3) = 4 - (8 + (-6) + 3)$
$= 4 - (2 + 3) = 4 - 5$
$= 4 + (-5) = -1$

53. $-8 + 8 = 0$

55. $-8 - 8 = -8 + (-8) = -16$

57. $-8 - (-8) = -8 + (+8) = 0$

59. $18 - 35 = 18 + (-35) = -17$

61. $-31 + 17 - 12 = -31 + 17 + (-12)$
$= -14 + (-12) = -26$

63. $26 - (-41) - 52 = 26 + 41 + (-52)$
$= 67 + (-52) = 15$

65. $-23 - 41 - 62 = -23 + (-41) + (-62)$
$= -64 + (-62) = -126$

67. $100 - 83 - 45 - 24 = 100 + (-83) + (-45) + (-24)$
$= 17 + (-45) + (-24)$
$= -28 + (-24) = -52$

69. $52 - (-38) - 67 = 52 + 38 + (-67)$
$= 90 + (-67) = 23$

71. $-246 - 327 - (-542) = -246 + (-327) + 542$
$= -573 + 542 = -31$

73. $9 - 5 \cdot 4 - 2 = 9 - 20 - 2$
$= 9 + (-20) + (-2)$
$= -11 + (-2) = -13$

75. $9 - 5(4 - 2) = 9 - 5 \cdot 2 = 9 - 10$
$= 9 + (-10) = -1$

77. $9 - (5 \cdot 4 - 2) = 9 - (20 - 2)$
$= 9 - (20 + (-2)) = 9 - 18$
$= 9 + (-18) = -9$

79. $|6-2| - |2-6| = |4| - |-4|$
    $= 4 - 4 = 0$

81. $|2-6| - (2-6) = |-4| - (-4) = 4 - (-4)$
    $= 4 + (+4) = 8$

83. $|-4-3+2| - 4 - 3 + 2 = |-4+(-3)+2| + (-4) + (-3) + 2$
    $= |-7+2| + (-4) + (-3) + 2$
    $= |-5| + (-4) + (-3) + 2 = 5 + (-4) + (-3) + 2$
    $= 1 + (-3) + 2 = -2 + 2 = 0$

85. $-6 - (-19) = -6 + (+19) = 13°C$

87. $\$643.47 - (-\$82.94) = \$643.47 + (+\$82.94) = \$726.41$

89. $29,028 - (-1,296) = 29,028 + (+1,296) = 30,324$ (feet)

91. $-1,296 - (-35,840) = -1,296 + (+35,840) = 34,544$ (feet)

93. $-8 - 2 - 7 = (-8) + (-2) + (-7)$. The terms are $-8, -2,$ and -7.

95. $-3 - m - n = (-3) + (-m) + (-n)$. The terms are $-3, -m,$ and $-n$.

97. $-15,000 - (-6,000) = -15,000 + (+6,000) = -9,000$. The reduced debt is $\$9,000$.

98. The left side is equal to 4, while the right side is equal to -4. In general, the absolute value of a difference of two integers is not the same as the difference of the absolute values of those integers.

99. 4 times -3 should mean "add -3 four times", giving $(-3) + (-3) + (-3) + (-3) = -12$.

100. We can interpret -4 times 3 to mean "subtract 3 four times", giving $-3 - 3 - 3 - 3 = -12$.
     Similarly, we can interpret -4 times -3 to mean "subtract -3 four times", giving
     $-(-3) - (-3) - (-3) - (-3) = +3 + (+3) + (+3) + (+3) = +12$.

Exercises 1.5

1. $(+7)(-4) = -(7 \cdot 4) = -28$

3. $(-7)(+4) = -(7 \cdot 4) = -28$

5. $(-7)(-4) = +(7 \cdot 4) = 28$

7. $-7 - 4 = -11$

9. $-(-7-4) = 11$

11. $\dfrac{30}{-6} = -\left(\dfrac{30}{6}\right) = -5$

13. $\dfrac{-30}{6} = -\left(\dfrac{30}{6}\right) = -5$

15. $\dfrac{-30}{-6} = +\left(\dfrac{30}{6}\right) = 5$

17. $\dfrac{0}{6} = 0$

19. Undefined (division by 0 is not allowed)

21. $5(-3) - 7 = -15 - 7 = -22$

23. $5 - 3 - 7 = 2 - 7 = -5$

25. $5 - (3-7) = 5 - (-4) = 9$

27. $5(-3-7) = 5(-10) = -50$

29. $45 - 72 - 18 = -27 - 18 = -45$

31. $105 - (28-81) = 105 - (-53) = 158$

33. $-4(-2)(6)(-5) = 8(6)(-5)$
    $= 48(-5) = -240$

35. $10 - 6(2-5) = 10 - 6(-3)$
    $= 10 + 18 = 28$

37. $12 - 4(3 - 8) = 12 - 4(-5)$
$= 12 + 20 = 32$

39. $12 - 4 \cdot 3 - 8 = 12 - 12 - 8$
$= 0 - 8 = -8$

41. $12 - (4 \cdot 3 - 8) = 12 - (12 - 8)$
$= 12 - 4 = 8$

43. $(12 - 4)(3 - 8) = (8)(-5)$
$= -40$

45. $12(-8) - (-15)(-9) = (-96) - (135)$
$= -231$

47. $\dfrac{-15 - 6 + 3}{-9} = \dfrac{-21 + 3}{-9} = \dfrac{-18}{-9} = 2$

49. $\dfrac{-20 - 8 - 12}{-4} = \dfrac{-28 - 12}{-4} = \dfrac{-40}{-4} = 10$

51. $\dfrac{-20 - (8 - 12)}{-4} = \dfrac{-20 - (-4)}{-4} = \dfrac{-20 + 4}{-4} = \dfrac{-16}{-4} = 4$

53. $\dfrac{-20 - 8(-12)}{-4} = \dfrac{-20 + 96}{-4} = \dfrac{76}{-4} = -19$

55. $\dfrac{18}{-9} - \dfrac{20}{-5} = -2 - (-4) = -2 + 4 = 2$

57. $\dfrac{-44 - 16 + 80}{-5} = \dfrac{-60 + 80}{-5} = \dfrac{20}{-5} = -4$

59. $\dfrac{9(-6)}{9 - 6} = \dfrac{-54}{3} = -18$

61. $\dfrac{6(4)(-8)}{6(4) - 8} = \dfrac{24(-8)}{24 - 8} = \dfrac{-192}{16} = -12$

63. $\dfrac{8(-6) - 2}{-3 - 2} = \dfrac{-48 - 2}{-5} = \dfrac{-50}{-5} = 10$

65. $\dfrac{9 - 5}{5 - 5} = \dfrac{4}{0}$, which is undefined, since division by 0 is not allowed.

67. $\dfrac{4(-10)}{-3 - 1} - \dfrac{4 - 10}{-3(-1)} = \dfrac{-40}{-4} - \dfrac{-6}{3}$
$= 10 - (-2) = 12$

69. $\dfrac{28(-12)}{36 - 42} = \dfrac{-336}{-6} = 56$

71. $-8[-4 - 6(4 - 7)] = -8[-4 - 6 \cdot -3]$
$= -8[-4 + 18]$
$= -8(14) = -112$

73. $12 - 4[7 - 3(6 - 2)] = 12 - 4[7 - 3 \cdot 4]]$
$= 12 - 4(7 - 12)$
$= 12 - 4(-5) = 19$
$= 12 + 20 = 32$

75. $56 - \dfrac{26 - 60}{2 - 4} = 56 - \dfrac{-34}{-2}$
$= 56 - 17 = 39$

77. $70 - \dfrac{40 + 5(-3)}{-5} = 70 - \dfrac{40 - 15}{-5} = 70 - \dfrac{25}{-5}$
$= 70 - (-5) = 70 + 5 = 75$

79. Even though the symbols are the same in both examples, the parentheses lead to different results. Without the parentheses, we have a subtraction example: $5 - 2 = 3$; with parentheses, we have a multiplication example: $5(-2) = -10$.

80. Again, the same symbols are used in both examples, but the location of the parentheses cause the difference this time. Our order of operations requires that we multiply before we subtract. Therefore, $6(-3) - 2 = -18 - 2 = -20$, while $6 - 3(-2) = 6 + 6 = 12$.

81. When we see $-7-8$, it must be interpreted as a subtraction, since no parentheses are present. Therefore, (a) and (b) are both wrong, since $-7-8 = -15$. (It would be correct to write $-7(-8) = +56$ or $(-7)(-8) = +56$.) Part (c) is wrong also. Since $2-4=-2$, it follows that $-3-(2-4) = -3-(-2) = -3+2 = -1$. In part (d), the order of operations has not been followed. Here, $-4-2(5-6) = -4-2(-1) = -4+2 = -2$. (The calculation $-4-2 = -6$ never legitimately enters into the computation of the expression.)

82. The statement "two negatives make a positive" is accurate for multiplication and division but **not** for addition and subtraction. For instance, $(-3) + (-2) = -5$ and $(-3) - (-2) = -1$, and neither of these answers is positive.

83. The pattern suggest that $-1(-2) = 2$ and $-2(-2) = 4$. This in turn suggests that the product of two negative numbers is positive.

Exercises 1.6

1. $4.2 + 5.9 = 10.1$
3. $4(5.1) = 20.4$
5. $\dfrac{12.8}{3.2} = 4$
7. $\dfrac{36.8}{1.5} = 24.5\overline{3}$
9. $\dfrac{8}{0.2} + \dfrac{12}{0.4} = 40 + 30 = 70$
11. $-2 - 5.3 = -7.3$
13. $-2(5.3) = -10.6$
15. $-2(-5.3) = 10.6$
17. $\dfrac{-8.4}{1.2} = -7$
19. $\dfrac{-6}{1.5} + \dfrac{21.6}{-1.2} = -4 - 18 = -22$
21. $0.831 - 0.746 - 0.294 = 0.085 - 0.294$
    $= -0.209$
23. $0.53(21) - 0.42(85) = 11.13 - 35.7$
    $= -24.57$
25. $12.4 - 20(0.8) + 4.7 = 12.4 - 16 + 4.7$
    $= 1.1$
27. $0.02(28.6 - 13.5) = 0.02(15.1)$
    $= 0.302$
29. $5.2 - 1.4(2.8 - 0.7) = 5.2 - 1.4(2.1)$
    $= 5.2 - 2.94$
    $= 2.26$
31. Between 4 and 5.
33. Between -3 and -2.
35. Between 7 and 8.
37. Between -1 and 0.
39. [number line]
41. [number line]
43. [number line]
45. [number line]
47. [number line]

In 49-53 there are other possible answers if examples are given.

49. Members: $1.25, \dfrac{3}{5}, \sqrt{2}$ Not members: 1, 2, 3

51. There are no members since every rational number is also real. Not members: 1, 2, 3

53. Members: $\dfrac{1}{2}, \dfrac{3}{4}, -\dfrac{5}{9}$ Not members: 1, 2, 3

55. (a) False. The rational number $\frac{2}{3}$ is not an integer.
    (b) True. Every integer can be written as itself over 1. For example, the integer 8 can be written as $\frac{8}{1}$.
    (c) False, because -15 is to the left of -10 on the number line.
    (d) True, because $|-15| = 15$ and $|-10| = 10$.
    (e) False. The rational number $-\frac{1}{4}$ is negative, not positive.
    (f) False. The integer -2 is negative, not positive.
    (g) False. The real number $\sqrt{5}$ is not a rational number.
    (h) True. Every rational number has a location on the number line and so is also a real number.

56. (a) Use the commutative property of addition to rewrite the problem as $12.9 - 1.7 - 8.3$. Then this equals $12.9 - 10 = 2.9$.
    (b) Use the commutative property of multiplication to rewrite the problem as $7(10)(-2.3)$. Then this becomes $7(-23) = -161$.
    (c) Use the associative property of multiplication to rewrite the problem as $-10\left(\frac{3.2}{-1.6}\right)$. Then this equals $-10(-2) = 20$.
    (d) First, use the commutative property of multiplication to rewrite the problem as $-30(-10)\left(\frac{1.6}{3}\right)$, which becomes $300\left(\frac{1.6}{3}\right)$. Then use the associative property of multiplication to write this as $\frac{300}{3}(1.6) = 100(1.6) = 160$.

57. Since $(1)(1) = 1$ and $(2)(2) = 4$, the square root of 3, to the nearest whole number, is 2. Since $(1.7)(1.7) = 2.89$ and $(1.8)(1.8) = 3.24$, the square root of 3, to the nearest tenth, is 1.7. Since $(1.73)(1.73) = 2.9929$ and $(1.74)(1.74) = 3.0276$, the square root of 3, to the nearest hundredth, is 1.73. Since $(1.732)(1.732) = 2.999824$ and $(1.733)(1.733) = 3.003289$ the square root of 3, to the nearest thousandth, is 1.732.

Chapter 1 Review Exercises

1. $\{2, 4, 6, 8, 10, 12, 14, 16, 18\}$
3. $\{2, 3, 5, 7, 11, 13, 17, 19\}$
5. $\{23, 29\}$
7. $2 \cdot 3 \cdot 5$
9. Prime
11. $2 \cdot 2 \cdot 5 \cdot 5$
13. $3 - 7 = -4$
15. $-7 - 5 = -12$
17. $-2 - (-6) = -2 + 6 = 4$
19. $-4 - 5 - 6 = -9 - 6 = -15$
21. $-7 + 12 - 5 = +5 - 5 = 0$
23. $7 - 4 + 3 - 9 = 3 + 3 - 9 = 6 - 9 = -3$
25. $8 - 5 - 6 = 3 - 6 = -3$
27. $8 - (5 - 6) = 8 - (-1) = 8 + 1 = 9$
29. $8(5 - 6) = 8(-1) = -8$
31. $8 - 3 - 6 = 5 - 6 = -1$

33. $8 - 3(-6) = 8 + 18 = 26$

35. $8 - (3 - 6) = 8 - (-3) = 8 + 3 = 11$

37. $8(-3)(-6) = -24(-6) = 144$

39. $9 - 4(3 - 7) = 9 - 4(-4) = 9 + 16 = 25$

41. $9 - 4 \cdot 3 - 7 = 9 - 12 - 7$
    $= -3 - 7 = -10$

43. $9 - (4 \cdot 3 - 7) = 9 - (12 - 7)$
    $= 9 - 5 = 4$

45. $(9 - 4)(3 - 7) = 5(-4) = -20$

47. $|4 - 9| - |3 - 7| = |-5| - |-4| = 5 - 4 = 1$

49. $\dfrac{-7 - 3}{-2(-5)} = \dfrac{-10}{+10} = -1$

51. $\dfrac{-4(-2)(-8)}{-4(2) - 8} = \dfrac{8(-8)}{-8 - 8} = \dfrac{-64}{-16} = 4$

53. $\dfrac{6 - 4(3 - 1)}{-2(-3) - 4} = \dfrac{6 - 4(2)}{6 - 4}$
    $= \dfrac{6 - 8}{2} = \dfrac{-2}{2} = -1$

55. $8 + 2[3 - 4(1 - 6)] = 8 + 2[3 - 4(-5)]$
    $= 8 + 2[3 + 20]$
    $= 8 + 2[23]$
    $= 8 + 46 = 54$

57. $28(-65) - 30(-8 - 7) = 28(-65) - 30(-15)$
    $= -1820 + 450 = -1370$

59. $\dfrac{-48 - 5(36)}{47 - 53} = \dfrac{-48 - 180}{-6}$
    $= \dfrac{-228}{-6} = 38$

61. $-3.4(6.85) - 2.1 = -23.29 - 2.1$
    $= -25.39$

63. $\dfrac{8.57 - 12.63}{-4.2} = \dfrac{-4.06}{-4.2}$
    $= 0.97$

Chapter 1 Practice Test

1. (a) True (b) False (c) False
   (d) True (both have eight elements) (e) $C = \emptyset$

3. $|3 - 8| - |1 - 6| = |-5| - |-5| = 5 - 5 = 0$

5. $-4(-3)(-2) = 12(-2) = -24$

7. $-3(-5) - 2(-1) = 15 - (-2) = 15 + 2 = 17$

9. $\dfrac{4(-8)}{4 - 8} - \dfrac{3 - 6}{-2 - 1} = \dfrac{-32}{-4} - \dfrac{-3}{-3} = 8 - 1 = 7$

11. $\dfrac{(-2)(-3)(-4)}{(-2)(-3) - 4} = \dfrac{6(-4)}{6 - 4} = \dfrac{-24}{2} = -12$

13. $8 - 5 \cdot 4 - 7 = 8 - 20 - 7 = -12 - 7 = -19$

15. $8 - 3[8 - 3(8 - 3)] = 8 - 3[8 - 3(5)]$
    $= 8 - 3[8 - 15] = 8 - 3(-7)$
    $= 8 - (-21) = 8 + 21 = 29$

17. $\dfrac{105(-42) - (-36)(-80)}{20(25) - 17(30)} = \dfrac{-4410 - 2880}{500 - 510} = \dfrac{-7290}{-10} = 729$

Chapter 2
Algebraic Expressions

Exercises 2.1

1. $x \cdot x \cdot x \cdot x \cdot x \cdot x$
3. $(-x)(-x)(-x)(-x)$
5. $-(x \cdot x \cdot x \cdot x)$
7. $x \cdot x \cdot y \cdot y \cdot y$
9. $x \cdot x + y \cdot y \cdot y$
11. $x \cdot y \cdot y \cdot y$
13. a^4
15. $x^2 y^3$
17. $-r^2 s^3$
19. $-x^2(-y)^3$
21. $x^3 x^5 = x^{3+5} = x^8$
23. $3^5 = 3 \cdot 3 \cdot 3 \cdot 3 \cdot 3 = 9 \cdot 3 \cdot 3 \cdot 3$
    $= 27 \cdot 3 \cdot 3 = 81 \cdot 3 = 243$
25. $-2^3 = -(2 \cdot 2 \cdot 2) = -(4 \cdot 2) = -8$
27. $(-2)^4 = (-2)(-2)(-2)(-2)$
    $= 4(-2)(-2) = -8(-2) = 16$
29. $3^2 + 3^3 - 3^4 = 3 \cdot 3 + 3 \cdot 3 \cdot 3 - 3 \cdot 3 \cdot 3 \cdot 3$
    $= 9 + 27 - 81 = -45$
31. $4 \cdot 3^2 - 2 \cdot 5^2 = 4 \cdot 3 \cdot 3 - 2 \cdot 5 \cdot 5$
    $= 36 - 50 = -14$
33. $(3-7)^2 - (4-5)^3 = (-4)^2 - (-1)^3$
    $= (-4)(-4) - (-1)(-1)(-1) = 16 - (-1) = 17$
35. $3^2 4^3 = 3 \cdot 3 \cdot 4 \cdot 4 \cdot 4 = 576$
37. $3^2 3^3 = 3^{2+3} = 3^5 = 3 \cdot 3 \cdot 3 \cdot 3 \cdot 3 = 243$
39. $x^3 x^5 = x^{3+5} = x^8$
41. $aa^2 a^4 = a^{1+2+4} = a^7$
43. $(3x)(5x)(4x) = (3 \cdot 5 \cdot 4)(x \cdot x \cdot x)$
    $= 60x^3$
45. $(3r^2)(2r^3) = (3 \cdot 2)(r^2 r^3)$
    $= 6r^{2+3} = 6r^5$
47. $(-3x^3)(5x^2) = (-3 \cdot 5)(x^3 x^2)$
    $= -15x^{3+2} = -15x^5$
49. $(-c^4)(2^3)(-5c) = [-2^3(-5)](c^4 c)$
    $= 40c^{4+1} = 40c^5$
51. $(8x^3 y^2)(4xy^5) = (8 \cdot 4)(x^3 x)(y^2 y^5)$
    $= 32x^{3+1} y^{2+5} = 32x^4 y^7$
53. $(-2xy)(x^2 y^2)(-3xy) = [-2(-3)](xx^2 x)(yy^2 y)$
    $= 6x^{1+2+1} y^{1+2+1} = 6x^4 y^4$
55. $(3a^4)^2 = (3a^4)(3a^4) = (3 \cdot 3)(a^4 a^4)$
    $= 9a^{4+4} = 9a^8$
57. $(-4n^2)^3 = (-4n^2)(-4n^2)(-4n^2)$
    $= [(-4)(-4)(-4)](n^2 n^2 n^2)$
    $= -64n^{2+2+2} = -64n^6$
59. $(x^2)^3 (x^4)^2 = (x^2 x^2 x^2)(x^4 x^4) = x^{2+2+2+4+4} = x^{14}$

61. 0.073

63. 14.758

65. 107.916

67. 48.337

69. In $3 \cdot 2^4$, the exponent of 4 applies only to the 2, not to the 3; in $(3 \cdot 2)^4$, the exponent of 4 applies to both the 2 and the 3. Put another way, we compute $3 \cdot 2^4$ by first raising 2 to the fourth power and then multiplying the result by 3. We compute $(3 \cdot 2)^4$ by first multiplying 2 by 3 and then raising the result to the fourth power.

70. (a) The exponent of 2 applies only to the 2, not to the 3.
    (b) Only x should be raised to the fourth power.
    (c) 5^4 is not the same as $5 \cdot 4$.
    (d) When we square 3, we get 9, not 6. Addition is performed in the exponents.
    (e) -3^4 is the opposite of 3^4, so it is equal to -81.
    (f) We add exponents when we multiply powers of x, not when we add powers of x.
    (g) When multiplying powers of x, we add exponents. So $x^2 x^7 = x^{2+7} = x^9$.
    (h) Here, we do not add exponents, since $(x^2)^4 = x^2 x^2 x^2 x^2 = x^{2+2+2+2} = x^8$.

71. $3 \cdot 5$ represents the product of 3 and 5. Its value is 15.
    5^3 means $5 \cdot 5 \cdot 5$, the product of three factors of 5. Its value is 125.

72. $3^5 = 3 \cdot 3 \cdot 3 \cdot 3 \cdot 3$, the product of five factors of 3. Its value is 243.
    $5^3 = 5 \cdot 5 \cdot 5$, the product of three factors of 5. Its value is 125.

73. $2^4 = 2 \cdot 2 \cdot 2 \cdot 2 = (2 \cdot 2) \cdot (2 \cdot 2) = 4 \cdot 4 = 4^2$. In general, if a and b are two different counting numbers, $a^b \neq b^a$ (see previous problem). The only distinct counting numbers for which $a^b = b^a$ are 2 and 4.

Exercises 2.2

1. $C = 22.75 + 0.12m$
    $= 22.75 + 0.12(86)$
    $= 22.75 + 10.32 = \$33.07$

3. $C = 18.95d + 0.14m$
    $= 18.95(5) + 0.14(264)$
    $= 94.75 + 36.96 = \$131.71$

5. $h = 80 + 40t - 16t^2$
    $= 80 + 40(3.6) - 16(3.6)^2$
    $= 80 + 144 - 207.36$
    $= 244 - 207.36 = 16.64$
    $= 16.6$ ft. (rounded to the nearest tenth)

7. $P = 12{,}600 + 6.35n - 0.002n^2$
    $= 12{,}600 + 6.35(4280) - 0.002(4280)^2$
    $= 12{,}600 + 27{,}178 - 36{,}636.8$
    $= 39{,}778 - 36{,}636.8 = \$3{,}141.20$

9. $I = 415 + 17.5h$
    $= 415 + 17.5(12)$
    $= 415 + 210 = \$625$

11. $C = 18.95 + 0.065n$
    $= 18.95 + 0.065(82)$
    $= 18.95 + 5.33 = \$24.28$

13. $A = s^2$
    $= (6.4)^2 = 40.96$
    $= 41.0$ sq. in. (rounded to the nearest tenth)

15. $V = LWH$
    $= (1.8)(2.3)(1.1) = (4.14)(1.1) = 4.554$
    $= 4.6$ cu. ft. (rounded to the nearest tenth)

17. $V = C\left(1 - \dfrac{n}{N}\right)$
    $= 18,000\left(1 - \dfrac{8}{15}\right) = 18,000\left(\dfrac{7}{15}\right)$
    $= \$8400$

19. $N = \dfrac{857,000}{0.01x^2 + 0.2x}$
    $= \dfrac{857,000}{0.01(550)^2 + 0.2(550)}$
    $= 273$ computers (rounded)

21. $R = -0.03x^2 + 0.9x + 51.4$

    (a)
x	3	4	5	6	7	8
R	53.8%	54.5%	55.2%	55.7%	56.2%	56.7%

    (b) Comparing the original data values with those computed using R we see that the formula seems to give a reasonable accurate approximation to the employment rate. The formula is most accurate for 1993 and least accurate for 1994 and 1998.

23. $N = 0.02x^3 - 0.36x^2 + 1.40x + 1.65$

    (a)
x	0	1	2	3	4	5
N	1.7	2.7	3.2	3.2	2.8	2.2

    (b) Comparing the original data values with those computed using N we see that the formula seems to give a reasonable accurate approximation to the unemployment rate. The formula is most accurate for 1990 and least accurate for 1992.

25. $-y = -(-3) = 3$

27. $-|y| = -|-3| = -3$

29. $x + y = (2) + (-3) = -1$

31. $x - y = (2) - (-3) = 2 + 3 = 5$

33. $|x - y| = |(2) - (-3)| = |5| = 5$

35. $|x| - y = |(2)| - (-3) = 2 + 3 = 5$

37. $|x| - |y| = |(2)| - |(-3)| = 2 - 3 = -1$

39. $x + y + z = (2) + (-3) + (-4)$
    $= 2 - 3 - 4 = -5$

41. $x + y - z = (2) + (-3) - (-4)$
    $= 2 - 3 + 4 = 3$

43. $xy - z = (2)(-3) - (-4)$
    $= -6 + 4 = -2$

45. $x(y - z) = (2)[(-3) - (-4)] = 2(1) = 2$

47. $x - (y - z) = (2) - [(-3) - (-4)]$
    $= 2 - 1 = 1$

49. $xy^2 = (2)(-3)^2 = 2(9) = 18$

51. $x + y^2 = (2) + (-3)^2 = 2 + 9 = 11$

53. $x^2 + y^2 = (2)^2 + (-3)^2 = 4 + 9 = 13$

55. $(x + y + z)^2 = [(2) + (-3) + (-4)]^2$
    $= (-5)^2 = 25$

57. $-y^2 = -(-3)^2 = -9$

59. $-x^2(-z) = -(2)^2[-(-4)] = -4(4) = -16$

61. $xyz^2 = (2)(-3)(-4)^2 = -6(16) = -96$

63. $x(y - z)^2 = (2)[(-3) - (-4)]^2$
    $= 2(1)^2 = 2(1) = 2$

65. $xy^2 - (xy)^2 = (2)(-3)^2 - [(2)(-3)]^2$
    $= 2(9) - (-6)^2$
    $= 18 - 36 = -18$

67. $(z - 3x)^2 = [(-4) - 3(2)]^2$
    $= (-4 - 6)^2 = (-10)^2 = 100$

69. $(5x + y)(3x - y) = [5(2) + (-3)][3(2) - (-3)]$
    $= (10 - 3)(6 + 3) = (7)(9) = 63$

71. $y^2 - 3y + 2 = (-3)^2 - 3(-3) + 2$
$= 9 + 9 + 2 = 20$

73. $3x^2 + 4x + 1 = 3(2)^2 + 4(2) + 1$
$= 3(4) + 8 + 1$
$= 12 + 8 + 1 = 21$

75. $x^2 + 3x^2y - 3xy^2 + y^3 = (2)^2 + 3(2)^2(-3) - 3(2)(-3)^2 + (-3)^3$
$= 4 + 3(4)(-3) - 3(2)(9) + (-27)$
$= 4 - 36 - 54 - 27 = -113$

77. $\dfrac{xy}{x+y} = \dfrac{(2)(-3)}{(2)+(-3)} = \dfrac{-6}{-1} = 6$

79. $\dfrac{2}{x} - \dfrac{y}{3} + \dfrac{z}{2} = \dfrac{2}{(2)} - \dfrac{(-3)}{3} + \dfrac{(-4)}{2}$
$= 1 - (-1) + (-2) = 0$

81. $3x - 4y$ has two terms. The first term has a coefficient of 3 and a literal part of x; the second term has a coefficient of -4 and a literal part of y.

83. $3x(-4y) = -12xy$ has one term with a coefficient of -12 and a literal part of xy.

85. $3x(z - y)$ has one term with a coefficient of 3 and a literal part of $x(z - y)$.

87. $4x^2 - 3x + 2$ has three terms. The first term has a coefficient of 4 and a literal part of x^2; the second term has a coefficient of -3 and a literal part of x; the third term has no literal part—it is just the constant 2.

89. $-x^2 + y - 13$ has three terms. The first term has a coefficient of -1 and a literal part of x^2; the second term has a coefficient of 1 and a literal part of y; the third term has no literal part—it is just the constant -13.

91. $3x(2x) + 4y(5y) = 6x^2 + 20y^2$ has two terms. The first term has a coefficient of 6 and a literal part of x^2; the second term has a coefficient of 20 and a literal part of y^2.

93. 2.031

95. -9.862

97. 0.046

99. 25.667

101. 2.44; 2.47

103. -6.26; -6.25

105. A term is an algebraic expression that is connected by multiplication (and/or division). If a term is formed by multiplying two or more expressions, each is called a factor of that term.

106. $x + xy + xyz$ is one of many possible answers.

Exercises 2.3

1. Essential
3. Nonessential
5. Essential
7. Nonessential
9. Essential
11. Essential (both)
13. The first is nonessential, the second is essential.
15. $x, 2x$: literal part is x, coefficients are 1 and 2.
$y, 3y$: literal part is y, coefficients are 1 and 3.

17. $2x^2$, $-x^2$: literal part is x^2, coefficients are 2 and -1.
    $-3x$, $-x$ literal part is x, coefficients are -3 and -1.
    $4x^3$: literal part is x^3, coefficient is 4.

19. 4, 5: constant terms.
    $4u$, $5u$: literal part is u, coefficients are 4 and 5.
    $4u^2$: literal part is u^2, coefficient is 4.

21. $5x^2$: literal part is x^2, coefficient is 5.
    $5x^2y$: literal part is x^2y, coefficient is 5.
    $5y^2$: literal part is y^2, coefficient is 5.
    $5xy^2$: literal part is xy^2, coefficient is 5.

23. $-x^2y$, $-2x^2y$: literal part is x^2y, coefficients are -1 and -2.
    $2xy^2$, $3xy^2$: literal part is xy^2, coefficients are 2 and 3.
    x^2y^2: literal part is x^2y^2, coefficient is 1.

25. $3(x+4) = 3 \cdot x + 3 \cdot 4 = 3x + 12$

27. $5(y-2) = 5 \cdot y - 5 \cdot 2 = 5y - 10$

29. $-2(x+7) = -2 \cdot x + (-2) \cdot 7 = -2x - 14$

31. $3(5x+2) = 3 \cdot 5x + 3 \cdot 2 = 15x + 6$

33. $-4(3x+1) = -4 \cdot 3x + (-4) \cdot 1 = -12x - 4$

35. $x(x+3) = x \cdot x + x \cdot 3 = x^2 + 3x$

37. $x(x^2+3x) = x \cdot x^2 + x \cdot 3x = x^3 + 3x^2$

39. $5x(2x-4) = 5x \cdot 2x - 5x \cdot 4 = 10x^2 - 20x$

41. $0.42(40-x) = 0.42 \cdot 40 - 0.42 \cdot x = 16.8 - 0.42x$

43. $1.28(3x+60) = 1.28 \cdot 3x + 1.28 \cdot 60 = 3.84x + 76.8$

45. $2x + 10 = 2 \cdot x + 2 \cdot 5 = 2(x+5)$

47. $5y - 20 = 5 \cdot y - 5 \cdot 4 = 5(y-4)$

49. $9x + 3y - 6 = 3 \cdot 3x + 3 \cdot y - 3 \cdot 2 = 3(3x + y - 2)$

51. $x^2 + xy = x \cdot x + x \cdot y = x(x+y)$

53. $2x + 5x = (2+5)x = 7x$

55. $2x^2 + 5x^2 = (2+5)x^2 = 7x^2$

57. $3a - 8a + 2a = (3 - 8 + 2)a = -3a$

59. $-3y + y - 2y = (-3 + 1 - 2)y = -4y$

61. $-x - 2x - 3x = (-1 - 2 - 3)x = -6x$

63. $2x - 3y - 7x + 5y = 2x - 7x - 3y + 5y$
    $= (2-7)x + (-3+5)y$
    $= -5x + 2y$

65. $3x + 5y + 2z$ cannot be simplified

67. $3x^2 + 7x + x^2 + 3x = 3x^2 + x^2 + 7x + 3x$
    $= (3+1)x^2 + (7+3)x$
    $= 4x^2 + 10x$

69. $x^2 - 2x + x^2 - x = x^2 + x^2 - 2x - x$
$= (1+1)x^2 + (-2-1)x = 2x^2 - 3x$

71. $5x^2y - 3x^2 + x^2y - x^2 = 5x^2y + x^2y - 3x^2 - x^2$
$= (5+1)x^2y + (-3-1)x^2$
$= 6x^2y - 4x^2$

73. $2x + 5x - 3y + y - 7x = 2x + 5x - 7x - 3y + y$
$= (2+5-7)x + (-3+1)y = 0x - 2y = -2y$

75. $-5s^2 + 3st - s^2 + 6s^2 = -5s^2 - s^2 + 6s^2 + 3st$
$= (-5-1+6)s^2 + 3st = 0s^2 + 3st = 3st$

77. $3a^2b + ab^2 - ab^2 - 2a^2b - ab^2 = 3a^2b - 2a^2b + ab^2 - ab^2 - ab^2$
$= (3-2)a^2b + (1-1-1)ab^2$
$= 1a^2b - 1ab^2 = a^2b - ab^2$

79. $0.8x - 5 - x + 2.7 = 0.8x - x - 5 + 2.7$
$= (0.8 - 1)x + (-5 + 2.7) = -0.2x - 2.3$

81. $0.28x + 5.48 + 0.54x = 0.28x + 0.54x + 5.48$
$= (0.28 + 0.54)x + 5.48 = 0.82x + 5.48$

83. $3(x+y) + 4x - y = 3x + 3y + 4x - y$
$= 3x + 4x + 3y - y$
$= (3+4)x + (3-1)y$
$= 7x + 2y$

85. $3(x+y) + 4(x-y) = 3x + 3y + 4x - 4y$
$= 3x + 4x + 3y - 4y$
$= (3+4)x + (3-4)y$
$= 7x - 1y = 7x - y$

87. $3(x+y) + 4x(-y) = 3x + 3y - 4xy$

89. $5x(x^2 + 3) + 2x(3 + x^2) = 5x^3 + 15x + 6x + 2x^3 = 5x^3 + 2x^3 + 15x + 6x$
$= (5+2)x^3 + (15+6)x = 7x^3 + 21x$

91. $5x(x^2 + 3) + 2x(3x^2) = 5x^3 + 15x + 6x^3 = 5x^3 + 6x^3 + 15x$
$= (5+6)x^3 + 15x = 11x^3 + 15x$

93. $0.06x + 0.09(x + 5000) = 0.06x + 0.09x + 0.09(5000)$
$= (0.06 + 0.09)x + 450 = 0.15x + 450$

95. $0.35x + 0.42(2x) + 0.54(80 - 3x) = 0.35x + 0.84x + 43.2 - 1.62x$
$= 0.35x + 0.84x - 1.62x + 43.2$
$= (0.35 + 0.84 - 1.62)x + 43.2 = -0.43x + 43.2$

97. The distribution property allows us to combine like terms.

98. (a) Our order of operations requires that multiplication be done before addition. Therefore, we should not add 5 and 3 first.

   (b) The distribution property requires that we multiply each term of $x - 4$ by 3.

   (c) We cannot add 5 and $3x$ to get $8x$, since these are not like terms.

   (d) This is correct as written.

99. (a) We can write $45(116) - 45(16)$ as $45(116 - 16)$. This equals $45(100)$ or $4,500$.
    (b) We can write $60(278) - 60(28)$ as $60(278 - 28)$. This equals $60(250)$ or $15,000$.

Exercises 2.4

1. $4x + y + 4(x + y) = 4x + y + 4x + 4y$
   $= 4x + 4x + y + 4y = 8x + 5y$

3. $5(m + 2n) + 3(m - n) = 5m + 10n + 3m - 3n$
   $= 5m + 3m + 10n - 3n = 8m + 7n$

5. $-2(x - 3y) + 5(y - x) = -2x + 6y + 5y - 5x$
   $= -2x - 5x + 6y + 5y = -7x + 11y$

7. $5 + 3(x - 2) = 5 + 3x - 6 = 3x + 5 - 6 = 3x - 1$

9. $5 - 3(x - 2) = 5 - 3x + 6 = -3x + 5 + 6 = -3x + 11$

11. $(5 - 3)(x - 2) = 2(x - 2) = 2x - 4$

13. $4(m - 3n) + 2(5m + 6n) = 4m - 12n + 10m + 12n$
    $= 4m + 10m - 12n + 12n = 14m$

15. $2(a - 2b) - 4(b - 2a) = 2a - 4b - 4b + 8a$
    $= 2a + 8a - 4b - 4b = 10a - 8b$

17. $3(2x^2 - 4y) + 4(5y - 3x^2) = 6x^2 - 12y + 20y - 12x^2$
    $= 6x^2 - 12x^2 - 12y + 20y = -6x^2 + 8y$

19. $8 - (3x - 4) = 8 - 3x + 4$
    $= -3x + 8 + 4 = -3x + 12$

21. $5y - (1 - 2y) = 5y - 1 + 2y$
    $= 5y + 2y - 1 = 7y - 1$

23. $5(x - 3y) - x - 3y = 5x - 15y - x - 3y$
    $= 5x - x - 15y - 3y$
    $= 4x - 18y$

25. $5(x - 3y) - (x - 3y) = 5x - 15y - x + 3y$
    $= 5x - x - 15y + 3y$
    $= 4x - 12y$

27. $5(x - 3y) - x(-3y) = 5x - 15y + 3xy$

29. $5x(-3y) - x(-3y) = -15xy + 3xy$
    $= -12xy$

31. $5x(-3y)(-x)(-3y) = [(5)(-3)(-1)(-3)](x \cdot x)(y \cdot y) = -45x^2y^2$

33. $2x^2(x - 2) + x(3x^2 - 4x) = 2x^3 - 4x^2 + 3x^3 - 4x^2$
    $= 2x^3 + 3x^3 - 4x^2 - 4x^2 = 5x^3 - 8x^2$

35. $3a(4a - 1) - a(4 - a) = 12a^2 - 3a - 4a + a^2$
    $= 12a^2 + a^2 - 3a - 4a = 13a^2 - 7a$

37. $4(x^2 + 7x) - (x^2 + 7x) = 4x^2 + 28x - x^2 - 7x$
    $= 4x^2 - x^2 + 28x - 7x = 3x^2 + 21x$

39. $3a(a^2 + 3b) + 4b^2(a^2 - b) = 3a^3 + 9ab + 4a^2b^2 - 4b^3$

41. $3x^2 - 7x + 4 - 8x^2 - 3 - x = 3x^2 - 8x^2 - 7x - x + 4 - 3$
$$= -5x^2 - 8x + 1$$

43. $x^2y(xy - x) - 5xy(x^2y - x^2) = x^3y^2 - x^3y - 5x^3y^2 + 5x^3y$
$$= x^3y^2 - 5x^3y^2 - x^3y + 5x^3y$$
$$= -4x^3y^2 + 4x^3y$$

45. $4u^2v(u - v) - (uv^3 + u^2v^2) = 4u^3v - 4u^2v^2 - uv^3 - u^2v^2$
$$= 4u^3v - 4u^2v^2 - u^2v^2 - uv^3$$
$$= 4u^3v - 5u^2v^2 - uv^3$$

47. $4(x + 3y) + (4x + 3y) + 4x(3y) = 4x + 12y + 4x + 3y + 12xy$
$$= 4x + 4x + 12y + 3y + 12xy$$
$$= 8x + 15y + 12xy$$

49. $6(m - 2n) + (6m - 2n) + 6m(-2n) = 6m - 12n + 6m - 2n - 12mn$
$$= 6m + 6m - 12n - 2n - 12mn$$
$$= 12m - 14n - 12mn$$

51. $3t^5(t^4 - 4) - (t^5 + t^4) - 2t^3(3t)(-t^5) = 3t^9 - 12t^5 - t^5 - t^4 + 6t^9$
$$= 3t^9 + 6t^9 - 12t^5 - t^5 - t^4$$
$$= 9t^9 - 13t^5 - t^4$$

53. $-3(-x + 2) + (8 - 5x) - (2 - 2x) = 3x - 6 + 8 - 5x - 2 + 2x$
$$= 3x - 5x + 2x - 6 + 8 - 2$$
$$= 0x + 0 = 0$$

55. $a - 2[a - 2(a - 2)] = a - 2[a - 2a + 4]$
$$= a - 2[-a + 4]$$
$$= a + 2a - 8 = 3a - 8$$

57. $x\{x - 4[x - (x - 4)]\} = x\{x - 4[x - x + 4]\}$
$$= x\{x - 4(4)\} = x\{x - 16\}$$
$$= x^2 - 16x$$

59. $3(x + 2) + 4[x - 3(2 - x)] = 3(x + 2) + 4[x - 6 + 3x]$
$$= 3(x + 2) + 4(x + 3x - 6) = 3(x + 2) + 4(4x - 6)$$
$$= 3x + 6 + 16x - 24$$
$$= 3x + 16x + 6 - 24 = 19x - 18$$

61. $4(y - 3) - 2[3y - 5(y - 1)] = 4(y - 3) - 2[3y - 5y + 5]$
$$= 4(y - 3) - 2(-2y + 5)$$
$$= 4y - 12 + 4y - 10$$
$$= 4y + 4y - 12 - 10 = 8y - 22$$

63. $t[3t - 4(t + 5)] - 2(t^2 - 4t) = t(3t - 4t - 20) - 2t^2 + 8t$
$$= t(-t - 20) - 2t^2 + 8t$$
$$= -t^2 - 20t - 2t^2 + 8t$$
$$= -3t^2 - 12t$$

65. $-a^2(3a-7) - 2a[a^2 - 4(a-2)] = -3a^3 + 7a^2 - 2a(a^2 - 4a + 8)$
$= -3a^3 + 7a^2 - 2a^3 + 8a^2 - 16a$
$= -5a^3 + 15a^2 - 16a$

67. (a) The solution is correct up to $3 + 2[5x + 12]$. At this point, it is wrong to add $3 + 2 = 5$. Instead, use the distributive property:
$3 + 2[5x + 12] = 3 + 10x + 24$
$= 10x + 3 + 24$
$= 10x + 27$

(b) Within the brackets, we must add x and $x + 3$, not multiply them. As a result, we get:
$3x + 5[x + (x + 3)] = 3x + 5[2x + 3]$
$= 3x + 10x + 15$
$= 13x + 15$

68. (a) Do not subtract $5 - 3$; distribute -3 instead.
(b) When -3 is distributed, it must multiply -4 as well as x.
(c) When we multiply -3 by -4, we get 12, not -12.
(d) This is done correctly.

Exercises 2.5

1. $n + 4$
3. $n - 4$
5. $n - 4$
7. $5n + 6$
9. $2n - 9$
11. $n(n + 7)$
13. $(n + 2)(n - 6)$
15. $2n - 8 = 14$
17. $5n + 4 = n - 2$
19. $r + s = rs$
21. $2(r + s) = rs - 3$
23. $x + (x + 1) = 2x + 1$, where $x =$ smaller integer
25. $x + (x + 2) = 2x + 2$, where $x =$ smaller even integer
27. $x(x + 2)(x + 4)$, where $x =$ smallest of the odd integers
29. $8x = 7(x + 2) - 4$, where $x =$ smaller even integer
31. $x^2 + (x + 1)^2 + (x + 2)^2 = 5$, where $x =$ smallest of the integer

33. $A = 8.65h$. When $h = 14$, we get:
$A = 8.65(14) = \$121.10$

35. $C = 225 + 0.045p$. When $p = 4800$, we get
$C = 225 + 0.045(4800)$
$= 225 + 216 = \$441$

37. $C = 42 + 9p$. When $p = 6$, we get:
$C = 42 + 9(6) = 42 + 54 = \$96$

39. $C = 0.65 + 0.42(m - 1)$. When $m = 12$, we get:
$C = 0.65 + 0.42[(12) - 1]$
$= 0.65 + 0.42(11)$
$= 0.65 + 4.62 = \$5.27$

41. $T = 8d + 10k$. When $d = 12$ and $k = 15$, we get:
$T = 8(12) + 10(15)$
$= 96 + 150 = 246$ km.

43. $T = 6L + 9D$. When $L = 5$ and $D = 8$, we get:
$T = 6(5) + 9(8)$
$= 30 + 72 = 102$ lawns

45. $0.30d$ dollars

47. $d + 0.04d = 1.04d$ dollars

49. $p - 0.20p = 0.80p$ dollars

51. $p = 380 + 0.28(380)$
$p = 380 + 106.40$
$= \$486.40$

53. (a) original price: $120

    less 20%: $120 - 0.20(120) = \$96$

    less 10%: $96 - 0.10(96) = \$86.40$

    (b) less 30%: $120 - 0.30(120) = \$84$

    They are not equivalent. 30% of the original price yields the better discount.

55. $285 + 0.52x$ dollars

57. $6h + 14(h - 7) = 6h + 14h - 98$
$= 20h - 98$ dollars

59. $48t + 54(t - 2) = 48t + 54t - 108$
$= 102t - 108$ miles

61. (a) 8 nickels
    (b) 40¢ ($= 8 \times 5$¢)
    (c) 12 dimes
    (d) 120¢ ($= 12 \times 10$¢) or $1.20
    (e) 9 quarters
    (f) 225¢ ($= 9 \times 25$¢) or $2.25
    (g) 29 coins
    (h) 40¢ + 120¢ + 225¢ = 385¢ or $3.85

63. (a) A grades 200 exams per minute.
    (b) A grades for 15 minutes.
    (c) A grades 3,000 exams ($= 200 \times 15$).
    (d) B grades 160 exams per minute.
    (e) B grades for 20 minutes.
    (f) B grades 3,200 exams ($= 160 \times 20$).
    (g) A and B grade 6,200 exams altogether ($= 3,000 + 3,200$).

65. (a) She walks at the rate of 100 meters per minute.
    (b) She walks for 25 minutes.
    (c) She walks 2,500 meters ($= 100 \times 25$).
    (d) She jogs at the rate of 220 meters per minute.
    (e) She jogs for 35 minutes.
    (f) She jogs 7,700 meters ($= 220 \times 35$).
    (g) She covers a distance of 10,200 meters altogether ($= 2,500 + 7,700$).

Chapter 2 Review Exercises

1. $x \cdot y \cdot y \cdot y$

3. $-(x \cdot x \cdot x \cdot x)$

5. $3 \cdot x \cdot x$

7. $x^2 y^3$

9. $a^2 - b^3$

11. $-z^4 = -(-2)^4 = -(-2)(-2)(-2)(-2)$
$= -16$

13. $xy^2 = (-3)(4)^2 = -3(16) = -48$

15. $xyz - (x + y + z) = (-3)(4)(-2) - [(-3) + (4) + (-2)]$
$= 24 - (-1) = 25$

17. $|xy| + z - |z| = |(-3)(4)| + (-2) - |(-2)|$
$= |-12| + (-2) - |-2|$
$= 12 - 2 - 2 = 8$

19. $2x^2 - (x + y)^2 = 2(-3)^2 - [(-3) + (4)]^2$
$= 2(9) - (1)^2$
$= 18 - 1 = 17$

21. $x^3 x^4 x = x^{3+4+1} = x^8$

23. $a^7 a^2 + a^3 a^6 = a^{7+2} + a^{3+6}$
$= a^9 + a^9 = 2a^9$

25. $4x^3 x^2 + 3x^2 x^4 = 4x^{3+2} + 3x^{2+4}$
$= 4x^5 + 3x^6$

27. $3x^2 - 7x + 7 - 5x^2 - x - 3 = 3x^2 - 5x^2 - 7x - x + 7 - 3$
$= -2x^2 - 8x + 4$

29. $2a^2 b(3ab^4) = (2 \cdot 3)(a^2 a)(bb^4)$
$= 6a^{2+1} b^{1+4} = 6a^3 b^5$

31. $2a^2 b(3a + b^4) = 2a^2 b(3a) + 2a^2 b(b^4)$
$= 6a^3 b + 2a^2 b^5$

33. $3x(2x + 4) + 5(x^2 - 3) = 6x^2 + 12x + 5x^2 - 15$
$= 6x^2 + 5x^2 + 12x - 15$
$= 11x^2 + 12x - 15$

35. $4y^2(y - 2) - y(y^2 - 5y) = 4y^3 - 8y^2 - y^3 + 5y^2$
$= 4y^3 - y^3 - 8y^2 + 5y^2 = 3y^3 - 3y^2$

37. $3xy(x^2 - 2y) + 4xy^2(y - x) = 3x^3 y - 6xy^2 + 4xy^3 - 4x^2 y^2$

39. $3(2x - 4y) - (x + 2y) - (x - 2y) = 6x - 12y - x - 2y - x + 2y$
$= 6x - x - x - 12y - 2y + 2y$
$= 4x - 12y$

41. $3x^4(x^3 - 2y^2) - 4x(x^2)(x^4) = 3x^7 - 6x^4 y^2 - 4x^7$
$= 3x^7 - 4x^7 - 6x^4 y^2$
$= -x^7 - 6x^4 y^2$

43. $3 - [x - 3 - (x - 3)] = 3 - [x - 3 - x + 3]$
$= 3 - [x - x - 3 + 3]$
$= 3 - 0 = 3$

45. $0.02x + 0.07(2x + 1250) = 0.02x + 0.07(2x) + 0.07(1250)$
$= 0.02x + 0.14x + 87.5 = 0.16x + 87.5$

47. $5x^2 + 10 = 5 \cdot x^2 + 5 \cdot 2 = 5(x^2 + 2)$

49. $3y - 6z + 9 = 3 \cdot y - 3 \cdot 2z + 3 \cdot 3$
$= 3(y - 2z + 3)$

51. $n + 7 = 3n - 4$

53. $n + (n + 2) = n - 5$

55. $d - 0.35d = 0.65d$ dollars

57. (a) He sells 12 newspaper subscriptions.
    (b) He earns $2 for each newspaper subscription.
    (c) He sells 9 magazine subscriptions.
    (d) He earns $5 for each magazine subscription.
    (e) He earns $24 ($= 12 \times 2$) for the newspaper subscriptions.
    (f) He earns $45 ($= 9 \times 5$) for the magazine subscriptions.
    (g) He earns $69 ($= 24 + 45$) altogether.

Chapter 2 Practice Test

1. $-3^4 = -(3 \cdot 3 \cdot 3 \cdot 3) = -81$

3. $(-3 - 4 + 6)^5 = (-7 + 6)^5 = (-1)^5$
   $= (-1)(-1)(-1)(-1)(-1)$
   $= -1$

5. $x - y - z = (-2) - (3) - (-4)$
   $= -2 - 3 + 4 = -5 + 4 = -1$

7. $x^3 - z^2 = (-2)^3 - (-4)^2$
   $= (-2)(-2)(-2) - (-4)(-4)$
   $= -8 - (+16) = -8 - 16 = -24$

9. $4x^2y - 5xy + y^2 - 3xy - 2y^2 - x^2y = 4x^2y - x^2y - 5xy - 3xy + y^2 - 2y^2$
   $= 3x^2y - 8xy - y^2$

11. $2x(x^2 - y) - 3(x - xy) - (2x^3 - 3x) = 2x^3 - 2xy - 3x + 3xy - 2x^3 + 3x$
    $= 2x^3 - 2x^3 - 2xy + 3xy - 3x + 3x$
    $= xy$

13. $4 - [x - 4(x - 4)] = 4 - [x - 4x + 16]$
    $= 4 - [-3x + 16]$
    $= 4 + 3x - 16 = 3x + 4 - 16$
    $= 3x - 12$

15. (a) $2n + 4$ (b) $5n - 20 = n$

17. $M = 99 + 0.15C$

Chapter 3
First Degree Equations and Inequalities

Exercises 3.1

1. $2(x-3) = 2x - 6$
    $2x - 6 = 2x - 6$
    identity

3. $5(x+2) = 5x + 2$
    $5x + 10 = 5x + 2$
    $10 = 2$
    contradiction

5. $2a + 4 + 3a = 6a + 4 - a$
    $5a + 4 = 5a + 4$
    identity

7. $2z^2 + 3z - 2z^2 - z = z + z + 1$
    $2z^2 - 2z^2 + 3z - z = z + z + 1$
    $2z = 2z + 1$
    $0 = 1$
    contradiction

9. $5u - 4(u-1) - u = u - 4 - (u-2)$
    $5u - 4u + 4 - u = u - 4 - u + 2$
    $5u - 4u - u + 4 = u - u - 4 + 2$
    $4 = -2$
    contradiction

11. $7 - 3(y-2) = y + 4(5-y) - 7$
    $7 - 3y + 6 = y + 20 - 4y - 7$
    $7 + 6 - 3y = 20 - 7 + y - 4y$
    $13 - 3y = 13 - 3y$
    identity

13. $w(w-2) - w^2 + 2w = 3(w+1) - (3w-1)$
    $w^2 - 2w - w^2 + 2w = 3w + 3 - 3w + 1$
    $w^2 - w^2 - 2w + 2w = 3w - 3w + 3 + 1$
    $0 = 4$
    contradiction

15. $2(x^2 - 3) - x(2x - 1) + x = 2 - x - (x+8) + 4x$
    $2x^2 - 6 - 2x^2 + x + x = 2 - x - x - 8 + 4x$
    $2x^2 - 2x^2 + x + x - 6 = -x - x + 4x + 2 - 8$
    $2x - 6 = 2x - 6$
    identity

17. CHECK $x = -3$:
    $x + 5 = -2$
    $(-3) + 5 \stackrel{?}{=} -2$
    $2 \neq -2$
    Therefore $x = -3$ does not satisfy the equation.

    CHECK $x = -7$:
    $x + 5 = -2$
    $(-7) + 5 \stackrel{?}{=} -2$
    $-2 \stackrel{\checkmark}{=} -2$
    Therefore $x = -7$ does satisfy the equation

19. CHECK $a = -5$:
    $2 - a = 3$
    $2 - (-5) \stackrel{?}{=} 3$
    $7 \neq 3$
    Therefore $a = -5$ does not satisfy the equation.

    CHECK $a = 5$:
    $2 - a = 3$
    $2 - (5) \stackrel{?}{=} 3$
    $-3 \neq 3$
    Therefore $a = 5$ does not satisfy the equation.

CHECK $a = 1$:
$$2 - a = 3$$
$$2 - (1) \stackrel{?}{=} 3$$
$$1 \neq 3$$
Therefore $a = 1$ does not satisfy the equation.

21. CHECK $y = 2$:
$$5y - 6 = y - 3$$
$$5(2) - 6 \stackrel{?}{=} (2) - 3$$
$$10 - 6 \stackrel{?}{=} 2 - 3$$
$$4 \neq -1$$
Therefore $y = 2$ does not satisfy the equation.

    CHECK $y = \dfrac{3}{4}$:
$$5y - 6 = y - 3$$
$$5\left(\dfrac{3}{4}\right) - 6 \stackrel{?}{=} \left(\dfrac{3}{4}\right) - 3$$
$$\dfrac{15}{4} - 6 \stackrel{?}{=} \dfrac{3}{4} - 3$$
$$\dfrac{-9}{4} \stackrel{\checkmark}{=} \dfrac{-9}{4}$$
Therefore $y = \dfrac{3}{4}$ does satisfy the equation.

23. CHECK $w = -4$:
$$6 - 2w = 10 - 3w$$
$$6 - 2(-4) \stackrel{?}{=} 10 - 3(-4)$$
$$6 + 8 \stackrel{?}{=} 10 + 12$$
$$14 \neq 22$$
Therefore $w = -4$ does not satisfy the equation.

    CHECK $w = 1$:
$$6 - 2w = 10 - 3w$$
$$6 - 2(1) \stackrel{?}{=} 10 - 3(1)$$
$$6 - 2 \stackrel{?}{=} 10 - 3$$
$$4 \neq 7$$
Therefore $w = 1$ does not satisfy the equation

25. CHECK $x = \dfrac{2}{5}$:
$$4(x - 7) - (x + 1) = 15 - x$$
$$4\left[\left(\dfrac{2}{5}\right) - 7\right] - \left[\left(\dfrac{2}{5}\right) + 1\right] \stackrel{?}{=} 15 - \left(\dfrac{2}{5}\right)$$
$$4\left(\dfrac{2}{5} - \dfrac{35}{5}\right) - \left(\dfrac{2}{5} + \dfrac{5}{5}\right) \stackrel{?}{=} \dfrac{75}{5} - \dfrac{2}{5}$$
$$4\left(\dfrac{-33}{5}\right) - \dfrac{7}{5} \stackrel{?}{=} \dfrac{73}{5}$$
$$\dfrac{-132}{5} - \dfrac{7}{5} \stackrel{?}{=} \dfrac{73}{5}$$
$$\dfrac{-139}{5} \neq \dfrac{73}{5}$$
Therefore $x = \dfrac{2}{5}$ does not satisfy the equation.

    CHECK $x = 7$:
$$4(x - 7) - (x + 1) = 15 - x$$
$$4[(7) - 7] - [(7) + 1] \stackrel{?}{=} 15 - (7)$$
$$4(0) - 8 \stackrel{?}{=} 15 - 7$$
$$0 - 8 \stackrel{?}{=} 15 - 7$$
$$-8 \neq 8$$
Therefore $x = 7$ does not satisfy the equation.

27. CHECK $z = -2$:
$$3z + 2(z - 1) = 4(z + 2) - (z + 5)$$
$$3(-2) + 2[(-2) - 1] \stackrel{?}{=} 4[(-2) + 2] - [(-2) + 5]$$
$$-6 + 2(-3) \stackrel{?}{=} 4(0) - 3$$
$$-6 - 6 \stackrel{?}{=} 0 - 3$$
$$-12 \neq -3$$
Therefore $z = -2$ does not satisfy the equation.

    CHECK $z = 1$:
$$3z + 2(z - 1) = 4(z + 2) - (z + 5)$$
$$3(1) + 2[(1) - 1] \stackrel{?}{=} 4[(1) + 2] - [(1) + 5]$$
$$3 + 2(0) \stackrel{?}{=} 4(3) - 6$$
$$3 + 0 \stackrel{?}{=} 12 - 6$$
$$3 \neq 6$$
Therefore $z = 1$ does not satisfy the equation.

29. CHECK $x = -2$:
$$x^2 - 3x = 2x - 6$$
$$(-2)^2 - 3(-2) \stackrel{?}{=} 2(-2) - 6$$
$$4 + 6 \stackrel{?}{=} -4 - 6$$
$$10 \neq -10$$
Therefore $x = -2$ does not satisfy the equation.

CHECK $x = 2$:
$$x^2 - 3x = 2x - 6$$
$$(2)^2 - 3(2) \stackrel{?}{=} 2(2) - 6$$
$$4 - 6 \stackrel{?}{=} 4 - 6$$
$$-2 \stackrel{\checkmark}{=} -2$$
Therefore $x = 2$ does satisfy the equation.

31. CHECK $a = -1$:
$$a^2 - 4a = 4 - a$$
$$(-1)^2 - 4(-1) \stackrel{?}{=} 4 - (-1)$$
$$1 + 4 \stackrel{?}{=} 4 + 1$$
$$5 \stackrel{\checkmark}{=} 5$$
Therefore $a = -1$ does satisfy the equation.

CHECK $a = 4$:
$$a^2 - 4a = 4 - a$$
$$(4)^2 - 4(4) \stackrel{?}{=} 4 - (4)$$
$$16 - 16 \stackrel{?}{=} 4 - 4$$
$$0 \stackrel{\checkmark}{=} 0$$
Therefore $a = 4$ does satisfy the equation.

33. CHECK $y = -2$:
$$y(y + 6) = (y + 2)^2$$
$$(-2)[(-2) + 6] \stackrel{?}{=} [(-2) + 2]^2$$
$$-2(4) \stackrel{?}{=} 0^2$$
$$-8 \neq 0$$
Therefore $y = -2$ does not satisfy the equation.

CHECK $y = 2$:
$$y(y + 6) = (y + 2)^2$$
$$(2)[(2) + 6] \stackrel{?}{=} [(2) + 2]^2$$
$$2(8) \stackrel{?}{=} 4^2$$
$$16 \stackrel{\checkmark}{=} 16$$
Therefore $y = 2$ does satisfy the equation.

35. A value satisfies or is a solution to an equation if both sides of the equation are equal when the value is substituted for the variable.

36. An identity is an equation that is always true, no matter what value is chosen for the variable.
A contradiction is an equation that is never true, no mater what value is chosen for the variable.
A conditional equation is one that is true for some values of the variables and false for the others.

37. 2   39. -3 41. -16 43. -3

45. $8 - 5(-3) = 8 + 15 = 23$ 47. $-6 + 3(-9) = -6 - 27 = -33$

Exercises 3.2

1. $x + 3 = 8$
$\; -3 \;\; -3$
$x \;\;\;\; = 5$

   CHECK $x = 5$:
   $(5) + 3 \stackrel{?}{=} 8$
   $8 \stackrel{\checkmark}{=} 8$

3. $y - 4 = 7$
$\; +4 \;\; +4$
$y \;\;\;\; = 11$

   CHECK $y = 11$:
   $(11) - 4 \stackrel{?}{=} 7$
   $7 \stackrel{\checkmark}{=} 7$

5. $a + 3 = 1$
$\; -3 \;\; -3$
$a \;\;\;\; = -2$

   CHECK $a = -2$:
   $(-2) + 3 \stackrel{?}{=} 1$
   $1 \stackrel{\checkmark}{=} 1$

7. $a - 5 = -8$
$\; +5 \;\; +5$
$a \;\;\;\; = -3$

   CHECK $a = -3$:
   $(-3) - 5 \stackrel{?}{=} -8$
   $-8 \stackrel{\checkmark}{=} -8$

9. $3x = 21$ CHECK $x = 7$:
   $\dfrac{\cancel{3}x}{\cancel{3}} = \dfrac{21}{3}$ $3(7) \stackrel{?}{=} 21$
   $x = 7$ $21 \stackrel{\checkmark}{=} 21$

11. $4x = 15$ CHECK $x = \dfrac{15}{4}$:
    $\dfrac{\cancel{4}x}{\cancel{4}} = \dfrac{15}{4}$ $4\left(\dfrac{15}{4}\right) \stackrel{?}{=} 15$
    $x = \dfrac{15}{4}$ $15 \stackrel{\checkmark}{=} 15$

13. $-4x = -12$ CHECK $x = 3$:
    $\dfrac{\cancel{-4}x}{\cancel{-4}} = \dfrac{-12}{-4}$ $-4(3) \stackrel{?}{=} -12$
    $x = 3$ $-12 \stackrel{\checkmark}{=} -12$

15. $3x + x = 8 - 12$ CHECK $x = -1$:
    $4x = -4$ $3(-1) + (-1) \stackrel{?}{=} 8 - 12$
    $\dfrac{\cancel{4}x}{\cancel{4}} = \dfrac{-4}{4}$ $-3 - 1 \stackrel{?}{=} -4$
    $x = -1$ $-4 \stackrel{\checkmark}{=} -4$

17. $4x - x = 2 - 8$ CHECK $x = -2$:
    $3x = -6$ $4(-2) - (-2) \stackrel{?}{=} 2 - 8$
    $\dfrac{\cancel{3}x}{\cancel{3}} = \dfrac{-6}{3}$ $-8 + 2 \stackrel{?}{=} 2 - 8$
    $x = -2$ $-6 \stackrel{\checkmark}{=} -6$

19. $2z - 3z - 11z = -4(6)$ CHECK $z = 2$:
    $-12z = -24$
    $\dfrac{\cancel{-12}z}{\cancel{-12}} = \dfrac{-24}{-12}$ $2(2) - 3(2) - 11(2) \stackrel{?}{=} -4(6)$
    $z = 2$ $4 - 6 - 22 \stackrel{?}{=} -24$
                         $-24 \stackrel{\checkmark}{=} -24$

21. $3w - 7w + 4w = 8 - 3$
    $0w = 5$
    $0 = 5$ (contradiction)
    No solutions

23. $2x + 1 = 7$ CHECK $x = 3$:
    $\dfrac{-1 \quad -1}{2x \quad\;\; = 6}$ $2(3) + 1 \stackrel{?}{=} 7$
    $\dfrac{\cancel{2}x}{\cancel{2}} = \dfrac{6}{2}$ $6 + 1 \stackrel{?}{=} 7$
    $x = 3$ $7 \stackrel{\checkmark}{=} 7$

25. $10x - 4 = 22$ CHECK $x = \dfrac{13}{5}$:
    $\dfrac{+4 \quad\; +4}{10x \quad\;\; = 26}$ $10\left(\dfrac{13}{5}\right) - 4 \stackrel{?}{=} 22$
    $\dfrac{\cancel{10}x}{\cancel{10}} = \dfrac{26}{10}$ $26 - 4 \stackrel{?}{=} 22$
    $x = \dfrac{13}{5}$ $22 \stackrel{\checkmark}{=} 22$

27. $2t = 3t + 5$ CHECK $t = -5$:
    $\dfrac{-3t \;\; -3t}{-t = \quad\;\; 5}$ $2(-5) \stackrel{?}{=} 3(-5) + 5$
    $t = -5$ $-10 \stackrel{?}{=} -15 + 5$
                         $-10 \stackrel{\checkmark}{=} -10$

29. $9 = 6 - 3a$ CHECK $a = -1$:
    $\dfrac{-6 \quad -6}{3 = \quad\; -3a}$ $9 \stackrel{?}{=} 6 - 3(-1)$
    $\dfrac{3}{-3} = \dfrac{\cancel{-3}a}{\cancel{-3}}$ $9 \stackrel{?}{=} 6 + 3$
    $-1 = a$ $9 \stackrel{\checkmark}{=} 9$

31. $20 = 3w - 1$ CHECK $w = 7$:
    $\dfrac{+1 \qquad +1}{21 = 3w}$ $20 \stackrel{?}{=} 3(7) - 1$
    $\dfrac{21}{3} = \dfrac{\cancel{3}w}{\cancel{3}}$ $20 \stackrel{?}{=} 21 - 1$
    $7 = w$ $20 \stackrel{\checkmark}{=} 20$

33. $8 - 3(x - 4)$
    $= 8 - 3x + 12$
    $= -3x + 20$

35. $4t - 2(t - 3) = 12$
    $4t - 2t + 6 = 12$
    $2t + 6 = 12$
    $ -6 -6$
    $\overline{2t = 6}$
    $\dfrac{2t}{2} = \dfrac{6}{2}$
    $t = 3$

37. $3a + 5(2a + 4) + a$
    $= 3a + 10a + 20 + a$
    $= 14a + 20$

39. Let $w =$ width.
    $6w = 40$
    $\dfrac{6w}{6} = \dfrac{40}{6}$
    $w = \dfrac{20}{3}$ or $w = 6\dfrac{2}{3}$
    The width of the rectangle is $6\dfrac{2}{3}$ in.

41. Let $l =$ length.
    $3.5l = 73.5$
    $\dfrac{3.5l}{3.5} = \dfrac{73.5}{3.5}$
    $l = 21$
    The length of the rectangle is 21 yd.

43. Let $t =$ time needed.
    $52t = 234$
    $\dfrac{52t}{52} = \dfrac{234}{52}$
    $t = 4.5$
    It would take someone 4.5 hours.

45. Let $m =$ number of miles driven.
    $0.17m + 3(29.95) = 123.17$
    $0.17m + 89.85 = 123.17$
    $ -89.85 -89.85$
    $\overline{0.17m = 33.32}$
    $\dfrac{0.17m}{0.17} = \dfrac{33.32}{0.17}$
    $m = 196$
    The car was driven 196 miles.

47. (a) $C = 0.003s + 0.21$
    $C = 0.003(55) + 0.21$
    $C = 0.375$
    The cost is $0.38 per mile..

    (b) $C = 0.003s + 0.21$
    $0.35 = 0.003s + 0.21$
    $-0.21 -0.21$
    $\overline{0.14 = 0.003s}$
    $\dfrac{0.14}{0.003} = \dfrac{0.003s}{0.003}$
    $\dfrac{140}{3}$ or $46\dfrac{2}{3} = s$
    The speed of the truck is $46\tfrac{2}{3}$ mph.

49. (a) $C = 4T - 160$
    $150 = 4T - 160$
    $+160 +160$
    $\overline{310 = 4T}$
    $\dfrac{310}{4} = \dfrac{4T}{4}$
    $\dfrac{155}{2}$ or $77.5 = T$
    The temperature is 77.5°F.

    (b) $C = 4(90) - 160$
    $C = 200$
    It will chirp 200 times per minute..

51. (a) $T = 46 - 0.0054A$
$T = 46 - 0.0054(5000)$
$T = 19$
The temperature is 19°F.

(b) $T = 46 - 0.0054A$
$32 = 46 - 0.0054A$
$\underline{-46 -46 }$
$-14 = -0.0054A$
$\dfrac{-14}{-0.0054} = \dfrac{-0.0054A}{-0.0054}$
$2593.6 = A$ (rounded to the nearest tenth)
The altitude is 2593.6 feet.

53. $N = 1.67s + 55$
$85 = 1.67s + 55$
$\underline{-55 -55}$
$30 = 1.67s$
$\dfrac{30}{1.67} = \dfrac{1.67s}{1.67}$
$18.0 = s$ (rounded to the nearest tenth)
The speed is 18.0 feet/sec.

55. Let $w =$ width.
Then $2w + 7 =$ length.
$2w + 2(2w + 7) = 50$
$2w + 4w + 14 = 50$
$6w + 14 = 50$
$\underline{-14 -14}$
$6w = 36$
$\dfrac{6w}{6} = \dfrac{36}{6}$
$w = 6$

The width is 6 cm. and the
length is $2(6) + 7 = 19$ cm.

57. Let $l =$ length.
Then $5l - 10 =$ width.
$2l + 2(5l - 10) = 100$
$2l + 10l - 20 = 100$
$12l - 20 = 100$
$\underline{+20 +20}$
$12l = 120$
$\dfrac{12l}{12} = \dfrac{120}{12}$
$l = 10$

The length is 10 yards and the
width is $5(10) - 10 = 40$ yards.

59. If two quantities are equal, then we will not disturb this equality if we change each quantity in exactly the same way.

60. Two equations are called equivalent if their solution sets are exactly the same.

61. $\begin{cases} x + 1 = 16 \\ x - 3 = 12 \\ x = 15 \\ 2x = 30 \end{cases}$ $\begin{cases} x = -1 \\ x + 7 = 6 \end{cases}$ $\begin{cases} 2x = 18 \\ x = 9 \\ x + 2 = 11 \\ x - 2 = 7 \end{cases}$ $\begin{cases} 2x = 10 \\ 5x = 25 \end{cases}$

62. To check your answer after you have solved an equation, write your answer in place of the variable in the original equation. If the result is a true statement, then your answer satisfies the equation.

63. (a) After subtracting 4 from both sides of the equation, we should get $2x = 4$, not $2x = 12$.
    (b) Subtracting 3 from $3x$ will not give x. (These are not like terms.)

(c) When we divide both sides of the equation by 3, we must remember to divide both terms on the left. We would get $x - 2 = 4$, not $x - 6 = 4$.

(d) We cannot divide both sides by x, since x might be equal to 0. (In this case, it is!)

(e) In the last step, we should divide both sides by -3. Then we get $\dfrac{-3x}{-3} = \dfrac{-6}{-3}$, which gives $x = 2$.

(f) We need to divide both sides of the final equation by -1 to obtain $x = -2$.

(g) This is correct as written.

65. $\dfrac{-15 - 5}{-2(-5)(-2)} = \dfrac{-20}{10(-2)}$
$= \dfrac{-20}{-20} = 1$

67. $7 - 5[7 - 5(7 - 5)] = 7 - 5[7 - 5(2)]$
$= 7 - 5(7 - 10)$
$= 7 - 5(-3)$
$= 7 + 15 = 22$

69. Prime

Exercises 3.3

1. $\begin{aligned} 8y + 4 &= 5y + 19 \\ -5y & -5y \\ \hline 3y + 4 &= 19 \\ -4 & -4 \\ \hline 3y &= 15 \end{aligned}$

$\dfrac{\cancel{3}y}{\cancel{3}} = \dfrac{15}{3}$

$y = 5$

CHECK $y = 5$:

$8(5) + 4 \stackrel{?}{=} 5(5) + 19$

$40 + 4 \stackrel{?}{=} 25 + 19$

$44 \stackrel{\checkmark}{=} 44$

3. $\begin{aligned} 2a + 5 &= 4a + 12 \\ -2a & -2a \\ \hline 5 &= 2a + 12 \\ -12 & -12 \\ \hline -7 &= 2a \end{aligned}$

$\dfrac{-7}{2} = \dfrac{\cancel{2}a}{\cancel{2}}$

$-\dfrac{7}{2} = a$

CHECK $a = -\dfrac{7}{2}$:

$2\left(-\dfrac{7}{2}\right) + 5 \stackrel{?}{=} 4\left(-\dfrac{7}{2}\right) + 12$

$-7 + 5 \stackrel{?}{=} -14 + 12$

$-2 \stackrel{\checkmark}{=} -2$

5. $\begin{aligned} 5r - 8 &= 3r - 20 \\ -3r & -3r \\ \hline 2r - 8 &= -20 \\ +8 & +8 \\ \hline 2r &= -12 \end{aligned}$

$\dfrac{\cancel{2}r}{\cancel{2}} = \dfrac{-12}{2}$

$r = -6$

CHECK $r = -6$:

$5(-6) - 8 \stackrel{?}{=} 3(-6) - 20$

$-30 - 8 \stackrel{?}{=} -18 - 20$

$-38 \stackrel{\checkmark}{=} -38$

7. $\begin{aligned} 10 - x &= 4 - 3x \\ +3x & +3x \\ \hline 10 + 2x &= 4 \\ -10 & -10 \\ \hline 2x &= -6 \\ \frac{2x}{2} &= \frac{-6}{2} \\ x &= -3 \end{aligned}$

CHECK $x = -3$:
$10 - (-3) \stackrel{?}{=} 4 - 3(-3)$
$10 + 3 \stackrel{?}{=} 4 + 9$
$13 \stackrel{\checkmark}{=} 13$

9. $\begin{aligned} -4 - 3u &= -2 - u \\ +3u & +3u \\ \hline -4 &= -2 + 2u \\ +2 & +2 \\ \hline -2 &= 2u \\ \frac{-2}{2} &= \frac{2u}{2} \\ -1 &= u \end{aligned}$

CHECK $u = -1$:
$-4 - 3(-1) \stackrel{?}{=} -2 - (-1)$
$-4 + 3 \stackrel{?}{=} -2 + 1$
$-1 \stackrel{\checkmark}{=} -1$

11. $\begin{aligned} x + 7 &= 7 - x \\ +x & +x \\ \hline 2x + 7 &= 7 \\ -7 & -7 \\ \hline 2x &= 0 \\ \frac{2x}{2} &= \frac{0}{2} \\ x &= 0 \end{aligned}$

CHECK $x = 0$:
$(0) + 7 \stackrel{?}{=} 7 - (0)$
$0 + 7 \stackrel{?}{=} 7 - 0$
$7 \stackrel{\checkmark}{=} 7$

13. $\begin{aligned} x + 7 &= 7 + x \\ -x & -x \\ \hline 7 &= 7 \quad \text{(identity)} \end{aligned}$
Always true

15. $\begin{aligned} x - 7 &= 7 + x \\ -x & -x \\ \hline -7 &= 7 \quad \text{(contradiction)} \end{aligned}$
No solutions

17. $\begin{aligned} x - 7 &= 7 - x \\ +x & +x \\ \hline 2x - 7 &= 7 \\ +7 & +7 \\ \hline 2x &= 14 \\ \frac{2x}{2} &= \frac{14}{2} \\ x &= 7 \end{aligned}$

CHECK $x = 7$:
$(7) - 7 \stackrel{?}{=} 7 - (7)$
$7 - 7 \stackrel{?}{=} 7 - 7$
$0 \stackrel{\checkmark}{=} 0$

19. $\begin{aligned} 2(t + 1) + 4t &= 29 \\ 2t + 2 + 4t &= 29 \\ 6t + 2 &= 29 \\ -2 & -2 \\ \hline 6t &= 27 \\ \frac{6t}{6} &= \frac{27}{6} \\ t &= \frac{9}{2} \end{aligned}$

CHECK $t = \frac{9}{2}$:
$2\left[\left(\frac{9}{2}\right) + 1\right] + 4\left(\frac{9}{2}\right) \stackrel{?}{=} 29$
$2\left(\frac{11}{2}\right) + 4\left(\frac{9}{2}\right) \stackrel{?}{=} 29$
$11 + 18 \stackrel{?}{=} 29$
$29 \stackrel{\checkmark}{=} 29$

21. $2(y+3) + 4(y-2) = 22$
$2y + 6 + 4y - 8 = 22$
$6y - 2 = 22$
$ +2 +2$
$6y = 24$
$\dfrac{6y}{6} = \dfrac{24}{6}$
$y = 4$

CHECK $y = 4$:
$2[(4) + 3] + 4[(4) - 2] \stackrel{?}{=} 22$
$2(7) + 4(2) \stackrel{?}{=} 22$
$14 + 8 \stackrel{?}{=} 22$
$22 \stackrel{\checkmark}{=} 22$

23. $4 + 3(3y - 5) = 2y - 11 + y$
$4 + 9y - 15 = 2y - 11 + y$
$9y - 11 = 3y - 11$
$ -3y -3y$
$6y - 11 = -11$
$ +11 +11$
$6y = 0$
$\dfrac{6y}{6} = \dfrac{0}{6}$
$y = 0$

CHECK $y = 0$:
$4 + 3[3(0) - 5] \stackrel{?}{=} 2(0) - 11 + (0)$
$4 + 3(0 - 5) \stackrel{?}{=} 0 - 11 + 0$
$4 + 3(-5) \stackrel{?}{=} -11$
$4 - 15 \stackrel{?}{=} -11$
$-11 \stackrel{\checkmark}{=} -11$

25. $3(a - 2) + 4(2 - a) = a + 2(a + 1)$
$3a - 6 + 8 - 4a = a + 2a + 2$
$-a + 2 = 3a + 2$
$+a +a$
$2 = 4a + 2$
$-2 -2$
$0 = 4a$
$\dfrac{0}{4} = \dfrac{4a}{4}$
$0 = a$

CHECK $a = 0$:
$3[(0) - 2] + 4[2 - (0)] \stackrel{?}{=} (0) + 2[(0) + 1]$
$3(-2) + 4(2) \stackrel{?}{=} 0 + 2(1)$
$-6 + 8 \stackrel{?}{=} 0 + 2$
$2 \stackrel{\checkmark}{=} 2$

27. $8z - 3(z - 3) = -9$
$8z - 3z + 9 = -9$
$5z + 9 = -9$
$ -9 -9$
$5z = -18$
$\dfrac{5z}{5} = \dfrac{-18}{5}$
$z = -\dfrac{18}{5}$

CHECK $z = -\dfrac{18}{5}$:
$8\left(-\dfrac{18}{5}\right) - 3\left[\left(-\dfrac{18}{5}\right) - 3\right] \stackrel{?}{=} -9$
$8\left(-\dfrac{18}{5}\right) - 3\left(-\dfrac{33}{5}\right) \stackrel{?}{=} -9$
$-\dfrac{144}{5} + \dfrac{99}{5} \stackrel{?}{=} -9$
$-\dfrac{45}{5} \stackrel{?}{=} -9$
$-9 \stackrel{\checkmark}{=} -9$

29.
$$20t - 5(t-1) = 0$$
$$20t - 5t + 5 = 0$$
$$15t + 5 = 0$$
$$\ \underline{-5-5}$$
$$15t = -5$$
$$\frac{\cancel{15}t}{\cancel{15}} = \frac{-5}{15}$$
$$t = -\frac{1}{3}$$

CHECK $t = -\frac{1}{3}$:
$$20\left(-\frac{1}{3}\right) - 5\left[\left(-\frac{1}{3}\right) - 1\right] \stackrel{?}{=} 0$$
$$20\left(-\frac{1}{3}\right) - 5\left(-\frac{4}{3}\right) \stackrel{?}{=} 0$$
$$-\frac{20}{3} + \frac{20}{3} \stackrel{?}{=} 0$$
$$0 \stackrel{\checkmark}{=} 0$$

31.
$$20 - 5(t-1) = 0$$
$$20 - 5t + 5 = 0$$
$$-5t + 25 = 0$$
$$\underline{\ -25\ \ -25}$$
$$-5t = -25$$
$$\frac{\cancel{-5}t}{\cancel{-5}} = \frac{-25}{-5}$$
$$t = 5$$

CHECK $t = 5$:
$$20 - 5[(5) - 1] \stackrel{?}{=} 0$$
$$20 - 5(4) \stackrel{?}{=} 0$$
$$20 - 20 \stackrel{?}{=} 0$$
$$0 \stackrel{\checkmark}{=} 0$$

33.
$$2(y-3) - 3(y-5) = 5y - 5(y-2)$$
$$2y - 6 - 3y + 15 = 5y - 5y + 10$$
$$-y + 9 = 10$$
$$\underline{\ -9\ \ -9}$$
$$-y = 1$$
$$\frac{-y}{-1} = \frac{1}{-1}$$
$$y = -1$$

CHECK $y = -1$:
$$2[(-1) - 3] - 3[(-1) - 5] \stackrel{?}{=} 5(-1) - 5[(-1) - 2]$$
$$2(-4) - 3(-6) \stackrel{?}{=} 5(-1) - 5(-3)$$
$$-8 + 18 \stackrel{?}{=} -5 + 15$$
$$10 \stackrel{\checkmark}{=} 10$$

35.
$$4x - 3(x + 8) = 5x - 2(x - 12) - 2x$$
$$4x - 3x - 24 = 5x - 2x + 24 - 2x$$
$$x - 24 = x + 24$$
$$\underline{-x-x}$$
$$-24 = 24 \quad \text{(contradiction)}$$
No solutions

37.
$$a - (5 - 3a) = 7a - (a - 3) - 8$$
$$a - 5 + 3a = 7a - a + 3 - 8$$
$$4a - 5 = 6a - 5$$
$$\underline{-4a-4a}$$
$$-5 = 2a - 5$$
$$\underline{+5+5}$$
$$0 = 2a$$
$$\frac{0}{2} = \frac{\cancel{2}a}{\cancel{2}}$$
$$0 = a$$

CHECK $a = 0$:
$$(0) - [5 - 3(0)] \stackrel{?}{=} 7(0) - [(0) - 3] - 8$$
$$0 - (5 - 0) \stackrel{?}{=} 0 - (-3) - 8$$
$$0 - 5 \stackrel{?}{=} 0 + 3 - 8$$
$$-5 \stackrel{\checkmark}{=} -5$$

39.
$$\begin{aligned} x^2 + 3x - 7 &= x^2 - 5x + 1 \\ -x^2 \qquad\quad &\quad -x^2 \\ \hline 3x - 7 &= -5x + 1 \\ +5x \quad &\quad +5x \\ \hline 8x - 7 &= 1 \\ +7 &\quad +7 \\ \hline 8x &= 8 \\ \frac{8x}{8} &= \frac{8}{8} \\ x &= 1 \end{aligned}$$

CHECK $x = 1$:
$$(1)^2 + 3(1) - 7 \stackrel{?}{=} (1)^2 - 5(1) + 1$$
$$1 + 3 - 7 \stackrel{?}{=} 1 - 5 + 1$$
$$-3 \stackrel{\checkmark}{=} -3$$

41.
$$\begin{aligned} x(x+2) + 3x &= x(x-11) - 12 \\ x^2 + 2x + 3x &= x^2 - x - 12 \\ x^2 + 5x &= x^2 - x - 12 \\ -x^2 \qquad &\quad -x^2 \\ \hline 5x &= -x - 12 \\ +x &\quad +x \\ \hline 6x &= -12 \\ \frac{6x}{6} &= \frac{-12}{6} \\ x &= -2 \end{aligned}$$

CHECK $x = -2$:
$$(-2)[(-2) + 2] + 3(-2) \stackrel{?}{=} (-2)[(-2) - 1] - 12$$
$$-2(0) + 3(-2) \stackrel{?}{=} -2(-3) - 12$$
$$0 - 6 \stackrel{?}{=} 6 - 12$$
$$-6 \stackrel{\checkmark}{=} -6$$

43.
$$\begin{aligned} 2z(z+1) + 3(z+2) &= 3z(z+2) - z^2 \\ 2z^2 + 2z + 3z + 6 &= 3z^2 + 6z - z^2 \\ 2z^2 + 5z + 6 &= 2z^2 + 6z \\ -2z^2 \qquad &\quad -2z^2 \\ \hline 5z + 6 &= 6z \\ -5z &\quad -5z \\ \hline 6 &= z \end{aligned}$$

CHECK $z = 6$:
$$2(6)[(6) + 1] + 3[(6) + 2] \stackrel{?}{=} 3(6)[(6) + 2] - (6)^2$$
$$12(7) + 3(8) \stackrel{?}{=} 18(8) - 36$$
$$84 + 24 \stackrel{?}{=} 144 - 36$$
$$108 \stackrel{\checkmark}{=} 108$$

45.
$$\begin{aligned} 0.3x - 0.82 &= 1.13 \\ + 0.82 &\quad + 0.82 \\ \hline 0.3x &= 1.95 \\ \frac{0.3x}{0.3} &= \frac{1.95}{0.3} \\ x &= 6.5 \end{aligned}$$

47.
$$\begin{aligned} 2.3t - 1.6(t + 0.1) &= -0.139 \\ 2.3t - 1.6t - 0.16 &= -0.139 \\ 0.7t - 0.16 &= -0.139 \\ + 0.16 &\quad + 0.16 \\ \hline 0.7t &= 0.021 \\ \frac{0.7x}{0.7} &= \frac{0.021}{0.7} \\ x &= 0.03 \end{aligned}$$

49.
$$3.4(t-8) = 10.6(t+3)$$
$$3.4t - 27.2 = 10.6t + 31.8$$
$$\underline{+27.2 +27.2}$$
$$3.4t = 10.6t + 59$$
$$\underline{-10.6t -10.6t}$$
$$-7.2t = 59$$
$$\frac{-7.2t}{-7.2} = \frac{59}{-7.2}$$
$$t = -8.19$$

51.
$$2.8\,x = 1.60x + 2100$$
$$\underline{-1.60x -1.60x}$$
$$1.2\,x = 2100$$
$$\frac{1.2x}{1.2} = \frac{2100}{1.2}$$
$$x = 1750$$

The company must sell 1,750 items in order to break even.

CHECK: When 1,750 items are sold, $R = 2.8(1750) = \$4900$.
When 1,750 items are produced,
$C = 1.60(1750) + 2100$
$ = 2800 + 2100 = \$4900$.
So revenue = cost.

53. Let x = length of the shorter piece (in feet).
Then $x + 8$ = length of the longer piece (in feet).
$$x + (x+8) = 30$$
$$2x + 8 = 30$$
$$\underline{-8 -8}$$
$$2x = 22$$
$$\frac{2x}{2} = \frac{22}{2}$$
$$x = 11$$

Then $x + 8 = 11 + 8 = 19$. Thus, the two pieces are 11 ft. and 19 ft.
CHECK: 19 is 8 more than 11, and $11 + 19 = 30$.

55. Let x = one of the numbers.
Then $3x + 4$ = the other number.
$$x + (3x + 4) = 24$$
$$4x + 4 = 24$$
$$\underline{-4 -4}$$
$$4x = 20$$
$$\frac{4x}{4} = \frac{20}{4}$$
$$x = 5$$

Then $3x + 4 = 3(5) + 4 = 19$. Thus, the two numbers are 5 and 19.
CHECK: 19 is 4 more than 3 times 5, and the sum of 5 and 19 is 24.

57. Let $x =$ one of the numbers.
Then $4x - 5 =$ the other number.
$$\begin{aligned} x + (4x - 5) &= 11 \\ 5x - 5 &= 11 \\ +5 +5& \\ \hline 5x &= 16 \\ \frac{\cancel{5}x}{\cancel{5}} &= \frac{16}{5} \\ x &= \frac{16}{5} \end{aligned}$$

Then $4x - 5 = 4\left(\dfrac{16}{5}\right) - 5 = \dfrac{39}{5}$. Thus, the two numbers are $\dfrac{16}{5}$ and $\dfrac{39}{5}$.

CHECK: $\dfrac{39}{5}$ is 5 less than 4 times $\dfrac{16}{5}$ and the sum of $\dfrac{16}{5}$ and $\dfrac{39}{5}$ is $\dfrac{55}{5} = 11$.

59. Let $x =$ the number.
$$\begin{aligned} x + (5x - 4) &= 27 \\ 6x - 4 &= 27 \\ +4 +4& \\ \hline 6x &= 31 \\ \frac{\cancel{6}x}{\cancel{6}} &= \frac{31}{6} \\ x &= \frac{31}{6} \end{aligned}$$

CHECK: 4 less than 5 times $\dfrac{31}{6}$ is $\dfrac{131}{6}$, and $\dfrac{31}{6} + \dfrac{131}{6} = \dfrac{162}{6} = 27$.

61. Let $x =$ the number.
$$\begin{aligned} (2x + 4) - x &= 12 \\ x + 4 &= 12 \\ -4 -4& \\ \hline x &= 8 \end{aligned}$$

Thus, the number is 8.
CHECK: 4 more than twice 8 is 20, and. $20 - 8 = 12$.

63. Let $x =$ the smallest number.
Then $2x - 5 =$ the middle number and $2x + 10 =$ the largest number.
$$\begin{aligned} x + (2x - 5) + (2x + 10) &= 80 \\ 5x + 5 &= 80 \\ -5 -5& \\ \hline 5x &= 75 \\ \frac{\cancel{5}x}{\cancel{5}} &= \frac{75}{5} \\ x &= 15 \end{aligned}$$

Then $2x - 5 = 2(15) - 5 = 25$ and $2x + 10 = 2(15) + 10 = 40$. Thus the numbers are 15, 25, and 40.
CHECK: 25 is 5 less than twice 15, and 40 is 10 more than twice 15. The sum of 15, 25, and 40 is 80.

65. Let $n =$ the first integer.
Then $n + 1 =$ the second integer and
$n + 2 =$ the third integer.
$$n + (n + 1) + (n + 2) = 45$$
$$3n + 3 = 45$$
$$\underline{-3 -3}$$
$$3n = 42$$
$$\frac{3n}{3} = \frac{42}{3}$$
$$n = 14$$

Then $n + 1 = 14 + 1 = 15$ and $n + 2 = 14 + 2 = 16$.
Thus, the consecutive integers are 14, 15, and 16.
CHECK: 14, 15, and 16 are three consecutive integers. The sum of 14, 15, and 16 is 45.

67. Let $n =$ the first odd integer.
Then $n + 2 =$ the second odd integer and
$n + 4 =$ the third odd integer.
$$n + (n + 2) + (n + 4) = 2(n + 4) + 29$$
$$3n + 6 = 2n + 8 + 29$$
$$3n + 6 = 2n + 37$$
$$\underline{-2n -2n}$$
$$n + 6 = 37$$
$$\underline{-6 -6}$$
$$n = 31$$

Then $n + 2 = 31 + 2 = 33$ and $n + 4 = 31 + 4 = 35$.
Thus, the consecutive odd integers are 31, 33, and 35.
CHECK: 31, 33, and 35 are three consecutive odd integers. Their sum is 99, which is 29 more than $2(35) = 70$.

69. Let $W =$ width of the rectangle.
$2W + 1 =$ length of the rectangle.
$$2W + 2(2W + 1) = 26$$
$$2W + 4W + 2 = 26$$
$$6W + 2 = 26$$
$$\underline{ -2 -2}$$
$$6W = 24$$
$$\frac{6W}{6} = \frac{24}{6}$$
$$W = 4$$

Then $2W + 1 = 2(4) + 1 = 9$ is the length of the rectangle. Thus, the rectangle has a width of 4 cm and a length of 9 cm.
CHECK: 9 is one more than twice 4, and $2(4) + 2(9) = 8 + 18 = 26$.

71. Let $x =$ length of the shortest side.
Then $x + 1 =$ length of the "middle" side
and $x + 2 =$ length of the longest side.
$$x + (x + 1) + (x + 2) = 24$$
$$3x + 3 = 24$$
$$-3 = -3$$
$$3x = 21$$
$$\frac{3x}{3} = \frac{21}{3}$$
$$x = 7$$

Then $x + 1 = 7 + 1 = 8$ and $x + 2 = 7 + 2 = 9$.
Thus, the sides of the triangle have lengths of 7 cm, 8 cm, and 9 cm.
CHECK: 7, 8, and 9 are three consecutive integers. A triangle whose sides are 7 cm, 8 cm, and 9 cm has a perimeter of $7 + 8 + 9 = 24$ cm.

73. Let $W =$ width of the rectangle.
Then $W + 6 =$ length of the rectangle.
$$2(W + 10) + 2[3(W + 6)] = 2W + 2(W + 6) + 56$$
$$2W + 20 + 6W + 36 = 2W + 2W + 12 + 56$$
$$8W + 56 = 4W + 68$$
$$-4W \qquad -4W$$
$$4W + 56 = +68$$
$$-56 \qquad -56$$
$$4W = 12$$
$$\frac{4W}{4} = \frac{12}{4}$$
$$W = 3$$

Then $W + 6 = 3 + 6 = 9$, so that the original rectangle has a width of 3 and a length of 9.
CHECK: 9 is 6 more than 3. The new rectangle will have a width of $3 + 10 = 13$ and a length of $3(9) = 27$, so that its perimeter is $2(13) + 2(27) = 80$. This is 56 more than $2(3) + 2(9) = 24$, the original perimeter.

75. Let $m =$ number of miles driven.
$$2(45) + 0.40m = 170$$
$$90 + 0.40m = 170$$
$$-90 \qquad -90$$
$$0.40m = 80$$
$$\frac{0.40m}{0.40} = \frac{80}{0.40}$$
$$m = 200$$

The truck was driven 200 miles.
CHECK: To rent the truck for two days at $45 per day costs $2(45) = \$90$. To drive 200 miles at $0.40 per mile costs $0.40(200) = \$80$. $\$90 + \$80 = \$170$.

77. Let d = the daily rental charge.
$$125 + 5d = 275$$
$$-125 \quad\quad -125$$
$$5d = 150$$
$$\frac{5d}{5} = \frac{150}{5}$$
$$d = 30$$

The daily rental charge is $30.
CHECK: To rent a computer for 5 days at $30 per day costs $5(30) = \$150$. With an installation charge of $125, the total five-day rental costs $150 + \$125 = \$275$.

79. Let q = the number of quarters.
Then $20 - q$ = the number of dimes.
$$25q + 10(20 - q) = 425$$
$$25q + 200 - 10q = 425$$
$$15q + 200 = 425$$
$$\quad\quad -200 \quad -200$$
$$15q = 225$$
$$\frac{15q}{15} = \frac{225}{15}$$
$$q = 15$$

Then $20 - q = 20 - 15 = 5$. Thus there are 15 quarters and 5 dimes in the collection.
CHECK: $15 + 5 = 20$. The values of 15 quarters is $15(0.25) = \$3.75$. The value of 5 dimes is $5(0.10) = \$0.50$.
$\$3.75 + \$0.50 = \$4.25$.

81. Let x = number of advanced-purchase tickets sold.
Then $150 - x$ = number of tickets sold at the door.
$$10x + 12(150 - x) = 1580$$
$$10x + 1800 - 12x = 1580$$
$$-2x + 1800 = 1580$$
$$\quad\quad -1800 \quad -1800$$
$$-2x = -220$$
$$\frac{-2x}{-2} = \frac{-220}{-2}$$
$$x = 110$$

110 advanced-purchase tickets were sold.
CHECK: If 110 tickets were sold in advance, then 40 were sold at the door. $110(10) = \$1100$ was collected from the advance sale and $40(12) = \$480$ was collected at the door.
$\$1100 + \$480 = \$1580$.

83. Let x = number of half-dollars.
Then $x + 11$ = number of quarters.
So $45 - x - (x + 11)$ = number of dimes.
$$50x + 25(x + 11) + 10[45 - x - (x + 11)] = 1110$$
$$50x + 25x + 275 + 10(34 - 2x) = 1110$$
$$75x + 275 + 340 - 20x = 1110$$
$$55x + 615 = 1110$$
$$\quad\quad -615 \quad -615$$
$$55x = 495$$
$$\frac{55x}{55} = \frac{495}{55}$$
$$x = 9$$

Then $x + 11 = 9 + 11 = 20$ and $45 - x - (x + 11) = 45 - 9 - (9 + 11) = 16$. So there are 9 half-dollars, 20 quarters, and 16 dimes in the collection.

CHECK: $9 + 20 + 16 = 45$. The value of 9 half-dollars is $4.50. The value of 20 quarters is $5.00. The value of 16 dimes is $1.60. Total value is $\$4.50 + \$5.00 + \$1.60 = \$11.10$.

85. Let t = number of hours the electrician worked.
Then $t + 4$ = number of hours her assistant worked

$$45t + 24(t + 4) = 464$$
$$45t + 24t + 96 = 464$$
$$69t + 96 = 464$$
$$\begin{array}{r}-96 -96\end{array}$$
$$69t = 368$$
$$\frac{\cancel{69}t}{\cancel{69}} = \frac{368}{69}$$
$$t = \frac{16}{3} \text{ or } 5\frac{1}{3}$$

The electrician worked for $5\frac{1}{3}$ hours.

CHECK: The electrician earns $\frac{16}{3}(\$45) = \$240$ for her time. Her assistant works $t + 4 = \frac{16}{3} + 4 = \frac{28}{3}$ hours and earns $\frac{28}{3}(\$24) = \$224$.
Together, they earn $\$240 + \$224 = \$464$.

87. Let t = the time (in hours) that each train travels up to the point that they pass by each other
Using $d = rt$, $60t$ = distance traveled by the slower train and $90t$ = distance traveled by the faster train.
$$60t + 90t = 300$$
$$150t = 300$$
$$\frac{\cancel{150}t}{\cancel{150}} = \frac{300}{150}$$
$$t = 2$$

Thus the trains pass by each other 2 hours later than 10:00 a.m., or at 12:00 noon.

CHECK: In 2 hours, the slower train travels $60(2) = 120$ miles and the faster train travels $90(2) = 180$ miles. Together they travel $120 + 180 = 300$ miles.

89. Let t = the time (in hours) traveled by the person who left earlier.
Then $t - 1$ = the time (in hours) traveled by the person who left 1 hour later.
$$55t + 45(t - 1) = 280$$
$$55t + 45t - 45 = 280$$
$$100t - 45 = 280$$
$$\begin{array}{r}+45 +45\end{array}$$
$$100t = 325$$
$$\frac{\cancel{100}t}{\cancel{100}} = \frac{325}{100}$$
$$t = \frac{13}{4} \text{ or } 3\frac{1}{4}$$

Thus, the people will be 280 km apart $3\frac{1}{4}$ hours later than 2:00 p.m., or at 5:15 p.m.

CHECK: The person who left earlier travels $\frac{13}{4}(55) = \frac{715}{4}$ km. The person who left later travels $t - 1 = \frac{13}{4} - 1 = \frac{9}{4}$ hours and covers $\frac{9}{4}(45) = \frac{405}{4}$ km. The distance between them is $\frac{715}{4} + \frac{405}{4} = \frac{1120}{4} = 280$ km.

91. Let t = number of hours needed to complete the running section.
Then $6 - t$ = number of hours needed to complete the bicycling section.

$$18t + 50(6-t) = 172$$
$$18t + 300 - 50t = 172$$
$$-32t + 300 = 172$$
$$ -300\ \ -300$$
$$\overline{-32t = -128}$$
$$\frac{-32t}{-32} = \frac{-128}{-32}$$
$$t = 4$$

Thus, it takes 4 hours to complete the running section of the course. Since the running rate is 18 kph, the running section of the course is $18(4) = 72$ km.

CHECK: The first person runs a distance of $18(4) = 72$ km. The second person bicycles a distance of $50(6-t) = 50(6-4) = 50(2) = 100$ km. The total distance covered is $72 + 100 = 172$ km.

93. Let r = slower rate.
Then $r + 15$ = faster rate.

$$5r = 4(r+15)$$
$$5r = 4r + 60$$
$$-4r\ \ -4r$$
$$\overline{r = 60}$$

A car that travels for 5 hours at the rate of 60 kph covers $5(60) = 300$ km, which is the distance between town A and town B.

CHECK: At the rate of 60 kph, the car travels $5(60) = 300$ km in 5 hours. At the rate of $60 + 15 = 75$ kph, the car travels $4(75) = 300$ km in 4 hours, which is the same distance.

95. Let t = number of hours the trainee works.
Then $t - 2$ = number of hours that the secretary works.

$$7t + 15(t-2) = 124$$
$$7t + 15t - 30 = 124$$
$$22t - 30 = 124$$
$$+30\ \ +30$$
$$\overline{22t = 154}$$
$$\frac{22t}{22} = \frac{154}{22}$$
$$t = 7$$

Thus, the pile of forms will be finished 7 hours after 9:00 a.m., or at 4:00 p.m.

CHECK: The trainee processes $7(7) = 49$ forms. The secretary processes $15(7-2) = 15(5) = 75$ forms. The total number of forms is $49 + 75 = 124$.

97. Let x = number of shares Margaret buys at a lower price.
Then $200 - x$ = number of shares Margaret buys at a higher price.

$$8.125x + 9.375(200-x) = 1725$$
$$8.125x + 1875 - 9.375x = 1725$$
$$-1.25x + 1875 = 1725$$
$$-1875\ \ -1875$$
$$\overline{-1.25x = -150}$$
$$\frac{-1.25x}{-1.25} = \frac{-150}{-1.25}$$
$$x = 120$$

Then $200 - x = 200 - 120 = 80$. Thus Margaret buys 120 shares at the lower price and 80 shares at the higher price.

CHECK: 120 shares at $8.125 per share cost $120(\$8.125) = \$975$.
80 shares at $9.375 per share cost $80(\$9.375) = \$750$.
Total cost of all shares is $\$975 + \$750 = \$1725$.

99. Let x = number of lighter boxes.
Then $x + 89$ = number of heavier boxes.
$$\begin{aligned} 6.58x + 9.32(x + 89) &= 1974.28 \\ 6.58x + 9.32x + 829.48 &= 1974.28 \\ 15.9x + 829.48 &= 1974.28 \\ -829.48 \quad\quad -829.48& \\ \hline 15.9x &= 1144.80 \\ \frac{15.9x}{15.9} &= \frac{1144.80}{15.9} \\ x &= 72 \end{aligned}$$

Then $x + 89 = 72 + 89 = 161$. Since there are 72 lighter boxes and 161 heavier boxes on the truck, there are $72 + 161 = 233$ boxes altogether.

CHECK: 72 boxes that weigh 6.58 kg each weigh 473.76 kg altogether; 161 boxes that weigh 9.32 kg each weigh 1500.52 kg altogether. The total weight of the boxes is $473.76 + 1500.52 = 1974.28$ kg.

101. Let x = number of basic web sites
Then $17 - x$ = number of deluxe sites.
$$\begin{aligned} 675x + 1850(17 - x) &= 17350 \\ 675x + 31450 - 1850x &= 17350 \\ -1175x + 31450 &= 17350 \\ -31450 \quad -31450& \\ \hline -1175x &= -14100 \\ \frac{-1175x}{-1175} &= \frac{-14100}{-1175} \\ x &= 12 \end{aligned}$$

Then $17 - x = 17 - 12 = 5$. There are 12 basic web sites and 5 deluxe web sites.

CHECK: 12 basic + 5 deluxe = 17 total web sites. The charge for the basic setups is 12($675) = $8100. The charge for the deluxe setups is 5($1850) = $9250. The total bill is $8100 + $9250 = $17,350.

Exercises 3.4

1. $x + 4 < 3$
    $-2 + 4 \stackrel{?}{<} 3$
    $2 \stackrel{\checkmark}{<} 3$
    Therefore, -2 satisfies the inequality.

3. $a - 2 > -1$
    $-3 - 2 \stackrel{?}{>} -1$
    $-5 \not> -1$
    Therefore, -3 does not satisfy the inequality.

5. $-y + 3 \leq 5$
    $-(-2) + 3 \stackrel{?}{\leq} 5$
    $2 + 3 \stackrel{?}{\leq} 5$
    $5 \stackrel{\checkmark}{\leq} 5$
    Therefore, -2 satisfies the inequality.

7. $-8 \leq 2 - x$
    $-8 \stackrel{?}{\leq} 2 - 6$
    $-8 \stackrel{\checkmark}{\leq} -4$
    Therefore, 6 satisfies the inequality.

9. $2z - 5 < -3$
    $2(1) - 5 \stackrel{?}{<} -3$
    $2 - 5 \stackrel{?}{<} -3$
    $-3 \not< -3$
    Therefore, 1 does not satisfy the inequality.

11. $12 < 5 + 2u$
    $12 \stackrel{?}{<} 5 + 2(3)$
    $12 \stackrel{?}{<} 5 + 6$
    $12 \not< 11$
    Therefore, 3 does not satisfy the inequality.

13. $\quad 7 - 4x < 8$
$\quad 7 - 4(-4) \overset{?}{<} 8$
$\quad 7 + 16 \overset{?}{<} 8$
$\quad 23 \not< 8$
Therefore, -4 does not satisfy the inequality.

15. $\quad -2 < 8 - x < 3$
$\quad -2 \overset{?}{<} 8 - 6 \overset{?}{<} 3$
$\quad -2 \overset{\checkmark}{<} 2 \overset{\checkmark}{<} 3$
Therefore, 6 satisfies the inequality.

17. $\quad 6 + 2(a - 3) < 1$
$\quad 6 + 2[(-2) - 3] \overset{?}{<} 1$
$\quad 6 + 2(-5) \overset{?}{<} 1$
$\quad 6 - 10 \overset{?}{<} 1$
$\quad -4 \overset{\checkmark}{<} 1$
Therefore, -2 satisfies the inequality.

19. $\quad -12 < 9 - 5(x+1) < -5$
$\quad -12 \overset{?}{<} 9 - 5[(3) + 1] \overset{?}{<} -5$
$\quad -12 \overset{?}{<} 9 - 5(4) \overset{?}{<} -5$
$\quad -12 \overset{?}{<} 9 - 20 \overset{?}{<} -5$
$\quad -12 \overset{\checkmark}{<} -11 \overset{\checkmark}{<} -5$
Therefore, 3 satisfies the inequality.

21. $\quad x - 3 < 2$
$\quad \underline{+3 \phantom{<} +3}$
$\quad x < 5$

23. $\quad a + 7 > 4$
$\quad \underline{-7 -7}$
$\quad a > -3$

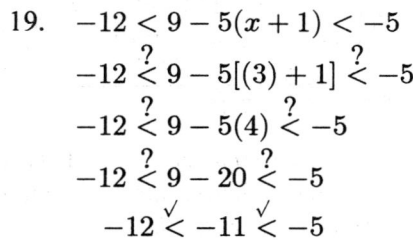

25. $\quad -4 > w + 2$
$\quad \underline{-2 -2}$
$\quad -6 > w$

27. $\quad z + 3 > 0$
$\quad \underline{-3 -3}$
$\quad z > -3$

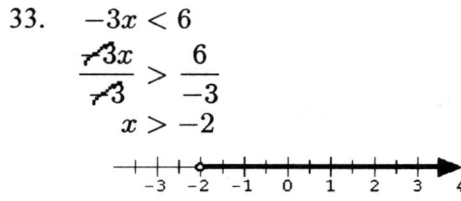

29. $\quad 3x \leq 12$
$\quad \dfrac{\cancel{3}x}{\cancel{3}} \leq \dfrac{12}{3}$
$\quad x \leq 4$

31. $\quad 4y > -8$
$\quad \dfrac{\cancel{4}y}{\cancel{4}} > \dfrac{-8}{4}$
$\quad y > -2$

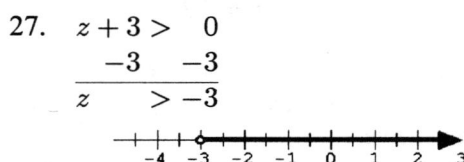

33. $\quad -3x < 6$
$\quad \dfrac{\cancel{-3}x}{\cancel{-3}} > \dfrac{6}{-3}$
$\quad x > -2$

35. $\quad -3x < -6$
$\quad \dfrac{\cancel{-3}x}{\cancel{-3}} > \dfrac{-6}{-3}$
$\quad x > 2$

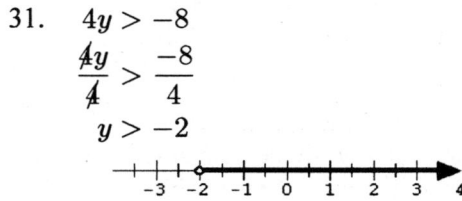

37. $\quad 7a > 0$
$\quad \dfrac{\cancel{7}a}{\cancel{7}} > \dfrac{0}{7}$
$\quad a > 0$

39. $\quad -7a \geq 0$
$\quad \dfrac{\cancel{-7}a}{\cancel{-7}} \leq \dfrac{0}{-7}$
$\quad a \leq 0$

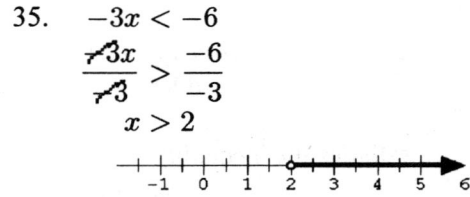

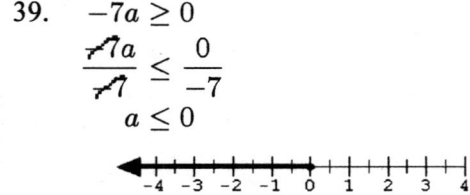

41. $\quad -x < 3$
$(-1)(-x) > (-1)3$
$\quad\quad x > -3$

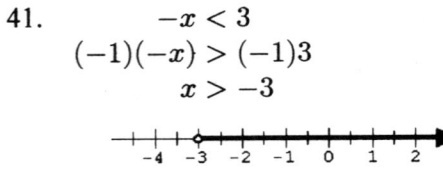

43. $\quad -x < -3$
$(-1)(-x) > (-1)(-3)$
$\quad\quad x > 3$

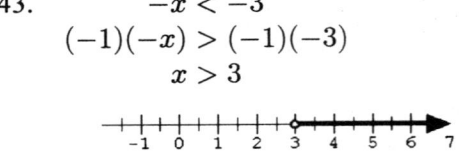

45. $\quad -5 < a - 4 \leq 2$
$\quad\quad +4 \quad\quad +4 \quad +4$
$\quad \overline{-1 < a \quad\quad\quad \leq 6}$

45. $\quad 1 \leq x + 3 \leq 5$
$\quad\quad -3 \quad\quad -3 \quad -3$
$\quad \overline{-2 \leq x \quad\quad \leq 2}$

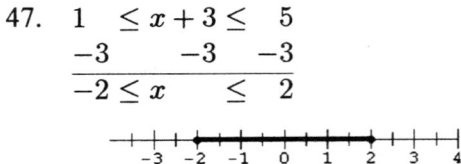

49. $\quad -6 < 3y < 3$
$\quad \dfrac{-6}{3} < \dfrac{3y}{3} < \dfrac{3}{3}$
$\quad -2 < y < 1$

51. $\quad 0 \leq -2x < 2$
$\quad \dfrac{0}{-2} \geq \dfrac{-2x}{-2} > \dfrac{2}{-2}$
$\quad 0 \geq x > -1 \quad \text{or} \quad -1 < x \leq 0$

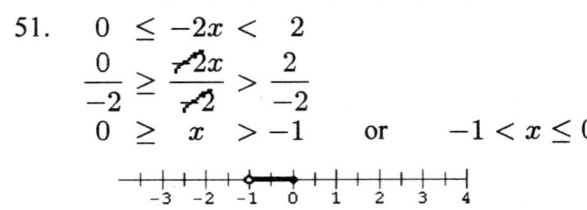

53. $\quad\quad -5 < -x < -1$
$(-1)(-5) > (-1)(-x) > (-1)(-1)$
$\quad\quad 5 > x > 1 \quad \text{or} \quad 1 < x < 5$

55. One number is less than a second number if the first number is located to the left of the second on the number line. (Equivalently, we could say that the second number is located to the right of the first on the number line.)

56. If we begin with an inequality, we may add the same quantity to both sides, subtract the same quantity from both sides, or multiply or divide both sides by the same positive quantity and get another inequality with the same symbol. If we multiply or divide both sides by the same negative quantity, we get another inequality with the reversed symbol.

57. When we multiply or divide both sides of an equality by the same nonzero quantity, it does not matter whether the quantity is positive or negative. That is not the case for inequalities (see Exercise 56).

58. $2 < x \leq 5$ means that x is between 2 and 5 on the number line, possibly equal to 5, but not equal to 2. That is, x is greater than 2 <u>and</u> less than or equal to 5.

59. $(2x^2)(3x^3) = (2 \cdot 3)(x^2 x^3) = 6x^5$

61. $4(x - 3y) + 2(3x - y) = 4x - 12y + 6x - 2y$
$\quad\quad\quad\quad\quad\quad\quad\quad = 4x + 6x - 12y - 2y$
$\quad\quad\quad\quad\quad\quad\quad\quad = 10x - 14y$

63. $x^2(x^3 - 4x) - (2x^5 - 4x^3) = x^5 - 4x^3 - 2x^5 + 4x^3$
$\quad\quad\quad\quad\quad\quad\quad\quad\quad\quad = x^5 - 2x^5 - 4x^3 + 4x^3$
$\quad\quad\quad\quad\quad\quad\quad\quad\quad\quad = -x^5$

Exercises 3.5

1. $x + 5 < 3$
   $\underline{-5 \phantom{<} -5}$
   $x < -2$

3. $a - 2 > -3$
   $\underline{+2 +2}$
   $a > -1$

5. $2y < 7$
   $\dfrac{\cancel{2}y}{\cancel{2}} < \dfrac{7}{2}$
   $y < \dfrac{7}{2}$

7. $2y > -7$
   $\dfrac{\cancel{2}y}{\cancel{2}} > \dfrac{-7}{2}$
   $y > -\dfrac{7}{2}$

9. $-2y < 7$
   $\dfrac{\cancel{-2}y}{\cancel{-2}} > \dfrac{7}{-2}$
   $y > -\dfrac{7}{2}$

11. $-2y > -7$
    $\dfrac{\cancel{-2}y}{\cancel{-2}} < \dfrac{-7}{-2}$
    $y < \dfrac{7}{2}$

13. $-x < 4$
    $\dfrac{-x}{-1} > \dfrac{4}{-1}$
    $x > -4$
    or
    $-x < 4$
    $(-1)(-x) > (-1)(4)$
    $x > -4$

15. $-1 > -y$
    $\dfrac{-1}{-1} < \dfrac{-y}{-1}$
    $1 < y$
    or
    $-1 > -y$
    $(-1)(-1) < (-1)(-y)$
    $1 < y$

17. $5x + 3 \leq 8$
    $\underline{-3 -3}$
    $5x \leq 5$
    $\dfrac{\cancel{5}x}{\cancel{5}} \leq \dfrac{5}{5}$
    $x \leq 1$

19. $2x - 9 \geq 16$
    $\underline{+9 +9}$
    $2x \geq 25$
    $\dfrac{\cancel{2}x}{\cancel{2}} \geq \dfrac{25}{2}$
    $x \geq \dfrac{25}{2}$

21. $2(z - 3) + 4 \geq -6$
    $2z - 6 + 4 \geq -6$
    $2z - 2 \geq -6$
    $\underline{+2 +2}$
    $2z \geq -4$
    $\dfrac{\cancel{2}z}{\cancel{2}} \geq \dfrac{-4}{2}$
    $z \geq -2$

23. $3(x + 4) + 2(x - 1) < 20$
    $3x + 12 + 2x - 2 < 20$
    $5x + 10 < 20$
    $\underline{-10 \phantom{<} -10}$
    $5x < 10$
    $\dfrac{\cancel{5}x}{\cancel{5}} < \dfrac{10}{5}$
    $x < 2$

25. $5(w+3) - 7w \leq 7$
$5w + 15 - 7w \leq 7$
$-2w + 15 \leq 7$
$ -15 -15$
$\overline{-2w \leq -8}$
$\dfrac{\cancel{-2}w}{\cancel{-2}} \geq \dfrac{-8}{-2}$
$w \geq 4$

27. $3(a+4) - 4(a-1) < 10$
$3a + 12 - 4a + 4 < 10$
$-a + 16 < 10$
$ -16 \phantom{<} -16$
$\overline{-a < -6}$
$\dfrac{-a}{-1} > \dfrac{-6}{-1}$
$a > 6$

29. $4(y-3) - (3y - 12) \geq 2$
$4y - 12 - 3y + 12 \geq 2$
$y \geq 2$

31. $2(u+2) - 2(u-1) < 5$
$2u + 4 - 2u + 2 < 5$
$6 < 5$
Contradiction

33. $4(x-2) - (4x - 3) < 6$
$4x - 8 - 4x + 3 < 6$
$-5 < 6$
Identity

35. $x + 3 < 2x + 7$
$-x \phantom{+3<2} -x$
$\overline{3 < x + 7}$
$\phantom{-x+3<} -7 -7$
$\overline{-4 < x}$

37. $5t - 3 \geq 3t + 10$
$-3t -3t$
$\overline{2t - 3 \geq 10}$
$+3 +3$
$\overline{2t \geq 13}$
$\dfrac{\cancel{2}t}{\cancel{2}} \geq \dfrac{13}{2}$
$t \geq \dfrac{13}{2}$

39. $2(a-5) + 3a > 6a - 6$
$2a - 10 + 3a > 6a - 6$
$5a - 10 > 6a - 6$
$-5a -5a$
$\overline{-10 > a - 6}$
$+6 +6$
$\overline{-4 > a}$

41. $4(w+2) - 3(w-1) > 5(w-1) - 5w$
$4w + 8 - 3w + 3 > 5w - 5 - 5w$
$w + 11 > -5$
$-11 -11$
$\overline{w > -16}$

43. $2y - 4(4y + 1) \leq 8 - (y + 2)$
$2y - 4y - 4 \leq 6 - y - 2$
$-2y - 4 \leq 6 - y$
$+2y +2y$
$\overline{-4 \leq 6 + y}$
$-6 -6$
$\overline{-10 \leq y}$

45. $2 < x + 7 < 10$
$-7 \phantom{<x} -7 \phantom{<} -7$
$\overline{-5 < x < 3}$

47. $3 < 2a + 5 < 7$
$-5 \phantom{<2a} -5 \phantom{<} -5$
$\overline{-2 < 2a < 2}$
$\dfrac{-2}{2} < \dfrac{\cancel{2}a}{\cancel{2}} < \dfrac{2}{2}$
$-1 < a < 1$

49. $\begin{aligned} -5 \leq -4y+3 &< 7 \\ -3 -3 &\phantom{<} -3 \\ \hline -8 \leq -4y &< 4 \\ \frac{-8}{-4} \geq \frac{-4y}{-4} &> \frac{4}{-4} \\ 2 \geq y &> -1 \\ \text{or} \\ -1 < y &\leq 2 \end{aligned}$

51. $\begin{aligned} 1 \leq 6-x &< 3 \\ -6 -6 &\phantom{<} -6 \\ \hline -5 \leq -x &< -3 \\ \frac{-5}{-1} \geq \frac{-x}{-1} &> \frac{-3}{-1} \\ 5 \geq x &> 3 \\ \text{or} \\ 3 < x &\leq 5 \end{aligned}$

53. $\begin{aligned} x+4 &< 2x-1 \\ -x &\phantom{<} -x \\ \hline 4 &< x-1 \\ +1 &\phantom{<} +1 \\ \hline 5 &< x \end{aligned}$

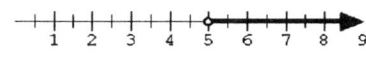

55. $\begin{aligned} 3(a+2) - 5a &\geq 2-a \\ 3a+6-5a &\geq 2-a \\ -2a+6 &\geq 2-a \\ +2a & +2a \\ \hline 6 &\geq 2+a \\ -2 & -2 \\ \hline 4 &\geq a \end{aligned}$

57. $\begin{aligned} -1 < x+3 &< 2 \\ -3 -3 &\phantom{<} -3 \\ \hline -4 < x &< -1 \end{aligned}$

59. $\begin{aligned} -1 \leq y-3 &\leq 2 \\ +3 +3 & +3 \\ \hline 2 \leq y &\leq 5 \end{aligned}$

61. $\begin{aligned} -3 \leq 4t+5 &< 9 \\ -5 -5 &\phantom{<} -5 \\ \hline -8 \leq 4t &< 4 \\ \frac{-8}{4} \leq \frac{4t}{4} &< \frac{4}{4} \\ -2 \leq t &< 1 \end{aligned}$

63. $\begin{aligned} -5 < 3-2x &\leq 9 \\ -3 -3 & -3 \\ \hline -8 < -2x &\leq 6 \\ \frac{-8}{-2} > \frac{-2x}{-2} &\geq \frac{6}{-2} \\ 4 > x &\geq -3 \\ \text{or} \\ -3 \leq x &< 4 \end{aligned}$

65. $\begin{aligned} 0.8 - 0.45(x-2) &\leq 0.26 \\ 0.8 - 0.45x + 0.9 &\leq 0.26 \\ -0.45x + 1.7 &\leq 0.26 \\ -1.7 & -1.7 \\ \hline -0.45x &\leq -1.44 \\ \frac{-0.45x}{-0.45} &\geq \frac{-1.44}{-0.45} \\ x &\geq 3.2 \end{aligned}$

67. $\begin{aligned} 5.468 < 2.9t - 12.86 &< 20.519 \\ +12.86 +12.86 &\phantom{<} +12.86 \\ \hline 18.328 < 2.9t &< 33.379 \\ \frac{18.328}{2.9} < \frac{2.9t}{2.9} &< \frac{33.379}{2.9} \\ 6.32 < t &< 11.51 \end{aligned}$

69. $7 > x$ 71. $t - 10$ 73. $x \geq 5$ 75. $y \leq 2$

77. Let $n =$ the number.
$$3n - 4 < 17$$
$$ +4 \phantom{<} +4$$
$$\overline{3n < 21}$$
$$\frac{\cancel{3}n}{\cancel{3}} < \frac{21}{3}$$
$$n < 7$$
All numbers less than 7 satisfy the condition.

79. Let $n =$ the number.
$$6n + 12 > 3n$$
$$-6n -6n$$
$$\overline{12 > -3n}$$
$$\frac{12}{-3} < \frac{\cancel{-}3n}{\cancel{-}3}$$
$$-4 < n$$
The number must be larger than -4.

81. Let $l =$ the length.
$$2l + 2(8) \geq 80$$
$$2l + 16 \geq 80$$
$$ -16 -16$$
$$\overline{2l \geq 64}$$
$$\frac{\cancel{2}l}{\cancel{2}} \geq \frac{64}{2}$$
$$l \geq 32$$
The length must be at least 32 cm.

83. Let $w =$ the width.
$$50 \leq 2w + 2(18) \leq 70$$
$$50 \leq 2w + 36 \leq 70$$
$$-36 -36 -36$$
$$\overline{14 \leq 2w \leq 34}$$
$$\frac{14}{2} \leq \frac{\cancel{2}w}{\cancel{2}} \leq \frac{34}{2}$$
$$7 \leq w \leq 17$$
The width is at least 7 in. and at most 17 in.

85. Let $d =$ the price of a ticket at the door. Then $d + 2 =$ price of a reserved seat.
$$150d + 300(d + 2) \geq 3750$$
$$150d + 300d + 600 \geq 3750$$
$$450d + 600 \geq 3750$$
$$ -600 -600$$
$$\overline{450d \geq 3150}$$
$$\frac{\cancel{450}d}{\cancel{450}} \geq \frac{3150}{450}$$
$$d \geq 7$$
So $d + 2 \geq 9$
The minimum price of a reserved seat is $9.

87. Let $s =$ annual sales (in dollars).
$$0.016(s - 82,000) \geq 1800$$
$$0.016s - 0.016(82,000) \geq 1800$$
$$0.016s - 1312 \geq 1800$$
$$+1312 +1312$$
$$\overline{0.016s \geq 3112}$$
$$\frac{\cancel{0.016}s}{\cancel{0.016}} \geq \frac{3112}{0.016}$$
$$s \geq 194,500$$
Marc's annual sales must be at least $194,500.

89. (a) Starting with $2 \leq -2x$, we must divide both sides of this inequality by -2, obtaining
$$\frac{2}{-2} \geq \frac{\cancel{-}2x}{\cancel{-}2} \text{ or } -1 \geq x.$$

(b) When we subtract 4 from both sides of the original inequality, the resulting inequality should read $2x < -2$, not $2x > -2$. (We only reverse the inequality sign when we multiply or divide by a negative quantity.)

(c) We must divide both sides of the inequality $-9 > 3x$ by 3 giving $\frac{-9}{3} > \frac{\cancel{3}x}{\cancel{3}}$ or $-3 > x$.

90. Only part (a) makes sense, since it requires that x be a number between -3 and 2. Parts (b), (c), (e) and (f) all lead to contradictions. For instance, the inequality $-5 < x < -8$ implies that $-5 < -8$, which is false. (the others are similar.) Part (d) makes no sense, since we cannot write double inequalities in which the inequality symbols point in opposite directions.

Chapter 3 Review Exercises

1. $5(x-4) - 3(x-3) = 3 - (14 - 2x)$
    $5x - 20 - 3x + 9 = 3 - 14 + 2x$
    $ 2x - 11 = 2x - 11$
    $\underline{-2x -2x}$
    $-11 = -11$
    Identity

3. $3a(a+3) - a(2a+4) = a^2 + 10$
    $3a^2 + 9a - 2a^2 - 4a = a^2 + 10$
    $a^2 + 5a = a^2 + 10$
    $\underline{-a^2 -a^2}$
    $5a = 10$
    $\dfrac{\cancel{5}a}{\cancel{5}} = \dfrac{10}{5}$
    $a = 2$
    Conditional equation

5. $2x - 5 = -7$

    CHECK $x = -6$:
    $2(-6) - 5 \stackrel{?}{=} -7$
    $-12 - 5 \stackrel{?}{=} -7$
    $-17 \neq -7$
    So -6 does not satisfy the equation.

    CHECK $x = -1$:
    $2(-1) - 5 \stackrel{?}{=} -7$
    $-2 - 5 \stackrel{?}{=} -7$
    $-7 \stackrel{\checkmark}{=} -7$
    So -1 satisfies the equation.

7. $4y + 3 \leq 10 + 2y$

    CHECK $y = 4$:
    $4(4) + 3 \stackrel{?}{\leq} 10 + 2(4)$
    $16 + 3 \stackrel{?}{\leq} 10 + 8$
    $19 \not\leq 18$
    So 4 does not satisfy the inequality.

    CHECK $y = \dfrac{5}{2}$:
    $4\left(\dfrac{5}{2}\right) + 3 \stackrel{?}{\leq} 10 + 2\left(\dfrac{5}{2}\right)$
    $10 + 3 \stackrel{?}{\leq} 10 + 5$
    $13 \stackrel{\checkmark}{\leq} 15$
    So $\dfrac{5}{2}$ satisfies the inequality.

9. $3t + 2(t + 1) = 3t + 3$

    CHECK $t = -2$:
    $3(-2) + 2[(-2) + 1] \stackrel{?}{=} 3(-2) + 3$
    $3(-2) + 2(-1) \stackrel{?}{=} 3(-2) + 3$
    $-6 - 2 \stackrel{?}{=} -6 + 3$
    $-8 \neq -3$
    So -2 does not satisfy the equation.

    CHECK $t = \dfrac{1}{2}$:
    $3\left(\dfrac{1}{2}\right) + 2\left[\left(\dfrac{1}{2}\right) + 1\right] \stackrel{?}{=} 3\left(\dfrac{1}{2}\right) + 3$
    $3\left(\dfrac{1}{2}\right) + 2\left(\dfrac{3}{2}\right) \stackrel{?}{=} 3\left(\dfrac{1}{2}\right) + 3$
    $\dfrac{3}{2} + 3 \stackrel{?}{=} \dfrac{3}{2} + 3$
    $\dfrac{9}{2} \stackrel{\checkmark}{=} \dfrac{9}{2}$
    So $\dfrac{1}{2}$ satisfies the equation.

11. $8 - 3(x - 2) > x - 4$
CHECK $x = -5$:
$$8 - 3[(-5) - 2] \stackrel{?}{>} (-5) - 4$$
$$8 - 3(-7) \stackrel{?}{>} -5 - 4$$
$$8 + 21 \stackrel{?}{>} -5 - 4$$
$$29 \stackrel{\checkmark}{>} -9$$
So -5 satisfies the inequality.

CHECK $x = 5$:
$$8 - 3[(5) - 2] \stackrel{?}{>} (5) - 4$$
$$8 - 3(3) \stackrel{?}{>} 5 - 4$$
$$8 - 9 \stackrel{?}{>} 5 - 4$$
$$-1 \not> 1$$
So 5 does not satisfy the inequality.

13. $a^2 + (a - 2)^2 = 20$
CHECK $a = -2$:
$$(-2)^2 + [(-2) - 2]^2 \stackrel{?}{=} 20$$
$$(-2)^2 + (-4)^2 \stackrel{?}{=} 20$$
$$4 + 16 \stackrel{?}{=} 20$$
$$20 \stackrel{\checkmark}{=} 20$$
So -2 satisfies the equation.

CHECK $a = 2$:
$$(2)^2 + [(2) - 2]^2 \stackrel{?}{=} 20$$
$$(2)^2 + (0)^2 \stackrel{?}{=} 20$$
$$4 + 0 \stackrel{?}{=} 20$$
$$4 \neq 20$$
So 2 does not satisfy the equation.

15.
$$\begin{aligned} 5x + 8 &= 2x - 7 \\ -2x & \quad\quad -2x \\ \hline 3x + 8 &= -7 \\ -8 & \quad\quad -8 \\ \hline 3x &= -15 \\ \frac{3x}{3} &= \frac{-15}{3} \\ x &= -5 \end{aligned}$$

17.
$$2(y + 4) - 2y = 8$$
$$2y + 8 - 2y = 8$$
$$8 = 8$$
Identity

19.
$$\begin{aligned} 2(3a + 4) + 8 &= 5(3a - 1) \\ 6a + 8 + 8 &= 15a - 5 \\ 6a + 16 &= 15a - 5 \\ -6a & \quad\quad -6a \\ \hline 16 &= 9a - 5 \\ +5 & \quad\quad +5 \\ \hline 21 &= 9a \\ \frac{21}{9} &= \frac{9a}{9} \\ \frac{7}{3} &= a \end{aligned}$$

21.
$$\begin{aligned} 8x - 3(x - 4) &= 4(x + 3) + 28 \\ 8x - 3x + 12 &= 4x + 12 + 28 \\ 5x + 12 &= 4x + 40 \\ -4x & \quad\quad -4x \\ \hline x + 12 &= 40 \\ -12 & \quad\quad -12 \\ \hline x &= 28 \end{aligned}$$

23.
$$a(a+3) - 2(a-1) = a(a-1) + 7$$
$$a^2 + 3a - 2a + 2 = a^2 - a + 7$$
$$a^2 + a + 2 = a^2 - a + 7$$
$$\underline{-a^2 \qquad\qquad -a^2}$$
$$a + 2 = -a + 7$$
$$\underline{+a \qquad\qquad +a}$$
$$2a + 2 = 7$$
$$\underline{-2 \qquad -2}$$
$$2a = 5$$
$$\frac{\cancel{2}a}{\cancel{2}} = \frac{5}{2}$$
$$a = \frac{5}{2}$$

25.
$$8 - 3(x-1) < 2$$
$$8 - 3x + 3 < 2$$
$$-3x + 11 < 2$$
$$\underline{\quad -11 \quad -11}$$
$$-3x < -9$$
$$\frac{\cancel{-3}x}{\cancel{-3}} > \frac{-9}{-3}$$
$$x > 3$$

27.
$$2(x-3) - 4(x-1) \geq 7 - x$$
$$2x - 6 - 4x + 4 \geq 7 - x$$
$$-2x - 2 \geq 7 - x$$
$$\underline{+2x \qquad\qquad +2x}$$
$$-2 \geq 7 + x$$
$$\underline{-7 \quad -7}$$
$$-9 \geq x$$

<--|--+--+--+--+--+--+--|-->
-10 -8 -6 -4 -2   0

29.
$$2 \leq 3a + 8 < 20$$
$$\underline{-8 \qquad -8 \quad -8}$$
$$-6 \leq 3a < 12$$
$$\frac{-6}{3} \leq \frac{\cancel{3}a}{\cancel{3}} < \frac{12}{3}$$
$$-2 \leq a < 4$$

<--+--+--+--+--+--+--o--+-->
 -2 -1   0   1   2   3   4   5

31. Let x = one of the numbers.
Then $2x - 3$ = the other number.
$$x + (2x - 3) = 18$$
$$3x - 3 = 18$$
$$\underline{\quad +3 \quad +3}$$
$$3x = 21$$
$$\frac{\cancel{3}x}{\cancel{3}} = \frac{21}{3}$$
$$x = 7$$

Then $2x - 3 = 2(7) - 3 = 14 - 3 = 11$.
Thus, the numbers are 7 and 11.

CHECK: 11 is 3 less than twice 7.
The sum of 7 and 11 is 18.

33. Let W = width of the rectangle.
Then $5W + 4$ = length of the rectangle.
$$2W + 2(5W + 4) = 80$$
$$2W + 10W + 8 = 80$$
$$12W + 8 = 80$$
$$\underline{\quad -8 \quad -8}$$
$$12W = 72$$
$$\frac{\cancel{12}W}{\cancel{12}} = \frac{72}{12}$$
$$W = 6$$

Then $5W + 4 = 5(6) + 4 = 30 + 4 = 34$.
Thus, the dimensions of the rectangle are 6 cm. and 34 cm.

CHECK: 34 is 4 more than 5 times 6.
$2(6) + 2(34) = 80$.

35. Let $x=$ number of first quality skirts
Then $150-x=$ number of irregular skirts
$$\begin{aligned} 15x+9(150-x) &= 2010 \\ 15x+1350-9x &= 2010 \\ 6x+1350 &= 2010 \\ -1350 & -1350 \\ \hline 6x &= 660 \\ \frac{\cancel{6}x}{\cancel{6}} &= \frac{660}{6} \\ x &= 110 \end{aligned}$$

Then $150-x = 150-(110) = 40$. Thus, the wholesaler bought 110 first quality skirts and 40 irregular skirts.

CHECK: 110 first quality skirts cost $15(110) = \$1650$. 40 irregular skirts cost $9(40) = \$360$. The total cost of the skirts is $\$1650 + \$360 = \$2010$.

37. Let $x=$ number of overtime hours the laborer must work.
$$\begin{aligned} 12(40)+18x &\geq 570 \\ 480+18x &\geq 570 \\ -480 & -480 \\ \hline 18x &\geq 90 \\ \frac{\cancel{18}x}{\cancel{18}} &\geq \frac{90}{18} \\ x &\geq 5 \end{aligned}$$

Thus, the minimum number of overtime hours that the laborer must work is 5.

Chapter 3 Practice Test

1. (a) $3x-5(x-2) = -2x+8$
$$\begin{aligned} 3x-5x+10 &= -2x+8 \\ -2x+10 &= -2x+8 \\ +2x & +2x \\ \hline 10 &= 8 \end{aligned}$$
Contradiction

(b) $3x-5(x-2) = -2x+10$
$$\begin{aligned} 3x-5x+10 &= -2x+10 \\ -2x+10 &= -2x+10 \\ +2x & +2x \\ \hline 10 &= 10 \end{aligned}$$
Identity

(c) $3x-5(x-2) = 2x-10$
$$\begin{aligned} 3x-5x+10 &= 2x-10 \\ -2x+10 &= 2x-10 \\ +2x & +2x \\ \hline 10 &= 4x-10 \\ +10 & +10 \\ \hline 20 &= 4x \\ \frac{20}{4} &= \frac{\cancel{4}x}{\cancel{4}} \\ 5 &= x \end{aligned}$$
Conditional

3. (a) $\quad\begin{aligned} 6 - 3x &= 3x - 10 \\ +3x &+3x \\ \hline 6 &= 6x - 10 \\ +10 &+10 \\ \hline 16 &= 6x \\ \frac{16}{6} &= \frac{\cancel{6}x}{\cancel{6}} \\ \tfrac{8}{3} &= x \end{aligned}$

(b) $\quad\begin{aligned} 2(3y - 5) - 4y &= 2 - (y + 12) \\ 6y - 10 - 4y &= 2 - y - 12 \\ 2y - 10 &= -y - 10 \\ +y &+y \\ \hline 3y - 10 &= -10 \\ +10 &+10 \\ \hline 3y &= 0 \\ \frac{\cancel{3}y}{\cancel{3}} &= \frac{0}{3} \\ y &= 0 \end{aligned}$

(c) $\quad\begin{aligned} 2a^2 - 3(a - 4) &= 2a(a - 6) + 3a \\ 2a^2 - 3a + 12 &= 2a^2 - 12a + 3a \\ 2a^2 - 3a + 12 &= 2a^2 - 9a \\ -2a^2 &-2a^2 \\ \hline -3a + 12 &= -9a \\ +3a &+3a \\ \hline 12 &= -6a \\ \frac{12}{-6} &= \frac{\cancel{-6}a}{\cancel{-6}} \\ -2 &= a \end{aligned}$

(d) $\quad\begin{aligned} 9 - 5(x - 2) &\geq 4 \\ 9 - 5x + 10 &\geq 4 \\ -5x + 19 &\geq 4 \\ -19 &-19 \\ \hline -5x &\geq -15 \\ \frac{\cancel{-5}x}{\cancel{-5}} &\leq \frac{-15}{-5} \\ x &\leq 3 \end{aligned}$

number line: arrow left from 3, closed at 3, marks -2 to 5

(e) $\quad\begin{aligned} 1 < 3 &- x \leq 5 \\ -3 \phantom{<} -3 & -3 \\ \hline -2 < & -x \leq 2 \\ \frac{-2}{-1} > \frac{-x}{-1} &\geq \frac{2}{-1} \\ 2 > x &\geq -2 \end{aligned}$

or

$-2 \leq x < 2$

number line: closed at -2, open at 2, marks -3 to 4

5. Let $x =$ number of used cassettes.
Then $20 - x =$ number of new cassettes.
$\begin{aligned} 1x + 3(20 - x) &= 46 \\ x + 60 - 3x &= 46 \\ -2x + 60 &= 46 \\ -60 &-60 \\ \hline -2x &= -14 \\ \frac{\cancel{-2}x}{\cancel{-2}} &= \frac{-14}{-2} \\ x &= 7 \end{aligned}$

Then $20 - x = 20 - 7 = 13$. Thus the person bought 7 used cassettes and 13 new cassettes.

CHECK: $7 + 13 \stackrel{\checkmark}{=} 20$.
7 cassettes at \$1 each cost \$7.
13 cassettes at \$3 each cost $13(3) = \$39$.
$\$7 + \$39 \stackrel{\checkmark}{=} \$46$

7. Let $x =$ sales (in dollars).
$\begin{aligned} 0.0125(x - 10{,}000) &\geq 1500 \\ 0.0125x - 125 &\geq 1500 \\ +125 &+125 \\ \hline 0.0125x &\geq 1625 \\ \frac{\cancel{0.0125}x}{\cancel{0.0125}} &\geq \frac{1625}{0.0125} \\ x &\geq 130{,}000 \end{aligned}$

Jan's sales must be at least \$130,000.

Chapters 1 - 3
Cumulative Review

1. $-8 - 5 - 7 = -13 - 7 = -20$

3. $12 - 4(3 - 5) = 12 - 4(-2)$
$= 12 + 8$
$= 20$

5. $-5^2 = -(5 \cdot 5) = -25$

7. $xx^2x^3 = x^{1+2+3} = x^6$

9. $x^2y - 2xy^2 - xy^2 - 3x^2y = x^2y - 3x^2y - 2xy^2 - xy^2$
$= (1 - 3)x^2y + (-2 - 1)xy^2$
$= -2x^2y - 3xy^2$

11. $2x(3x^2 - 4y) = 2x(3x^2) - 2x(4y)$
$= 6x^3 - 8xy$

13. $-3u^2(u^3)(-5v) = [(-3)(-5)](u^2u^3)v$
$= 15u^5v$

15. $4(m - 3n) + 3(2m - n) = 4m - 12n + 6m - 3n$
$= 4m + 6m - 12n - 3n$
$= 10m - 15n$

17. $2ab(a^2 - ab) - 4a^2(ab - b^2) = 2a^3b - 2a^2b^2 - 4a^3b + 4a^2b^2$
$= 2a^3b - 4a^3b - 2a^2b^2 + 4a^2b^2 = (2 - 4)a^3b + (-2 + 4)a^2b^2$
$= -2a^3b + 2a^2b^2$

19. $x^2y - xy^2 - (xy^2 - x^2y) = x^2y - xy^2 - xy^2 + x^2y$
$= x^2y + x^2y - xy^2 - xy^2 = (1 + 1)x^2y + (-1 - 1)xy^2$
$= 2x^2y - 2xy^2$

21. $x - 3[x - 4(x - 5)] = x - 3[x - 4x + 20]$
$= x - 3[-3x + 20] = x + 9x - 60$
$= 10x - 60$

23. $3xy(4x^3y - 2y) - 2x(3y^2)(2x^3) = 12x^4y^2 - 6xy^2 - 12x^4y^2$
$= 12x^4y^2 - 12x^4y^2 - 6xy^2 = (12 - 12)x^4y^2 - 6xy^2$
$= -6xy^2$

25. $x^2 = (-2)^2 = 4$

27. $xy^2 - (xy)^2 = (-2)(-3)^2 - [(-2)(-3)]^2$
$= -2(9) - (6)^2 = -18 - 36$
$= -54$

29. $|x - y - z| = |(-2) - (-3) - (5)|$
$= |-2 + 3 - 5| = |-4|$
$= 4$

31. $2x - 4y^2 = 2(-2) - 4(-3)^2$
$= 2(-2) - 4(9) = -4 - 36$
$= -40$

33. $$\begin{aligned} 2x + 11 &= 5x + 10 \\ -2x & -2x \\ \hline 11 &= 3x + 10 \\ -10 & -10 \\ \hline 1 &= 3x \\ \frac{1}{3} &= \frac{3x}{3} \\ \frac{1}{3} &= x \end{aligned}$$

35. $$\begin{aligned} 9(a+1) - 3(2a-2) &= 12 \\ 9a + 9 - 6a + 6 &= 12 \\ 3a + 15 &= 12 \\ -15 & -15 \\ \hline 3a &= -3 \\ \frac{3a}{3} &= \frac{-3}{3} \\ a &= -1 \end{aligned}$$

37. $$\begin{aligned} 4(5 - x) - 2(6 - 2x) &= 8 \\ 20 - 4x - 12 + 4x &= 8 \\ 8 &= 8 \end{aligned}$$
Identity

39. $$\begin{aligned} 2(s+4) + 3(2s+2) &> 6s \\ 2s + 8 + 6s + 6 &> 6s \\ 8s + 14 &> 6s \\ -8s & -8s \\ \hline 14 &> -2s \\ \frac{14}{-2} &< \frac{-2s}{-2} \\ -7 &< s \end{aligned}$$

[number line from -8 to 0 with open circle at -7, shaded right]

41. $$\begin{aligned} 1 < 2y - 5 &\leq 3 \\ +5 \phantom{<} +5 & +5 \\ \hline 6 < 2y &\leq 8 \\ \frac{6}{2} < \frac{2y}{2} &\leq \frac{8}{2} \\ 3 < y &\leq 4 \end{aligned}$$

[number line from -1 to 6 with open circle at 3, closed circle at 4]

43. $$\begin{aligned} 4(3d - 2) + 6(8 - d) &= 9d - 4(d - 10) \\ 12d - 8 + 48 - 6d &= 9d - 4d + 40 \\ 6d + 40 &= 5d + 40 \\ -5d & -5d \\ \hline d + 40 &= 40 \\ -40 & -40 \\ \hline d &= 0 \end{aligned}$$

45. Let x = the number
$$\begin{aligned} 2x - 8 &\geq x + 4 \\ -x & -x \\ \hline x - 8 &\geq 4 \\ +8 & +8 \\ \hline x &\geq 12 \end{aligned}$$
The number is at least 12.

47. Let x = number of danishes that Louise bought.
$18 - x$ = number of pastries that Louise bought.
$$\begin{aligned} 40x + 55(18 - x) &= 825 \\ 40x + 990 - 55x &= 825 \\ -15x + 990 &= 825 \\ -990 & -990 \\ \hline -15x &= -165 \\ \frac{-15x}{-15} &= \frac{-165}{-15} \\ x &= 11 \end{aligned}$$

Then $18 - x = 18 - 11 = 7$.
Thus, Louise bought 11 danishes and 7 pastries.

CHECK: 11 danishes cost $11(\$0.40) = \$4.40$. 7 pastries cost $7(\$0.55) = \$3.85$. The total cost of the assortment is $\$4.40 + \$3.85 = \$8.25$

Chapters 1 - 3
Cumulative Practice Test

1. (a) $-3^4 + 3(-2)^3 = -(3 \cdot 3 \cdot 3 \cdot 3) + 3((-2)(-2)(-2))$
   $= -81 + 3(-8) = -81 - 24$
   $= -105$

   (b) $(3 - 7 - 2)^2 = (-4 - 2)^2 = (-6)^2$
   $= (-6)(-6) = 36$

3. (a) $5x^2 - 4x - 8 - 7x^2 - x + 11 = 5x^2 - 7x^2 - 4x - x - 8 + 11$
   $= -2x^2 - 5x + 3$

   (b) $2(x - 3y) + 5(y - x) = 2x - 6y + 5y - 5x = 2x - 5x - 6y + 5y$
   $= -3x - y$

   (c) $3a(a^2 - 2b) - 5(a^3 - ab) = 3a^3 - 6ab - 5a^3 + 5ab = 3a^3 - 5a^3 - 6ab + 5ab$
   $= -2a^3 - ab$

   (d) $2(u - 4v) - 3(v - 2u) - (8u - 11v) = 2u - 8v - 3v + 6u - 8u + 11v$
   $= 2u + 6u - 8u - 8v - 3v + 11v$
   $= 0u + 0v = 0$

   (e) $4x^2y^3(x - 5y) - 2xy(5y^2)(-2xy) = 4x^3y^3 - 20x^2y^4 + 20x^2y^4$
   $= 4x^3y^3$

   (f) $6 - a[6 - a(6 - a)] = 6 - a[6 - 6a + a^2]$
   $= 6 - 6a + 6a^2 - a^3$

5. (a) $7 - 3a \geq 13$
   $-7 -7$
   $\overline{-3a \geq 6}$
   $\dfrac{\cancel{-3}a}{\cancel{-3}} \leq \dfrac{6}{-3}$
   $a \leq -2$

   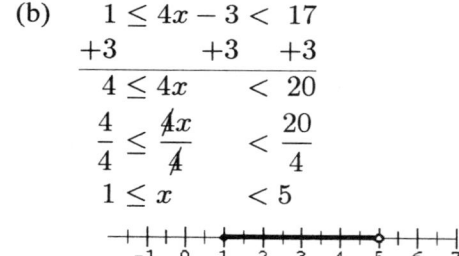

   (b) $1 \leq 4x - 3 < 17$
   $+3 +3 +3$
   $\overline{4 \leq 4x < 20}$
   $\dfrac{4}{4} \leq \dfrac{\cancel{4}x}{\cancel{4}} < \dfrac{20}{4}$
   $1 \leq x < 5$

Chapter 4
Rational Expressions

Exercises 4.1

1. $\dfrac{18}{30} = \dfrac{3 \cdot \cancel{6}}{5 \cdot \cancel{6}} = \dfrac{3}{5}$

3. $\dfrac{-9}{21} = \dfrac{-3 \cdot \cancel{3}}{7 \cdot \cancel{3}} = \dfrac{-3}{7} = -\dfrac{3}{7}$

5. $\dfrac{-15}{-6} = \dfrac{5(\cancel{-3})}{2(\cancel{-3})} = \dfrac{5}{2}$

7. $\dfrac{-5+8}{10-4} = \dfrac{3}{6} = \dfrac{1 \cdot \cancel{3}}{2 \cdot \cancel{3}} = \dfrac{1}{2}$

9. $\dfrac{8-5(2)}{6-4(3)} = \dfrac{8-10}{6-12} = \dfrac{-2}{-6}$
$= \dfrac{1(\cancel{-2})}{3(\cancel{-2})} = \dfrac{1}{3}$

11. $\dfrac{x^3}{x} = \dfrac{x^2 \cdot \cancel{x}}{1 \cdot \cancel{x}} = \dfrac{x^2}{1} = x^2$

13. $\dfrac{x}{x^3} = \dfrac{1 \cdot \cancel{x}}{x^2 \cdot \cancel{x}} = \dfrac{1}{x^2}$

15. $\dfrac{10x}{4x^2} = \dfrac{5 \cdot \cancel{2x}}{2x \cdot \cancel{2x}} = \dfrac{5}{2x}$

17. $\dfrac{-3z^6}{5z^2} = \dfrac{-3z^4 \cdot \cancel{z^2}}{5 \cdot \cancel{z^2}} = \dfrac{-3z^4}{5} = -\dfrac{3z^4}{5}$

19. $\dfrac{12t^5}{30t^{10}} = \dfrac{2 \cdot \cancel{6t^5}}{5t^5 \cdot \cancel{6t^5}} = \dfrac{2}{5t^5}$

21. $\dfrac{6ab^5}{-2a^3b^2} = \dfrac{3b^3 \cdot \cancel{2ab^2}}{-a^2 \cdot \cancel{2ab^2}} = \dfrac{3b^3}{-a^2} = -\dfrac{3b^3}{a^2}$

23. $\dfrac{(2x^3)(6x^2)}{(4x)(3x^4)} = \dfrac{12x^5}{12x^5} = \dfrac{1 \cdot \cancel{12x^5}}{1 \cdot \cancel{12x^5}} = \dfrac{1}{1} = 1$

25. $\dfrac{(r^3t^2)(-rt^3)}{2r^2t^7} = \dfrac{-r^4t^5}{2r^2t^7} = \dfrac{-r^2 \cdot \cancel{r^2t^5}}{2t^2 \cdot \cancel{r^2t^5}}$
$= \dfrac{-r^2}{2t^2} = -\dfrac{r^2}{2t^2}$

27. $\dfrac{3a(5b)(-4ab^3)}{6ab(2a^2b^2)} = \dfrac{-60a^2b^4}{12a^3b^3}$
$= \dfrac{-5b \cdot \cancel{12a^2b^3}}{a \cdot \cancel{12a^2b^3}}$
$= \dfrac{-5b}{a} = -\dfrac{5b}{a}$

29. $\dfrac{(2x)^5}{(4x)^3} = \dfrac{32x^5}{64x^3} = \dfrac{x^2 \cdot \cancel{32x^3}}{2 \cdot \cancel{32x^3}} = \dfrac{x^2}{2}$

31. $\dfrac{(-4x^3)^2}{(-2x^4)^3} = \dfrac{16x^6}{-8x^{12}} = \dfrac{2 \cdot \cancel{8x^6}}{-x^6 \cdot \cancel{8x^6}}$
$= \dfrac{2}{-x^6} = -\dfrac{2}{x^6}$

33. $\dfrac{(2xy^2)^3}{(4x^2y^3)^3} = \dfrac{8x^3y^6}{64x^6y^9} = \dfrac{1 \cdot \cancel{8x^3y^6}}{8x^3y^3 \cdot \cancel{8x^3y^6}}$
$= \dfrac{1}{8x^3y^3}$

35. $\dfrac{5x-2x}{10x-4x} = \dfrac{3x}{6x} = \dfrac{1 \cdot \cancel{3x}}{2 \cdot \cancel{3x}} = \dfrac{1}{2}$

37. $\dfrac{5a(2x)}{15a(8x)} = \dfrac{10ax}{120ax} = \dfrac{1 \cdot \cancel{10ax}}{12 \cdot \cancel{10ax}} = \dfrac{1}{12}$

39. $\dfrac{4s-3t}{8s-9t}$ cannot be reduced.

41. $\dfrac{7a^2 - 5a^2 - 6a^2}{4a - 8a} = \dfrac{-4a^2}{-4a} = \dfrac{a(\cancel{-4a})}{1(\cancel{-4a})}$
$= \dfrac{a}{1} = a$

43. $\dfrac{5x^2 - 3x - x^2 + 3x}{6x^2 - 5x - 2x^2 + 5x} = \dfrac{4x^2}{4x^2}$
$= \dfrac{1 \cdot \cancel{4x^2}}{1 \cdot \cancel{4x^2}} = \dfrac{1}{1} = 1$

45. The value of a fraction is unchanged when its numerator and denominator are both either multiplied or divided by the same nonzero quantity.

46. We would conclude that $2 = 4$. It was incorrect to write
$\dfrac{4 + \cancel{2}}{1 + \cancel{2}} = \dfrac{4}{1}$.
The 2's cannot be crossed out, since 2 is a common term, not a common factor.

47. (a) The factor of x in the reduced fraction should be in its numerator.
    (b) When both factors in the numerator are crossed out, a factor of 1 remains.
    (c) As in part (b), a factor of 1 remains in the numerator, so that the reduced fraction is $\dfrac{1}{3x^2 y}$.
    (d) Crossing out x's amounts to crossing out terms rather than factors.

48. The fraction in (a), (c), (d), and (e) are equivalent. So are the fractions in (b), (f), and (g).

49. Let $w =$ width of the rectangle. Then $2w + 5 =$ length of the rectangle.
$$\begin{aligned} 2w + 2(2w + 5) &= 34 \\ 2w + 4w + 10 &= 34 \\ 6w + 10 &= 34 \\ -10 &\quad -10 \\ \hline 6w &= 24 \\ \dfrac{\cancel{6}w}{\cancel{6}} &= \dfrac{24}{6} \\ w &= 4 \end{aligned}$$
Then $2w + 5 = 2(4) + 5 = 13$. Thus, the rectangle has a width of 4 cm and a length of 13 cm.

Exercises 4.2

1. $\dfrac{-4}{9} \cdot \dfrac{-2}{3} = \dfrac{(-4)(-2)}{9 \cdot 3} = \dfrac{8}{27}$

3. $\dfrac{\overset{-2}{\cancel{-6}}}{\underset{2}{\cancel{10}}} \cdot \dfrac{\overset{3}{\cancel{15}}}{\underset{3}{\cancel{9}}} = \dfrac{\overset{-1}{\cancel{-2}}}{\cancel{2}} \cdot \dfrac{\overset{1}{\cancel{3}}}{\cancel{3}} = \dfrac{-1}{1} = -1$

5. $\dfrac{2}{3y} \cdot \dfrac{x}{5} = \dfrac{2x}{3y \cdot 5} = \dfrac{2x}{15y}$

7. $\dfrac{x^2}{4y} \cdot \dfrac{5x}{3y} = \dfrac{x^2 \cdot 5x}{4y \cdot 3y} = \dfrac{5x^3}{12y^2}$

9. $\dfrac{4}{5} \div \dfrac{5}{4} = \dfrac{4}{5} \cdot \dfrac{4}{5} = \dfrac{4 \cdot 4}{5 \cdot 5} = \dfrac{16}{25}$

11. $\dfrac{6x}{y} \div \dfrac{y^2}{2x^2} = \dfrac{6x}{y} \cdot \dfrac{2x^2}{y^2} = \dfrac{6x \cdot 2x^2}{y \cdot y^2} = \dfrac{12x^3}{y^3}$

13. $\dfrac{\overset{1}{\cancel{3}}}{2\cancel{t}} \cdot \dfrac{\cancel{t}w}{\underset{2}{\cancel{6}}} = \dfrac{1 \cdot w}{2 \cdot 2} = \dfrac{w}{4}$

15. $4 \cdot \dfrac{x}{12} = \dfrac{\overset{1}{\cancel{4}}}{1} \cdot \dfrac{x}{\underset{3}{\cancel{12}}} = \dfrac{1 \cdot x}{1 \cdot 3} = \dfrac{x}{3}$

17. $4 \div \dfrac{x}{12} = \dfrac{4}{1} \div \dfrac{x}{12} = \dfrac{4}{1} \cdot \dfrac{12}{x}$
$= \dfrac{4 \cdot 12}{1 \cdot x} = \dfrac{48}{x}$

19. $\dfrac{x}{12} \div 4 = \dfrac{x}{12} \div \dfrac{4}{1} = \dfrac{x}{12} \cdot \dfrac{1}{4}$
$= \dfrac{x \cdot 1}{12 \cdot 4} = \dfrac{x}{48}$

21. $\dfrac{\cancel{2x}^{-1}}{\cancel{3y^2}_{y}} \cdot \dfrac{\cancel{9y}^{-3}}{\cancel{4x}_{2}} = \dfrac{(-1)(-3)}{2y} = \dfrac{3}{2y}$

23. $\dfrac{\cancel{m^3 n^2}^{m^2}}{\cancel{2m}_{1}} \cdot \dfrac{\cancel{6}^{3}}{\cancel{n^3}_{n}} = \dfrac{m^2 \cdot 3}{1 \cdot n} = \dfrac{3m^2}{n}$

25. $\dfrac{3uv^2}{5w} \div \dfrac{6u^2 v}{15w} = \dfrac{\cancel{3uv^2}^{1v}}{\cancel{5w}_{1}} \cdot \dfrac{\cancel{15w}^{3}}{\cancel{6u^2 v}_{2u}}$
$= \dfrac{v \cdot 3}{2 \cdot u} = \dfrac{3v}{2u}$

27. $6xy \cdot \dfrac{2x}{3y} = \dfrac{\cancel{6xy}^{2}}{1} \cdot \dfrac{2x}{\cancel{3y}_{1}} = \dfrac{2x \cdot 2x}{1 \cdot 1}$
$= \dfrac{4x^2}{1} = 4x^2$

29. $6xy \div \dfrac{2x}{3y} = \dfrac{\cancel{6xy}^{3}}{1} \cdot \dfrac{3y}{\cancel{2x}_{1}} = \dfrac{3y \cdot 3y}{1 \cdot 1}$
$= \dfrac{9y^2}{1} = 9y^2$

31. $\dfrac{2x}{3y} \div (6xy) = \dfrac{2x}{3y} \div \dfrac{6xy}{1} = \dfrac{\cancel{2x}^{1}}{3y} \cdot \dfrac{1}{\cancel{6xy}_{3}}$
$= \dfrac{1 \cdot 1}{3y \cdot 3y} = \dfrac{1}{9y^2}$

33. $\dfrac{\cancel{4x}^{-2}}{\cancel{9y}_{3}} \cdot \dfrac{x^2}{y^2} \cdot \dfrac{\cancel{3y}^{1}}{\cancel{2x}_{1}} = \dfrac{-2 \cdot x^2}{3 \cdot y^2}$
$= -\dfrac{2x^2}{3y^2}$

35. $\dfrac{9}{a^2} \cdot \left(\dfrac{a}{3} \div \dfrac{3}{a}\right) = \dfrac{9}{a^2} \cdot \left(\dfrac{a}{3} \cdot \dfrac{a}{3}\right)$
$= \dfrac{\cancel{9}^{1}}{\cancel{a^2}_{1}} \cdot \dfrac{\cancel{a^2}^{1}}{\cancel{9}_{1}} = \dfrac{1 \cdot 1}{1 \cdot 1} = \dfrac{1}{1} = 1$

37. $\dfrac{9}{a^2} \div \left(\dfrac{\cancel{a}^{1}}{\cancel{3}_{1}} \cdot \dfrac{\cancel{3}^{1}}{\cancel{a}_{1}}\right) = \dfrac{9}{a^2} \div \dfrac{1}{1} = \dfrac{9}{a^2} \cdot \dfrac{1}{1}$
$= \dfrac{9 \cdot 1}{a^2 \cdot 1} = \dfrac{9}{a^2}$

39. $\dfrac{9}{a^2} \div \left(\dfrac{a}{3} \div \dfrac{3}{a}\right) = \dfrac{9}{a^2} \div \left(\dfrac{a}{3} \cdot \dfrac{a}{3}\right)$
$= \dfrac{9}{a^2} \div \dfrac{a^2}{9} = \dfrac{9}{a^2} \cdot \dfrac{9}{a^2}$
$= \dfrac{9 \cdot 9}{a^2 \cdot a^2} = \dfrac{81}{a^4}$

41. $\left(\dfrac{9}{a^2} \div \dfrac{a}{3}\right) \div \dfrac{a}{3} = \left(\dfrac{9}{a^2} \cdot \dfrac{3}{a}\right) \div \dfrac{a}{3}$
$= \dfrac{27}{a^3} \div \dfrac{a}{3} = \dfrac{27}{a^3} \cdot \dfrac{3}{a}$
$= \dfrac{27 \cdot 3}{a^3 \cdot a} = \dfrac{81}{a^4}$

43. $\dfrac{\frac{-10}{27}}{\frac{12}{35}} = \dfrac{-10}{27} \div \dfrac{12}{35} = \dfrac{\cancel{-10}^{-5}}{27} \cdot \dfrac{35}{\cancel{12}_{6}}$
$= \dfrac{-5 \cdot 35}{27 \cdot 6} = \dfrac{-175}{162}$

45. $\dfrac{\frac{-18}{49}}{14} = \dfrac{-18}{49} \div \dfrac{14}{1} = \dfrac{-18}{49} \cdot \dfrac{1}{14}$

$\phantom{\dfrac{\frac{-18}{49}}{14}} = \dfrac{-9 \cdot 1}{49 \cdot 7} = \dfrac{-9}{343}$

47. $\dfrac{\frac{2x}{3}}{\frac{10x}{9}} = \dfrac{2x}{3} \div \dfrac{10x}{9} = \dfrac{2x}{3} \cdot \dfrac{9}{10x}$

$\phantom{\dfrac{\frac{2x}{3}}{\frac{10x}{9}}} = \dfrac{1 \cdot 3}{1 \cdot 5} = \dfrac{3}{5}$

49. $\dfrac{\frac{x^2}{3}}{\frac{x}{6}} = \dfrac{x^2}{3} \div \dfrac{x}{6} = \dfrac{x^2}{3} \cdot \dfrac{6}{x}$

$\phantom{\dfrac{\frac{x^2}{3}}{\frac{x}{6}}} = \dfrac{x \cdot 2}{1 \cdot 1} = \dfrac{2x}{1} = 2x$

51. $\dfrac{\frac{x}{y^2}}{\frac{y}{x^2}} = \dfrac{x}{y^2} \div \dfrac{y}{x^2} = \dfrac{x}{y^2} \cdot \dfrac{x^2}{y}$

$\phantom{\dfrac{\frac{x}{y^2}}{\frac{y}{x^2}}} = \dfrac{x \cdot x^2}{y^2 \cdot y} = \dfrac{x^3}{y^3}$

53. $\dfrac{\frac{2u}{z^2}}{\frac{4z}{u}} = \dfrac{2u}{z^2} \div \dfrac{4z}{u} = \dfrac{2u}{z^2} \cdot \dfrac{u}{4z}$

$\phantom{\dfrac{\frac{2u}{z^2}}{\frac{4z}{u}}} = \dfrac{u \cdot u}{z^2 \cdot 2z} = \dfrac{u^2}{2z^3}$

55. $\dfrac{3x^2 - x^2}{4y^2 - y^2} \cdot \dfrac{2y + y}{x^2 + x^2} = \dfrac{2x^2}{3y^2} \cdot \dfrac{3y}{2x^2}$

$\phantom{\dfrac{3x^2 - x^2}{4y^2 - y^2}} = \dfrac{1 \cdot 1}{y \cdot 1} = \dfrac{1}{y}$

57. $\dfrac{3x^2 \cdot x^2}{4y^2 \cdot y^2} \cdot \dfrac{2y \cdot y}{x^2 \cdot x^2} = \dfrac{3x^4}{4y^4} \cdot \dfrac{2y^2}{x^4}$

$\phantom{\dfrac{3x^2 \cdot x^2}{4y^2 \cdot y^2}} = \dfrac{3}{2y^2}$

59. $\dfrac{6x^2y - x^2y}{2x^3 - 5x^3} \div \dfrac{5xy^2 - 2xy^2}{6x - x} = \dfrac{5x^2y}{-3x^3} \div \dfrac{3xy^2}{5x}$

$\phantom{\dfrac{6x^2y - x^2y}{2x^3 - 5x^3}} = \dfrac{5x^2y}{-3x^3} \cdot \dfrac{5x}{3xy^2}$

$\phantom{\dfrac{6x^2y - x^2y}{2x^3 - 5x^3}} = \dfrac{25x^3y}{-9x^4y^2}$

$\phantom{\dfrac{6x^2y - x^2y}{2x^3 - 5x^3}} = -\dfrac{25}{9xy}$

61. $\dfrac{6x^2y \cdot x^2y}{2x^3 - 5x^3} \div \dfrac{5xy^2 - 2xy^2}{6x \cdot x} = \dfrac{6x^4y^2}{-3x^3} \div \dfrac{3xy^2}{6x^2}$

$\phantom{\dfrac{6x^2y \cdot x^2y}{2x^3 - 5x^3}} = \dfrac{6x^4y^2}{-3x^3} \cdot \dfrac{6x^2}{3xy^2}$

$\phantom{\dfrac{6x^2y \cdot x^2y}{2x^3 - 5x^3}} = \dfrac{36 \, x^6 y^2}{-9 x^4 y^2}$

$\phantom{\dfrac{6x^2y \cdot x^2y}{2x^3 - 5x^3}} = -4x^2$

63. Let x = number of miles.

$x = 100 \div \dfrac{8}{5} = 100 \cdot \dfrac{5}{8}$

$ = \dfrac{100}{1} \cdot \dfrac{5}{8} = \dfrac{25 \cdot 5}{1 \cdot 2}$

$ = \dfrac{125}{2}$ or 62.5

So 100 kilometers is equivalent to 62.5 miles.

65. Let $x =$ part of the cake each person receives.
$$x = \frac{3}{8} \div 4 = \frac{3}{8} \cdot \frac{1}{4}$$
$$= \frac{3 \cdot 1}{8 \cdot 4} = \frac{3}{32}$$
Each person will receive $\frac{3}{32}$ of the cake.

67. Let $x =$ number of gallons
$$x = \left(7\frac{1}{2}\right)\left(5\frac{1}{4}\right) = \frac{15}{2} \cdot \frac{21}{4}$$
$$= \frac{15 \cdot 21}{2 \cdot 4} = \frac{315}{8} = 39\frac{3}{8}$$
There are $39\frac{3}{8}$ gallons.

69. Let $x =$ number of hors doeuvres.
$$x = 6\frac{1}{4} \div \frac{3}{4} = \frac{25}{\cancel{4}_1} \cdot \frac{\cancel{4}^1}{3}$$
$$= \frac{25}{3} = 8\frac{1}{3}$$
$8\frac{1}{3}$ hors doeuvres can be prepared.

71. Let $x =$ number of miles
$$x = \left(8\frac{1}{3}\right)\left(21\frac{4}{5}\right) = \frac{\cancel{25}^5}{3} \cdot \frac{109}{\cancel{5}_1}$$
$$= \frac{5 \cdot 109}{3 \cdot 1} = \frac{545}{3} = 181\frac{2}{3}$$
$181\frac{2}{3}$ miles can be driven.

73. Let $x =$ number of sheets.
$$x = 76 \div 12\frac{2}{3} = 76 \div \frac{38}{3}$$
$$= \frac{\cancel{76}^2}{1} \cdot \frac{3}{\cancel{38}_1} = 6$$
6 sheets will be needed.

75. 0.48   77. 0.45   79. 0.556

81. To divide by a fraction, multiply by its reciprocal. This rule works because division is defined to be the inverse of multiplication.

82. (a) The fraction $\frac{2y}{3x}$ must be inverted, leading to $\frac{3x}{2y} \cdot \frac{3x}{2y} = \frac{9x^2}{4y^2}$.

    (b) The factor 5 is $\frac{5}{1}$, not $\frac{5}{5}$. Thus, $5 \cdot \frac{3x}{2} = \frac{5}{1} \cdot \frac{3x}{2} = \frac{5 \cdot 3x}{1 \cdot 2} = \frac{15x}{2}$.

83. A collection of dimes and quarters has a total value of \$2.75. If there are three more dimes than quarters, how many of each type of coin are in the collection?

    Let $x =$ the number of quarters.

85. Let $x =$ smallest angle. Then $2x =$ largest angle and $x + 20 =$ third angle.
$$\begin{aligned} x + 2x + (x+20) &= 180 \\ 4x + 20 &= 180 \\ -20 -20& \\ \hline 4x &= 160 \\ \frac{4x}{4} &= \frac{160}{4} \\ x &= 40 \end{aligned}$$
Then $2x = 2(40) = 80$ and $x + 20 = 40 + 20 = 60$. So the three angles are 40°, 60°, and 80°.

Exercises 4.3

1. $\dfrac{5}{3} + \dfrac{4}{3} = \dfrac{5+4}{3} = \dfrac{\cancel{9}^3}{\cancel{3}_1} = \dfrac{3}{1} = 3$

3. $\dfrac{3}{5} - \dfrac{7}{5} = \dfrac{3-7}{5} = \dfrac{-4}{5} = -\dfrac{4}{5}$

5. $\dfrac{7}{9} - \dfrac{5}{9} - \dfrac{8}{9} = \dfrac{7-5-8}{9} = \dfrac{\cancel{-6}^{-2}}{\cancel{9}_3} = -\dfrac{2}{3}$

7. $\dfrac{2}{3} + \dfrac{4}{5} = \dfrac{2(5)}{3(5)} + \dfrac{4(3)}{5(3)} = \dfrac{10}{15} + \dfrac{12}{15}$
$= \dfrac{10+12}{15} = \dfrac{22}{15}$

9. $\dfrac{2}{3} \cdot \dfrac{4}{5} = \dfrac{2 \cdot 4}{3 \cdot 5} = \dfrac{8}{15}$

11. $\dfrac{2}{3} - \dfrac{5}{6} = \dfrac{2(2)}{3(2)} - \dfrac{5}{6} = \dfrac{4}{6} - \dfrac{5}{6}$
$= \dfrac{4-5}{6} = \dfrac{-1}{6} = -\dfrac{1}{6}$

13. $3 + \dfrac{3}{4} - \dfrac{3}{8} = \dfrac{3}{1} + \dfrac{3}{4} - \dfrac{3}{8}$
$= \dfrac{3(8)}{1(8)} + \dfrac{3(2)}{4(2)} - \dfrac{3}{8}$
$= \dfrac{24}{8} + \dfrac{6}{8} - \dfrac{3}{8}$
$= \dfrac{24 + 6 - 3}{8} = \dfrac{27}{8}$

15. $\dfrac{3}{2} - \dfrac{4}{5} + \dfrac{7}{10} = \dfrac{3(5)}{2(5)} - \dfrac{4(2)}{5(2)} + \dfrac{7}{10}$
$= \dfrac{15}{10} - \dfrac{8}{10} + \dfrac{7}{10}$
$= \dfrac{15 - 8 + 7}{10} = \dfrac{\cancel{14}^7}{\cancel{10}_5} = \dfrac{7}{5}$

17. $\dfrac{5}{6} - \dfrac{3}{8} + \dfrac{1}{4} = \dfrac{5(4)}{6(4)} - \dfrac{3(3)}{8(3)} + \dfrac{1(6)}{4(6)}$
$= \dfrac{20}{24} - \dfrac{9}{24} + \dfrac{6}{24}$
$= \dfrac{20 - 9 + 6}{24} = \dfrac{17}{24}$

19. $2 + \dfrac{1}{3} - \dfrac{1}{2} = \dfrac{2}{1} + \dfrac{1}{3} - \dfrac{1}{2}$
$= \dfrac{2(6)}{1(6)} + \dfrac{1(2)}{3(2)} - \dfrac{1(3)}{2(3)}$
$= \dfrac{12}{6} + \dfrac{2}{6} - \dfrac{3}{6}$
$= \dfrac{12 + 2 - 3}{6} = \dfrac{11}{6}$

21. $\dfrac{8}{3x} + \dfrac{4}{3x} = \dfrac{8+4}{3x} = \dfrac{\cancel{12}^4}{\cancel{3x}_1} = \dfrac{4}{x}$

23. $\dfrac{8}{3x} \cdot \dfrac{4}{3x} = \dfrac{8 \cdot 4}{3x \cdot 3x} = \dfrac{32}{9x^2}$

25. $\dfrac{7}{6x} + \dfrac{13}{6x} - \dfrac{11}{6x} = \dfrac{7 + 13 - 11}{6x}$
$= \dfrac{\cancel{9}^3}{\cancel{6x}_2} = \dfrac{3}{2x}$

27. $\dfrac{3y}{7x} - \dfrac{5y}{7x} + \dfrac{4y}{7x} = \dfrac{3y - 5y + 4y}{7x} = \dfrac{2y}{7x}$

29. $\dfrac{w}{9z} - \dfrac{5w}{9z} + \dfrac{4w}{9z} = \dfrac{w - 5w + 4w}{9z}$
$= \dfrac{0}{9z} = 0$

31. $\dfrac{-5a}{7b} + \dfrac{3a}{7b} - \dfrac{12a}{7b} = \dfrac{-5a + 3a - 12a}{7b}$
$= \dfrac{\cancel{-14a}^{-2}}{\cancel{7b}_{1}}$
$= \dfrac{-2a}{b} = -\dfrac{2a}{b}$

33. $\dfrac{x+3}{3x} + \dfrac{x-6}{3x} = \dfrac{x+3+x-6}{3x}$
$= \dfrac{2x-3}{3x}$

35. $\dfrac{3y^2-5}{4y} + \dfrac{5-4y^2}{4y} = \dfrac{3y^2-5+5-4y^2}{4y}$
$= \dfrac{\cancel{-y^2}^{-y}}{\cancel{4y}} = \dfrac{-y}{4} = -\dfrac{y}{4}$

37. $\dfrac{5x+2}{10x} - \dfrac{x+2}{10x} = \dfrac{5x+2-(x+2)}{10x}$
$= \dfrac{5x+2-x-2}{10x}$
$= \dfrac{\cancel{4x}^{2}}{\cancel{10x}_{5}} = \dfrac{2}{5}$

39. $\dfrac{2a-1}{3a} - \dfrac{5a-1}{3a} = \dfrac{2a-1-(5a-1)}{3a}$
$= \dfrac{2a-1-5a+1}{3a}$
$= \dfrac{\cancel{-3a}^{-1}}{\cancel{3a}_{1}} = \dfrac{-1}{1} = -1$

41. $\dfrac{w-4}{6w} - \dfrac{w-3}{6w} + \dfrac{5}{6w} = \dfrac{w-4-(w-3)+5}{6w} = \dfrac{w-4-w+3+5}{6w}$
$= \dfrac{\cancel{4}^{2}}{\cancel{6w}_{3}} = \dfrac{2}{3w}$

43. $\dfrac{t^2-3t+2}{5t^2} - \dfrac{7t+t^2}{5t^2} = \dfrac{t^2-3t+2-(7t+t^2)}{5t^2}$
$= \dfrac{t^2-3t+2-7t-t^2}{5t^2} = \dfrac{-10t+2}{5t^2}$

45. $\dfrac{3}{x} + \dfrac{2}{y} = \dfrac{3(y)}{x(y)} + \dfrac{2(x)}{y(x)} = \dfrac{3y}{xy} + \dfrac{2x}{xy}$
$= \dfrac{3y+2x}{xy}$

47. $\dfrac{3}{x} \cdot \dfrac{2}{y} = \dfrac{3 \cdot 2}{x \cdot y} = \dfrac{6}{xy}$

49. $\dfrac{5}{3x} - \dfrac{7}{2} = \dfrac{5(2)}{3x(2)} - \dfrac{7(3x)}{2(3x)} = \dfrac{10}{6x} - \dfrac{21x}{6x}$
$= \dfrac{10-21x}{6x}$

51. $\dfrac{5}{48x} + \dfrac{3}{24y} = \dfrac{5(y)}{48x(y)} + \dfrac{3(2x)}{24y(2x)}$
$= \dfrac{5y}{48xy} + \dfrac{6x}{48xy} = \dfrac{5y+6x}{48xy}$

53. $\dfrac{4}{x^2} - \dfrac{3}{2x} = \dfrac{4(2)}{x^2(2)} - \dfrac{3(x)}{2x(x)}$
$= \dfrac{8}{2x^2} - \dfrac{3x}{2x^2} = \dfrac{8-3x}{2x^2}$

55. $\dfrac{\overset{2}{\cancel{4}}}{x^2} \cdot \dfrac{3}{\cancel{2}x} = \dfrac{2 \cdot 3}{x^2 \cdot x} = \dfrac{6}{x^3}$

57. $\dfrac{2}{3x^2} + \dfrac{3}{2x^2} = \dfrac{2(2)}{3x^2(2)} + \dfrac{3(3)}{2x^2(3)}$
$= \dfrac{4}{6x^2} + \dfrac{9}{6x^2} = \dfrac{4+9}{6x^2} = \dfrac{13}{6x^2}$

59. $\dfrac{7}{12a^2} - \dfrac{9}{84a} = \dfrac{7(7)}{12a^2(7)} - \dfrac{9(a)}{84a(a)}$
$= \dfrac{49}{84a^2} - \dfrac{9a}{84a^2} = \dfrac{49-9a}{84a^2}$

61. $\dfrac{1}{x} + 2 = \dfrac{1}{x} + \dfrac{2}{1} = \dfrac{1}{x} + \dfrac{2(x)}{1(x)}$
$= \dfrac{1}{x} + \dfrac{2x}{x} = \dfrac{1+2x}{x}$

63. $\dfrac{5}{3xy} + \dfrac{1}{6y^2} = \dfrac{5(2y)}{3xy(2y)} + \dfrac{1(x)}{6y^2(x)}$
$= \dfrac{10y}{6xy^2} + \dfrac{x}{6xy^2} = \dfrac{10y+x}{6xy^2}$

65. $\dfrac{7}{6a^2b} + \dfrac{3}{4ab^3} = \dfrac{7(2b^2)}{6a^2b(2b^2)} + \dfrac{3(3a)}{4ab^3(3a)}$
$= \dfrac{14b^2}{12a^2b^3} + \dfrac{9a}{12a^2b^3}$
$= \dfrac{14b^2+9a}{12a^2b^3}$

67. $\dfrac{7}{\underset{2}{\cancel{6}}a^2b} \cdot \dfrac{\overset{1}{\cancel{3}}}{4ab^3} = \dfrac{7 \cdot 1}{2a^2b \cdot 4ab^3} = \dfrac{7}{8a^3b^4}$

69. $\dfrac{7}{24x^2} + \dfrac{1}{60x} = \dfrac{7(5)}{24x^2(5)} + \dfrac{1(2x)}{60x(2x)}$
$= \dfrac{35}{120x^2} + \dfrac{2x}{120x^2}$
$= \dfrac{35+2x}{120x^2}$

71. $\dfrac{15}{48rt^2} - \dfrac{7}{72t^3} = \dfrac{15(3t)}{48rt^2(3t)} - \dfrac{7(2r)}{72t^3(2r)}$
$= \dfrac{45t}{144rt^3} - \dfrac{14r}{144rt^3}$
$= \dfrac{45t-14r}{144rt^3}$

73. $\dfrac{3y}{20x^2} + \dfrac{9x}{50y} = \dfrac{3y(5y)}{20x^2(5y)} + \dfrac{9x(2x^2)}{50y(2x^2)}$
$= \dfrac{15y^2}{100x^2y} + \dfrac{18x^3}{100x^2y}$
$= \dfrac{15y^2+18x^3}{100x^2y}$

75. $\dfrac{4y}{3x^2} - \dfrac{3}{2x} + \dfrac{y}{x^2}$
$= \dfrac{4y(2)}{3x^2(2)} - \dfrac{3(3x)}{2x(3x)} + \dfrac{y(6)}{x^2(6)}$
$= \dfrac{8y}{6x^2} - \dfrac{9x}{6x^2} + \dfrac{6y}{6x^2} = \dfrac{14y-9x}{6x^2}$

77. $\dfrac{3}{4m^2n} - \dfrac{5}{6mn^3} + \dfrac{1}{8n^2} = \dfrac{3(6n^2)}{4m^2n(6n^2)} - \dfrac{5(4m)}{6mn^3(4m)} + \dfrac{1(3m^2n)}{8n^2(3m^2n)}$
$= \dfrac{18n^2}{24m^2n^3} - \dfrac{20m}{24m^2n^3} + \dfrac{3m^2n}{24m^2n^3} = \dfrac{18n^2-20m+3m^2n}{24m^2n^3}$

79. $\dfrac{x}{y} + \dfrac{y}{x} + \dfrac{3x}{2y} = \dfrac{x(2x)}{y(2x)} + \dfrac{y(2y)}{x(2y)} + \dfrac{3x(x)}{2y(x)} = \dfrac{2x^2}{2xy} + \dfrac{2y^2}{2xy} + \dfrac{3x^2}{2xy}$

$= \dfrac{2x^2 + 2y^2 + 3x^2}{2xy} = \dfrac{5x^2 + 2y^2}{2xy}$

81. $t - \dfrac{3}{t} = \dfrac{t}{1} - \dfrac{3}{t} = \dfrac{t(t)}{1(t)} - \dfrac{3}{t} = \dfrac{t^2}{t} - \dfrac{3}{t} = \dfrac{t^2 - 3}{t}$

83. $3x^2 + \dfrac{1}{x} - \dfrac{2}{x^2} = \dfrac{3x^2}{1} + \dfrac{1}{x} - \dfrac{2}{x^2} = \dfrac{3x^2(x^2)}{1(x^2)} + \dfrac{1(x)}{x(x)} - \dfrac{2}{x^2} = \dfrac{3x^4}{x^2} + \dfrac{x}{x^2} - \dfrac{2}{x^2} = \dfrac{3x^4 + x - 2}{x^2}$

85. $\dfrac{2x+3}{24} + \dfrac{x}{18} = \dfrac{(2x+3)(3)}{24(3)} + \dfrac{x(4)}{18(4)} = \dfrac{6x+9}{72} + \dfrac{4x}{72} = \dfrac{6x+9+4x}{72} = \dfrac{10x+9}{72}$

87. $\dfrac{2x+3}{x} + \dfrac{x}{2} = \dfrac{(2x+3)(2)}{x(2)} + \dfrac{x(x)}{2(x)}$

$= \dfrac{4x+6}{2x} + \dfrac{x^2}{2x}$

$= \dfrac{4x+6+x^2}{2x}$

89. $\dfrac{a-5}{2} + \dfrac{3}{a} = \dfrac{(a-5)(a)}{2(a)} + \dfrac{3(2)}{a(2)}$

$= \dfrac{a^2 - 5a}{2a} + \dfrac{6}{2a}$

$= \dfrac{a^2 - 5a + 6}{2a}$

91. A least common denominator (LCD) is the "smallest" expression that is exactly divisible by each of the denomiantors in a problem. We need the LCD in order to add or subtract two or more fractions. (Actually, a common denominator is really all that is needed, but the LCD is the most efficient one to use.)

92. Each of the original denominators is a factor of the LCD, so this is a common denominator. Since each factor of the LCD was chosen the maximum number of times that it appears as a factor in any one of the denominators, there are no extra factors. Thus, this must be the smallest common denominators possible.

93. The LCD for $\dfrac{3}{10}$ and $\dfrac{7}{9}$ is 90. The LCD for $\dfrac{5}{6}$ and $\dfrac{3}{4}$ is 12. The LCD of two fractions will simply be the product of the two denominators when these denominators have no prime factor in common.

94. (a) We must subtract the entire expression $(5-x)$ from $(x+3)$ in the numerator. This gives $\dfrac{x+3-(5-x)}{x} = \dfrac{x+3-5+x}{x} = \dfrac{2x-2}{x}$.

   (b) We cannot cancel when performing addition.

   (c) When building fractions, we must multiply both numerator and denominator by the same quantity. So $\dfrac{5x}{2y} = \dfrac{5x(3x)}{2y(3x)} = \dfrac{15x^2}{6xy}$ and $\dfrac{7y}{6x} = \dfrac{7y(y)}{6x(y)} = \dfrac{7y^2}{6xy}$. Then $\dfrac{5x}{2y} + \dfrac{7y}{6x} = \dfrac{15x^2}{6xy} + \dfrac{7y^2}{6xy} = \dfrac{15x^2 + 7y^2}{6xy}$.

   (d) The cancellation step undoes the building step and brings the problem back to its original form. The final step is incorrect, since we cannot add two fractions with unlike denominators.

   (e) The cancellation is not allowed, since there are no common factors to be crossed out.

95. Reducing these fractions would reverse the building process and bring us back to the original problem.

97. Let $p = $ weight of package in pounds.
$$10 + 4p = 28$$
$$\underline{-10 \qquad -10}$$
$$4p = 18$$
$$\frac{\cancel{4}p}{\cancel{4}} = \frac{18}{4}$$
$$p = 4\frac{1}{2}$$

The package weighed $4\frac{1}{2}$ pounds.

Exercises 4.4

1. $\quad \dfrac{x}{3} = 9 \qquad$ LCD $= 3$

   $3\left(\dfrac{x}{3}\right) = 3 \cdot 9$

   $\dfrac{\overset{1}{\cancel{3}}}{1} \cdot \dfrac{x}{\underset{1}{\cancel{3}}} = 27$

   $x = 27$

3. $\quad \dfrac{a}{4} = -6 \qquad$ LCD $= 4$

   $4\left(\dfrac{a}{4}\right) = 4 \cdot (-6)$

   $\dfrac{\overset{1}{\cancel{4}}}{1} \cdot \dfrac{a}{\underset{1}{\cancel{4}}} = -24$

   $a = -24$

5. $\quad \dfrac{y}{6} = \dfrac{5}{4} \qquad$ LCD $= 12$

   $12\left(\dfrac{y}{6}\right) = 12\left(\dfrac{5}{4}\right)$

   $\dfrac{\overset{2}{\cancel{12}}}{1} \cdot \dfrac{y}{\underset{1}{\cancel{6}}} = \dfrac{\overset{3}{\cancel{12}}}{1} \cdot \dfrac{5}{\underset{1}{\cancel{4}}}$

   $2y = 15$

   $y = \dfrac{15}{2}$

7. $\quad \dfrac{w}{8} = \dfrac{-7}{6} \qquad$ LCD $= 24$

   $24\left(\dfrac{w}{8}\right) = 24\left(\dfrac{-7}{6}\right)$

   $\dfrac{\overset{3}{\cancel{24}}}{1} \cdot \dfrac{w}{\underset{1}{\cancel{8}}} = \dfrac{\overset{4}{\cancel{24}}}{1} \cdot \dfrac{-7}{\underset{1}{\cancel{6}}}$

   $3w = -28$

   $w = \dfrac{-28}{3} = -\dfrac{28}{3}$

9. $\quad \dfrac{3x}{2} = -18 \qquad$ LCD $= 2$

   $2\left(\dfrac{3x}{2}\right) = 2 \cdot (-18)$

   $\dfrac{\overset{1}{\cancel{2}}}{1} \cdot \dfrac{3x}{\underset{1}{\cancel{2}}} = -36$

   $3x = -36$

   $x = -12$

11. $\quad 16 = \dfrac{4}{5}x \qquad$ LCD $= 5$

    $5 \cdot 16 = 5\left(\dfrac{4}{5}x\right)$

    $80 = \dfrac{\overset{1}{\cancel{5}}}{1} \cdot \dfrac{4}{\underset{1}{\cancel{5}}}x$

    $80 = 4x$

    $20 = x$

13. $\dfrac{x}{3} - 2 = \dfrac{2}{3}$ LCD = 3

$3\left(\dfrac{x}{3} - 2\right) = 3\left(\dfrac{2}{3}\right)$

$\dfrac{\cancel{3}^1}{1} \cdot \dfrac{x}{\cancel{3}_1} - 3 \cdot 2 = \dfrac{\cancel{3}^1}{1} \cdot \dfrac{2}{\cancel{3}_1}$

$x - 6 = 2$

$x = 8$

15. $\dfrac{3a}{4} + 2 = \dfrac{5}{4}$ LCD = 4

$4\left(\dfrac{3a}{4} + 2\right) = 4\left(\dfrac{5}{4}\right)$

$\dfrac{\cancel{4}^1}{1} \cdot \dfrac{3a}{\cancel{4}_1} + 4 \cdot 2 = \dfrac{\cancel{4}^1}{1} \cdot \dfrac{5}{\cancel{4}_1}$

$3a + 8 = 5$

$3a = -3$

$a = -1$

17. $\dfrac{u}{2} - \dfrac{u}{4} = 2$ LCD = 4

$4\left(\dfrac{u}{2} - \dfrac{u}{4}\right) = 4 \cdot 2$

$\dfrac{\cancel{4}^2}{1} \cdot \dfrac{u}{\cancel{2}_1} - \dfrac{\cancel{4}^1}{1} \cdot \dfrac{u}{\cancel{4}_1} = 8$

$2u - u = 8$

$u = 8$

19. $\dfrac{y}{3} + \dfrac{y}{5} < \dfrac{6}{5}$ LCD = 15

$15\left(\dfrac{y}{3} + \dfrac{y}{5}\right) < 15 \cdot \dfrac{6}{5}$

$\dfrac{\cancel{15}^5}{1} \cdot \dfrac{y}{\cancel{3}_1} + \dfrac{\cancel{15}^3}{1} \cdot \dfrac{y}{\cancel{5}_1} < \dfrac{\cancel{15}^3}{1} \cdot \dfrac{6}{\cancel{5}_1}$

$5y + 3y < 18$

$8y < 18$

$y < \dfrac{9}{4}$

21. $3x - \dfrac{2}{3}x = \dfrac{4}{3}$ LCD = 3

$3\left(3x - \dfrac{2}{3}x\right) = 3 \cdot \dfrac{4}{3}$

$3 \cdot 3x - \dfrac{\cancel{3}^1}{1} \cdot \dfrac{2x}{\cancel{3}_1} = \dfrac{\cancel{3}^1}{1} \cdot \dfrac{4}{\cancel{3}_1}$

$9x - 2x = 4$

$7x = 4$

$x = \dfrac{4}{7}$

23. $\dfrac{a}{4} - \dfrac{a}{3} \geq \dfrac{5}{2}$ LCD = 12

$12\left(\dfrac{a}{4} - \dfrac{a}{3}\right) \geq 12 \cdot \dfrac{5}{2}$

$\dfrac{\cancel{12}^3}{1} \cdot \dfrac{a}{\cancel{4}_1} - \dfrac{\cancel{12}^4}{1} \cdot \dfrac{a}{\cancel{3}_1} \geq \dfrac{\cancel{12}^6}{1} \cdot \dfrac{5}{\cancel{2}_1}$

$3a - 4a \geq 30$

$-a \geq 30$

$a \leq -30$

(Remember to reverse the inequality symbol when multiplying or dividing by a negative quantity.)

25. $0.7x + 0.4x = 5.5$

LCD = 10

$10(0.7x + 0.4x) = 10(5.5)$

$10(0.7x) + 10(0.4x) = 10(5.5)$

$7x + 4x = 55$

$11x = 55$

$x = 5$

27. $0.3x - 0.25x = 2$ LCD = 100

$100(0.3x - 0.25x) = 100(2)$

$100(0.3x) - 100(0.25x) = 100(2)$

$30x - 25x = 200$

$5x = 200$

$x = 40$

29.
$$0.8m + 0.05m = 0.34 \quad \text{LCD} = 100$$
$$100(0.8m + 0.05m) = 100(0.34)$$
$$100(0.8m) + 100(0.05m) = 100(0.34)$$
$$80m + 5m = 34$$
$$85m = 34$$
$$m = \frac{34}{85} = \frac{2 \cdot \cancel{17}}{5 \cdot \cancel{17}} = \frac{2}{5}$$

31.
$$\frac{w+3}{4} = \frac{w+4}{3} \quad \text{LCD} = 12$$
$$12\left(\frac{w+3}{4}\right) = 12\left(\frac{w+4}{3}\right)$$
$$\frac{\cancel{12}^3}{1} \cdot \frac{w+3}{\cancel{4}_1} = \frac{\cancel{12}^4}{1} \cdot \frac{w+4}{\cancel{3}_1}$$
$$3(w+3) = 4(w+4)$$
$$3w + 9 = 4w + 16$$
$$9 = w + 16$$
$$-7 = w$$

33.
$$\frac{w+3}{4} + 1 = \frac{w+4}{3} \quad \text{LCD} = 12$$
$$12\left(\frac{w+3}{4} + 1\right) = 12\left(\frac{w+4}{3}\right)$$
$$\frac{\cancel{12}^3}{1} \cdot \frac{w+3}{\cancel{4}_1} + 12 \cdot 1 = \frac{\cancel{12}^4}{1} \cdot \frac{w+4}{\cancel{3}_1}$$
$$3(w+3) + 12 = 4(w+4)$$
$$3w + 9 + 12 = 4w + 16$$
$$3w + 21 = 4w + 16$$
$$21 = w + 16$$
$$5 = w$$

35.
$$\frac{x+1}{2} + x = 6 \quad \text{LCD} = 2$$
$$2\left(\frac{x+1}{2} + x\right) = 2 \cdot 6$$
$$\frac{\cancel{2}^1}{1} \cdot \frac{x+1}{\cancel{2}_1} + 2 \cdot x = 12$$
$$x + 1 + 2x = 12$$
$$3x + 1 = 12$$
$$3x = 11$$
$$x = \frac{11}{3}$$

37.
$$\frac{y}{6} - \frac{y-2}{4} > 1 \quad \text{LCD} = 12$$
$$12\left(\frac{y}{6} - \frac{y-2}{4}\right) > 12 \cdot 1$$
$$\frac{\cancel{12}^2}{1} \cdot \frac{y}{\cancel{6}_1} - \frac{\cancel{12}^3}{1} \cdot \frac{y-2}{\cancel{4}_1} > 12 \cdot 1$$
$$2y - 3(y-2) > 12$$
$$2y - 3y + 6 > 12$$
$$-y + 6 > 12$$
$$-y > 6$$
$$y < -6$$

39. $$3 - \frac{a+1}{4} = \frac{a+4}{2} \quad \text{LCD} = 4$$

$$4\left(3 - \frac{a+1}{4}\right) = 4\left(\frac{a+4}{2}\right)$$

$$4 \cdot 3 - \frac{\cancel{4}^1}{1} \cdot \frac{a+1}{\cancel{4}_1} = \frac{\cancel{4}^2}{1} \cdot \frac{a+4}{\cancel{2}_1}$$

$$12 - (a+1) = 2(a+4)$$
$$12 - a - 1 = 2a + 8$$
$$11 - a = 2a + 8$$
$$11 = 3a + 8$$
$$3 = 3a$$
$$1 = a$$

41. $$\frac{2y-3}{2} - \frac{y-5}{3} = \frac{1}{6} \quad \text{LCD} = 6$$

$$6\left(\frac{2y-3}{2} - \frac{y-5}{3}\right) = 6 \cdot \frac{1}{6}$$

$$\frac{\cancel{6}^3}{1} \cdot \frac{2y-3}{\cancel{2}_1} - \frac{\cancel{6}^2}{1} \cdot \frac{y-5}{\cancel{3}_1} = \frac{\cancel{6}^1}{1} \cdot \frac{1}{\cancel{6}_1}$$

$$3(2y-3) - 2(y-5) = 1$$
$$6y - 9 - 2y + 10 = 1$$
$$4y + 1 = 1$$
$$4y = 0$$
$$y = 0$$

43. $$\frac{x+2}{3} - \frac{2x+3}{4} = \frac{x+4}{8} \quad \text{LCD} = 24$$

$$24\left(\frac{x+2}{3} - \frac{2x+3}{4}\right) = 24\left(\frac{x+4}{8}\right)$$

$$\frac{\cancel{24}^8}{1} \cdot \frac{x+2}{\cancel{3}_1} - \frac{\cancel{24}^6}{1} \cdot \frac{2x+3}{\cancel{4}_1} = \frac{\cancel{24}^3}{1} \cdot \frac{x+4}{\cancel{8}_1}$$

$$8(x+2) - 6(2x+3) = 3(x+4)$$
$$8x + 16 - 12x - 18 = 3x + 12$$
$$-4x - 2 = 3x + 12$$
$$-2 = 7x + 12$$
$$-14 = 7x$$
$$-2 = x$$

45. $$\frac{t}{2} + \frac{t-1}{3} + \frac{t-6}{4} = t - 2 \quad \text{LCD} = 12$$

$$12\left(\frac{t}{2} + \frac{t-1}{3} + \frac{t-6}{4}\right) = 12(t-2)$$

$$\frac{\cancel{12}^6}{1} \cdot \frac{t}{\cancel{2}_1} + \frac{\cancel{12}^4}{1} \cdot \frac{t-1}{\cancel{3}_1} + \frac{\cancel{12}^3}{1} \cdot \frac{t-6}{\cancel{4}_1} = 12(t-2)$$

$$6t + 4(t-1) + 3(t-6) = 12(t-2)$$
$$6t + 4t - 4 + 3t - 18 = 12t - 24$$
$$13t - 22 = 12t - 24$$
$$t - 22 = -24$$
$$t = -2$$

47. $$0.5(x+2) - 0.3(x-4) = 3$$
$$\text{LCD} = 10$$
$$10[0.5(x+2) - 0.3(x-4)] = 10 \cdot 3$$
$$10[0.5(x+2)] - 10[0.3(x-4)] = 10 \cdot 3$$
$$5(x+2) - 3(x-4) = 30$$
$$5x + 10 - 3x + 12 = 30$$
$$2x + 22 = 30$$
$$2x = 8$$
$$x = 4$$

49. $$0.06x + 0.04(5000 - x) = 320$$
$$\text{LCD} = 100$$
$$100[0.06x + 0.04(5000 - x)] = 100 \cdot 320$$
$$100(0.06x) + 100[0.04(5000 - x)] = 100 \cdot 320$$
$$6x + 4(5000 - x) = 32000$$
$$6x + 20000 - 4x = 32000$$
$$2x + 20000 = 32000$$
$$2x = 12000$$
$$x = 6000$$

51. $0.035x + 0.052(x + 5000) = 1130$
    LCD = 1000

$$1000[0.035x + 0.052(x + 5000)] = 1000 \cdot 1130$$
$$1000(0.035x) + 1000[0.052(x + 5000)] = 1000 \cdot 1130$$
$$35x + 52(x + 5000) = 1130000$$
$$35x + 52x + 260000 = 1130000$$
$$87x + 260000 = 1130000$$
$$87x = 870000$$
$$x = 10,000$$

53. $3(y+2) + \dfrac{y+3}{5} = \dfrac{9y+8}{2}$
    LCD = 10

$$10\left[3(y+2) + \dfrac{y+3}{5}\right] = 10\left(\dfrac{9y+8}{2}\right)$$

$$10[3(y+2)] + \dfrac{\overset{2}{\cancel{10}}}{1} \cdot \dfrac{y+3}{\underset{1}{\cancel{5}}} = \dfrac{\overset{5}{\cancel{10}}}{1} \cdot \dfrac{9y+8}{\underset{1}{\cancel{2}}}$$

$$30(y+2) + 2(y+3) = 5(9y+8)$$
$$30y + 60 + 2y + 6 = 45y + 40$$
$$32y + 66 = 45y + 40$$
$$66 = 13y + 40$$
$$26 = 13y$$
$$2 = y$$

55. $z + \dfrac{z+5}{3} - \dfrac{z-2}{6} = \dfrac{z+4}{4} + 1$
    LCD = 12

$$12\left(z + \dfrac{z+5}{3} - \dfrac{z-2}{6}\right) = 12\left(\dfrac{z+4}{4} + 1\right)$$

$$12 \cdot z + \dfrac{\overset{4}{\cancel{12}}}{1} \cdot \dfrac{z+5}{\underset{1}{\cancel{3}}} - \dfrac{\overset{2}{\cancel{12}}}{1} \cdot \dfrac{z-2}{\underset{1}{\cancel{6}}} = \dfrac{\overset{3}{\cancel{12}}}{1} \cdot \dfrac{z+4}{\underset{1}{\cancel{4}}} + 12 \cdot 1$$

$$12z + 4(z+5) - 2(z-2) = 3(z+4) + 12$$
$$12z + 4z + 20 - 2z + 4 = 3z + 12 + 12$$
$$14z + 24 = 3z + 24$$
$$11z + 24 = 24$$
$$11z = 0$$
$$z = 0$$

57. $3 \leq \dfrac{x}{3} - \dfrac{x+1}{2} \leq 6$
    LCD = 6

$$6 \cdot 3 \leq \quad 6\left(\dfrac{x}{3} - \dfrac{x+1}{2}\right) \quad \leq 6 \cdot 6$$

$$18 \leq \dfrac{\overset{2}{\cancel{6}}}{1} \cdot \dfrac{x}{\underset{1}{\cancel{3}}} - \dfrac{\overset{3}{\cancel{6}}}{1} \cdot \dfrac{x+1}{\underset{1}{\cancel{2}}} \leq 36$$

$$18 \leq \quad 2x - 3(x+1) \quad \leq 36$$
$$18 \leq \quad 2x - 3x - 3 \quad \leq 36$$
$$18 \leq \quad -x - 3 \quad \leq 36$$
$$21 \leq \quad -x \quad \leq 39$$
$$-21 \geq \quad x \quad \geq -39$$
$$\text{or} \quad -39 \leq x \leq -21$$

59. The LCD of 3, 2, and 5 is 30. So

$$\dfrac{x}{3} + \dfrac{x}{2} + \dfrac{x}{5} = \dfrac{x(10)}{3(10)} + \dfrac{x(15)}{2(15)} + \dfrac{x(6)}{5(6)}$$
$$= \dfrac{10x}{30} + \dfrac{15x}{30} + \dfrac{6x}{30} = \dfrac{31x}{30}$$

61. $\dfrac{x}{3} + \dfrac{x}{2} + \dfrac{x}{5} = 62 \quad$ LCD = 30

$$30\left(\dfrac{x}{3} + \dfrac{x}{2} + \dfrac{x}{5}\right) = 30 \cdot 62$$

$$\dfrac{\overset{10}{\cancel{30}}}{1} \cdot \dfrac{x}{\underset{1}{\cancel{3}}} + \dfrac{\overset{15}{\cancel{30}}}{1} \cdot \dfrac{x}{\underset{1}{\cancel{2}}} + \dfrac{\overset{6}{\cancel{30}}}{1} \cdot \dfrac{x}{\underset{1}{\cancel{5}}} = 1860$$

$$10x + 15x + 6x = 1860$$
$$31x = 1860$$
$$x = 60$$

63. $$\frac{x+5}{2} - \frac{x-1}{4} = 2 \quad \text{LCD} = 4$$
$$4\left(\frac{x+5}{2} - \frac{x-1}{4}\right) = 4 \cdot 2$$
$$\frac{\overset{2}{\cancel{4}}}{1} \cdot \frac{x+5}{\underset{1}{\cancel{2}}} - \frac{\overset{1}{\cancel{4}}}{1} \cdot \frac{x-1}{\underset{1}{\cancel{4}}} = 4 \cdot 2$$
$$2(x+5) - (x-1) = 8$$
$$2x + 10 - x + 1 = 8$$
$$x + 11 = 8$$
$$x = -3$$

65. The LCD of 2 and 4 is 4. So
$$\frac{x+5}{2} - \frac{x-1}{4} = \frac{(x+5)(2)}{2(2)} - \frac{x-1}{4}$$
$$= \frac{2(x+5)}{4} - \frac{x-1}{4}$$
$$= \frac{2(x+5) - (x-1)}{4}$$
$$= \frac{2x + 10 - x + 1}{4}$$
$$= \frac{x+11}{4}$$

67. The shipment of wooden boxes contains 80 pieces altogether. If oak boxes sell for $30 apiece and mahogany boxes sell for $50 apiece, how many of each type are in the shipment if its total value is $3600?
Let x = number of oak boxes.

69. Let x = weight of the package.
$$7 + 4(x-1) = 43$$
$$7 + 4x - 4 = 43$$
$$4x + 3 = 43$$
$$4x = 40$$
$$x = 10$$
The package weighs 10 lbs.

Exercises 4.5

1. $\dfrac{\#\text{ red}}{\#\text{ black}} = \dfrac{7}{5}$

3. $\dfrac{\#\text{ black}}{\#\text{ red}} = \dfrac{5}{7}$

5. $\dfrac{\#\text{ with}}{\#\text{ without}} = \dfrac{11}{5}$

7. $\dfrac{\overset{1}{\cancel{x}}}{3\cancel{x}} = \dfrac{1}{3}$

9. $\dfrac{a}{b+c}$

11. $$\frac{x}{5} = \frac{12}{3}$$
$$15 \cdot \frac{x}{5} = 15 \cdot \frac{12}{3}$$
$$\frac{\overset{3}{\cancel{15}}}{1} \cdot \frac{x}{\underset{1}{\cancel{5}}} = \frac{\overset{5}{\cancel{15}}}{1} \cdot \frac{12}{\underset{1}{\cancel{3}}}$$
$$3x = 60$$
$$x = 20$$

13. $$\frac{a}{6} = \frac{5}{3}$$
$$6 \cdot \frac{a}{6} = 6 \cdot \frac{5}{3}$$
$$\frac{\overset{1}{\cancel{6}}}{1} \cdot \frac{a}{\underset{1}{\cancel{6}}} = \frac{\overset{2}{\cancel{6}}}{1} \cdot \frac{5}{\underset{1}{\cancel{3}}}$$
$$a = 10$$

15. $$\frac{y}{15} = \frac{20}{6}$$
$$30 \cdot \frac{y}{15} = 30 \cdot \frac{20}{6}$$
$$\frac{\overset{2}{\cancel{30}}}{1} \cdot \frac{y}{\underset{1}{\cancel{15}}} = \frac{\overset{5}{\cancel{30}}}{1} \cdot \frac{20}{\underset{1}{\cancel{6}}}$$
$$2y = 100$$
$$y = 50$$

17. $$\frac{y}{9} = \frac{4}{3}$$
$$9 \cdot \frac{y}{9} = 9 \cdot \frac{4}{3}$$
$$\frac{\overset{1}{\cancel{9}}}{1} \cdot \frac{y}{\underset{1}{\cancel{9}}} = \frac{\overset{3}{\cancel{9}}}{1} \cdot \frac{4}{\underset{1}{\cancel{3}}}$$
$$y = 12$$

19. $x = 0.12$

21. $t = 11.59$

23. $$\frac{y + 0.3}{0.7} = \frac{y - 0.2}{0.9}$$
$$63 \cdot \frac{y + 0.3}{0.7} = 63 \cdot \frac{y - 0.2}{0.9}$$
$$\frac{\overset{90}{\cancel{63}}}{1} \cdot \frac{y + 0.3}{\underset{1}{\cancel{0.7}}} = \frac{\overset{70}{\cancel{63}}}{1} \cdot \frac{y - 0.2}{\underset{1}{\cancel{0.9}}}$$
$$90y + 27 = 70y - 14$$
$$20y + 27 = -14$$
$$20y = -41$$
$$\frac{\cancel{20}y}{\underset{1}{\cancel{20}}} = \frac{-41}{20}$$
$$y = -2.05$$

25. Let $x =$ number of red marbles in the jar.
$$\frac{x}{210} = \frac{7}{5}$$
$$210 \cdot \frac{x}{210} = 210 \cdot \frac{7}{5}$$
$$\frac{\overset{1}{\cancel{210}}}{1} \cdot \frac{x}{\underset{1}{\cancel{210}}} = \frac{\overset{42}{\cancel{210}}}{1} \cdot \frac{7}{\underset{1}{\cancel{5}}}$$
$$x = 294$$
There are 294 red marbles in the jar.

27. Let $x =$ number of students who got 90 or above.
$$\frac{x}{24} = \frac{3}{8}$$
$$24 \cdot \frac{x}{24} = 24 \cdot \frac{3}{8}$$
$$\frac{\overset{1}{\cancel{24}}}{1} \cdot \frac{x}{\underset{1}{\cancel{24}}} = \frac{\overset{3}{\cancel{24}}}{1} \cdot \frac{3}{\underset{1}{\cancel{8}}}$$
$$x = 9$$
There were 9 students who got 90 or above.

29. Let $W =$ width of the rectangle.
$$\frac{18}{W} = \frac{9}{4}$$
$$4W \cdot \frac{18}{W} = 4W \cdot \frac{9}{4}$$
$$\frac{4W}{1} \cdot \frac{18}{\underset{1}{\cancel{W}}} = \frac{\cancel{4}W}{1} \cdot \frac{9}{\underset{1}{\cancel{4}}}$$
$$72 = 9W$$
$$8 = W$$
The width of the rectangle is 8 cm.

31. $$\frac{\text{shorter side}}{\text{longer side}} = \frac{\overset{2}{\cancel{6x}}}{\underset{5}{\cancel{15x}}} = \frac{2}{5}$$

33. Let $x =$ length of the scale drawing of the 20-meter wall (in cm).
$$\frac{x}{20} = \frac{5}{12}$$
$$60 \cdot \frac{x}{20} = 60 \cdot \frac{5}{12}$$
$$\frac{\cancel{60}^{3}}{1} \cdot \frac{x}{\cancel{20}_{1}} = \frac{\cancel{60}^{5}}{1} \cdot \frac{5}{\cancel{12}_{1}}$$
$$3x = 25$$
$$x = \frac{25}{3} \approx 8.33$$
The scale drawing of the 20-meter wall is 8.33 cm.

35. Let $x =$ number of kilograms in 10 lbs.
$$\frac{10}{x} = \frac{2.2}{1}$$
$$x \cdot \frac{10}{x} = x \cdot \frac{2.2}{1}$$
$$\frac{\cancel{x}}{1} \cdot \frac{10}{\cancel{x}} = 2.2x$$
$$10 = 2.2x$$
$$\frac{10}{2.2} = x$$
$$4.55 \approx x$$
There are about 4.55 kilograms in 10 lbs.

37. Let $x =$ number of yards in 100 meters.
$$\frac{100}{x} = \frac{0.92}{1}$$
$$x \cdot \frac{100}{x} = x \cdot \frac{0.92}{1}$$
$$\frac{\cancel{x}}{1} \cdot \frac{100}{\cancel{x}} = 0.92x$$
$$100 = 0.92x$$
$$\frac{100}{0.92} = x$$
$$108.70 \approx x$$
There are about 108.70 yards in 100 meters.

39. Let $x =$ distance between the two cities (in miles).
$$\frac{12.6}{x} = \frac{5}{10}$$
$$10x \cdot \frac{12.6}{x} = 10x \cdot \frac{5}{10}$$
$$\frac{10\cancel{x}}{1} \cdot \frac{12.6}{\cancel{x}} = \frac{\cancel{10}x}{1} \cdot \frac{5}{\cancel{10}_{1}}$$
$$126 = 5x$$
$$25.2 = x$$
The distance between the two cities is 25.2 miles.

41. Let $x =$ Rolfe's share of the profits (in dollars).
$$\frac{18,500}{2,800} = \frac{12,800}{x}$$
$$\frac{185}{28} = \frac{12,800}{x}$$
$$\cancel{28}x \cdot \frac{185}{\cancel{28}} = 28\cancel{x} \cdot \frac{12,800}{\cancel{x}}$$
$$185x = 358,400$$
$$x = 1,937.30$$
Rolfe's share of the profits was $1,937.30.

43. Let $x =$ number of teaspoons of garlic needed.

$$\frac{2}{6} = \frac{x}{20}$$

$$\overset{10}{\cancel{60}} \cdot \frac{2}{\cancel{6}} = \overset{3}{\cancel{60}} \cdot \frac{x}{\underset{1}{\cancel{20}}}$$

$$20 = 3x$$

$$\frac{20}{3} = x$$

$$6.67 \approx x$$

Let $y =$ number of tablespoons of olive oil needed.

$$\frac{5}{6} = \frac{y}{20}$$

$$\overset{10}{\cancel{60}} \cdot \frac{5}{\cancel{6}} = \overset{3}{\cancel{60}} \cdot \frac{y}{\underset{1}{\cancel{20}}}$$

$$50 = 3y$$

$$\frac{50}{3} = y$$

$$16.67 \approx y$$

So 6.67 teaspoons of garlic and 16.67 tablespoons of olive oil are needed for 20 servings.

45. Let $x =$ tax due (in dollars).

$$\frac{788}{126,400} = \frac{x}{98,200}$$

$$98,200 \cdot \frac{788}{126,400} = \cancel{98,200} \cdot \frac{x}{\cancel{98,200}}$$

$$612.20 \approx x$$

So the tax due on the property is $612.20.

47. Let $x =$ ounces of copper in the alloy. Then $52 - x =$ ounces of tin in the alloy.

$$\frac{x}{52-x} = \frac{11}{2}$$

$$2(52-x) \cdot \frac{x}{52-x} = \cancel{2}(52-x) \cdot \frac{11}{\cancel{2}}$$

$$2x = 572 - 11x$$

$$13x = 572$$

$$x = \frac{572}{13} = 44$$

Then $52 - x = 52 - 44 = 8$. So the alloy contains 44 ounces of copper and 8 ounces of tin.

49. Let $x =$ Samantha's share of the profits. Then $128,650 - x =$ Greg's share of the profits.

$$\frac{x}{128,650 - x} = \frac{7}{5}$$

$$5(128,650-x) \cdot \frac{x}{128,650-x} = \cancel{5}(128,650-x) \cdot \frac{7}{\cancel{5}}$$

$$5x = 7(128,650 - x)$$

$$5x = 900,550 - 7x$$

$$12x = 900,550$$

$$x = \frac{900,550}{12} \approx 75,045.83$$

Then $128,650 - x = 128,650 - 75,045.83 = 53,604.17$. So Samantha's share of the profits is $75,045.83 and Greg's share of the profits is $53,604.17.

51. Let x = number of losers.
$$\frac{8}{3} = \frac{976}{x}$$
$$\not{3}x \cdot \frac{8}{\not{3}} = 3\not{x} \cdot \frac{976}{\not{x}}$$
$$8x = 2928$$
$$x = \frac{2928}{8} = 366$$
There were 366 losers.

53. Let x = total value of the estate.
$$\frac{7}{11} = \frac{56,300}{x}$$
$$\not{11}x \cdot \frac{7}{\not{11}} = 11\not{x} \cdot \frac{56,300}{\not{x}}$$
$$7x = 619,300$$
$$x = \frac{619,300}{7} \approx 88,471.43$$
The total value of the estate was $88,471.43.

55. Let x = estimated number of deer in the entire area.
$$\frac{48}{x} = \frac{18}{60}$$
$$60\not{x} \cdot \frac{48}{\not{x}} = \not{60}x \cdot \frac{18}{\not{60}\not{x}}$$
$$2880 = 18x$$
$$160 = x$$
The estimated number of deer in the entire area is 160.

57.
$$\frac{x}{6} = \frac{18}{9} \qquad \frac{y}{9} = \frac{8}{6}$$
$$\overset{3}{\not{18}} \cdot \frac{x}{\not{6}} = \overset{2}{\not{18}} \cdot \frac{18}{\not{9}} \qquad \overset{2}{\not{18}} \cdot \frac{y}{\not{9}} = \overset{3}{\not{18}} \cdot \frac{8}{\not{6}}$$
$$3x = 36 \qquad 2y = 24$$
$$x = 12 \qquad y = 12$$

59.
$$\frac{x}{12.5} = \frac{5.2}{8.4} \qquad \frac{y}{8.4} = \frac{9.6}{12.5}$$
$$(8.4)(\not{12.5}) \cdot \frac{x}{\not{12.5}} = (\not{8.4})(12.5) \cdot \frac{5.2}{\not{8.4}} \qquad (12.5)(\not{8.4}) \cdot \frac{y}{\not{8.4}} = (\not{12.5})(8.4) \cdot \frac{9.6}{\not{12.5}}$$
$$8.4x = 65 \qquad 12.5y = 80.64$$
$$x = \frac{65}{8.4} \approx 7.74 \qquad y = \frac{80.64}{12.5} \approx 6.45$$

61. Let x = height of the street light (in feet).
$$\frac{6}{8} = \frac{x}{23}$$
$$(23)(\not{8}) \cdot \frac{6}{\not{8}} = (\not{23})(8) \cdot \frac{x}{\not{23}}$$
$$138 = 8x$$
$$17.25 = x$$
The height of the street light is 17.25 feet.

63. Let w = width of the river (in meters).
$$\frac{w}{1800} = \frac{500}{750}$$
$$(\not{1800}) \cdot \frac{w}{\not{1800}} = \frac{\overset{2}{\not{500}}}{\underset{1}{\not{750}}} \cdot \overset{600}{\not{1800}}$$
$$w = 1200$$
The width of the river is 1200 meters.

65. Let $x =$ Cherise's share (in dollars).
Then $85,600 - x =$ Jerome's share (in dollars).
$$\frac{x}{85,600 - x} = \frac{3}{2}$$
$$2(85,600 - x) \cdot \frac{x}{85,600 - x} = 2(85,600 - x) \cdot \frac{3}{2}$$
$$2x = 3(85,600 - x)$$
$$2x = 256,800 - 3x$$
$$5x = 256,800$$
$$x = 51,360$$
Then $85,600 - x = 85,600 - 51,360 = 34,240$.
So Cherise's share is $51,360 and Jerome's share is $34,240.

Exercises 4.6

1. Let $x =$ the number.
$$\frac{2}{3}x + 5 = 9$$
$$3\left(\frac{2}{3}x + 5\right) = 3 \cdot 9$$
$$3 \cdot \frac{2}{3}x + 3 \cdot 5 = 3 \cdot 9$$
$$2x + 15 = 27$$
$$2x = 12$$
$$x = 6$$
Thus, the number is 6.
CHECK: $\frac{2}{3}$ of 6 is 4, and 5 more than 4 is 9.

3. Let $x =$ the number.
$$\frac{3}{4}x - 2 = \frac{1}{8}x - 7$$
$$8\left(\frac{3}{4}x - 2\right) = 8\left(\frac{1}{8}x - 7\right)$$
$$8 \cdot \frac{3}{4}x - 8 \cdot 2 = 8 \cdot \frac{1}{8}x - 8 \cdot 7$$
$$6x - 16 = x - 56$$
$$5x - 16 = -56$$
$$5x = -40$$
$$x = -8$$
Thus, the number is -8.
CHECK: 2 less than $\frac{3}{4}$ of -8 is 2 less than -6 or -8. $\frac{1}{8}$ of -8 is -1, and -8 is 7 less than -1.

5. Let $L =$ length of the rectangle (in meters).
Then $\frac{1}{2}L =$ width of the rectangle (in meters).
$$2(L) + 2\left(\frac{1}{2}L\right) = 36$$
$$2L + 2 \cdot \frac{1}{2}L = 36$$
$$2L + L = 36$$
$$3L = 36$$
$$L = 12$$
Then $\frac{1}{2}L = \frac{1}{2}(12) = 6$.
Thus, the dimensions of the rectangle are 12 meters and 6 meters.

CHECK: The width of the 6 m × 12 m rectangle is half its length. The perimeter of this rectangle is $2(6) + 2(12) = 12 + 24 = 36$ meters.

7. Let x = length of the longest side of the triangle (in inches).
Then $\frac{3}{4}x$ = length of the medium side of the triangle (in inches),
and $\frac{1}{2}\left(\frac{3}{4}x\right)$ = length of the shortest side of the triangle (in inches).

$$x + \frac{3}{4}x + \frac{1}{2}\left(\frac{3}{4}x\right) = 17$$
$$x + \frac{3}{4}x + \frac{3}{8}x = 17$$
$$8\left(x + \frac{3}{4}x + \frac{3}{8}x\right) = 8 \cdot 17$$
$$8 \cdot x + \overset{2}{\cancel{8}} \cdot \frac{3}{\cancel{4}}x + \overset{1}{\cancel{8}} \cdot \frac{3}{\cancel{8}}x = 8 \cdot 17$$
$$8x + 6x + 3x = 136$$
$$17x = 136$$
$$x = 8$$

Then $\frac{3}{4}x = \frac{3}{4}(8) = 6$, and
$$\frac{1}{2}\left(\frac{3}{4}x\right) = \frac{1}{2}\left(\frac{3}{4}(8)\right) = 3.$$
Thus, the sides of the triangle are 8 inches, 6 inches, and 3 inches.

CHECK: 6 is $\frac{3}{4}$ of 8,
and 3 is $\frac{1}{2}$ of 6.
The sum of 8, 6, and 3 is 17.

9. Let x = number of regular tickets sold.
Then $350 - x$ = number of combination tickets.
$$15x + 22(350 - x) = 6895$$
$$15x + 7700 - 22x = 6895$$
$$-7x + 7700 = 6895$$
$$-7x = -805$$
$$x = 115$$

Then $350 - x = 350 - (115) = 235$.
Thus, there were 115 regular tickets and 235 combination tickets sold.

CHECK: 115 regular tickets at $15 each gives $1725. 235 combination tickets at $22 each gives $5170. The total amount collected is $1725 + $5170 = $6985.

11. Let x = number of quarters in the collection.
Then $2x + 3$ = number of dimes.
$$25x + 10(2x + 3) = 255$$
$$25x + 20x + 30 = 255$$
$$45x + 30 = 255$$
$$45x = 225$$
$$x = 5$$

Then $2x + 3 = 2(5) + 3 = 13$.
Thus, there are 5 quarters and 13 dimes in the collection.

CHECK: 13 is 3 more than twice 5. 5 quarters are worth $5(\$0.25) = \$1.25$, while 13 dimes are worth $13(\$0.10) = \$1.30$.
Total value = $\$1.25 + \$1.30 = \$2.55$.

13. Let t = number of minutes that the older machine works.
Then $t - 15$ = number of minutes that the newer machine works.
$$175t + 250(t - 15) = 13675$$
$$175t + 250t - 3750 = 13675$$
$$425t - 3750 = 13675$$
$$425t = 17425$$
$$t = 41$$

Since the older machine began the sorting process at 10:00 a.m. and worked for 41 minutes, the sorting is completed at 10:41 a.m.

CHECK: The older machine sorts $175(41) = 7175$ screws, while the newer machine sorts $250(41 - 15) = 250(26) = 6500$ screws. In all, $7175 + 6500 = 13,675$ screws are sorted.

15. Let x = amount of money invested 6.35%.
    Then $x + 4000$ = amount of money invested at 7.28%.
    $$0.0635x + 0.0728(x + 4000) = 972.70$$
    $$10000[0.0635x + 0.0728(x + 4000)] = 10000(972.70)$$
    $$10000(0.0635x) + 10000[0.0728(x + 4000)] = 10000(972.70)$$
    $$635x + 728(x + 4000) = 9727000$$
    $$635x + 728x + 2912000 = 9727000$$
    $$1363x + 2912000 = 9727000$$
    $$1363x = 6815000$$
    $$x = 5000$$

    Thus, $5000 was invested at 6.35%.
    CHECK: $5000 at 6.35% yields $317.50 in interest. $5000 + $4000 = $9000 at 7.28% yields $655.20 in interest. The total interest earned is $317.50 + $655.20 = $972.70.

17. Let x = amount of money invested 9%.
    Then $800 - x$ = amount of money invested at 6%.
    $$0.09x + 0.06(800 - x) = 67.50$$
    $$100[0.09x + 0.06(800 - x)] = 100(67.50)$$
    $$100(0.09x) + 100[0.06(800 - x)] = 100(67.50)$$
    $$9x + 6(800 - x) = 6750$$
    $$9x + 4800 - 6x = 6750$$
    $$3x + 4800 = 6750$$
    $$3x = 1950$$
    $$x = 650$$

    Then $800 - x = 800 - 650 = 150$.
    Thus, $650 was invested at 9% and $150 was invested at 6%.

    CHECK: The interest on the 9% investment is $0.09(\$650) = \$58.50$. The interest on the 6% investment is $0.06(\$150) = \$9.00$. Total interest = $\$58.50 + \$9.00 = \$67.50$.

19. Let x = amount of money invested 8%.
    Then $6000 - x$ = amount of money invested at 12%.
    $$0.08x + 0.12(6000 - x) = 0.09(6000)$$
    $$100[0.08x + 0.12(6000 - x)] = 100[0.09(6000)]$$
    $$100(0.08x) + 100[0.12(6000 - x)] = 100[0.09(6000)]$$
    $$8x + 12(6000 - x) = 9(6000)$$
    $$8x + 72000 - 12x = 54000$$
    $$-4x + 72000 = 54000$$
    $$-4x = -18000$$
    $$x = 4500$$

    Then $6000 - x = 6000 - 4500 = 1500$.
    Thus, $1500 should be invested at 12%.

    CHECK: The interest on the 8% investment is $0.08(\$4500) = \$360$. The interest on the 12% investment is $0.12(\$1500) = \$180$. The total interest $\$360 + \$180 = \$540$ is 9% of $6000.

21. Let $x =$ number of ml of 30% hydrochloric acid solution.
$$0.30x + 0.50(30) = 0.45(x + 30)$$
$$100[0.30x + 0.50(30)] = 100[0.45(x + 30)]$$
$$100(0.30x) + 100[0.50(30)] = 100[0.45(x + 30)]$$
$$30x + 50(30) = 45(x + 30)$$
$$30x + 1500 = 45x + 1350$$
$$1500 = 15x + 1350$$
$$150 = 15x$$
$$10 = x$$

Thus, 10 ml of 30% hydrochloric acid solution should be used in the mixture.

CHECK: In 10 ml of a 30% solution, there are 3 ml of acid. In 30 ml of a 50% solution, there are 15 ml of acid. So there are $3 + 15 = 18$ ml of acid in the mixture, which is 45% of 40 ml.

23. Let $x =$ number of liters of 2.4% salt solution. Then $90 - x =$ number of liters of 4.6% salt solution.
$$0.024x + 0.046(90 - x) = 0.03(90)$$
$$1000[0.024x + 0.046(90 - x)] = 1000[0.03(90)]$$
$$1000(0.024x) + 1000[0.046(90 - x)] = 1000[0.03(90)]$$
$$24x + 46(90 - x) = 30(90)$$
$$24x + 4140 - 46x = 2700$$
$$-22x + 4140 = 2700$$
$$-22x = -1440$$
$$x = 65.5$$

Then $90 - x = 90 - 65.5 = 24.5$. Thus, 65.5 liters of the 2.4% solution should be mixed with 24.5 liters of the 4.6% solution.

CHECK: In 65.5 liters of a 2.4% salt solution, there are $0.024(65.5) = 1.57$ liters of salt. In 24.5 liters of a 4.6% salt solution, there are $0.046(24.5) = 1.13$ liters of salt. In the mixture, there are $1.57 + 1.13 = 2.70$ liters of salt, which is 3% of 90 liters.

25. Let $x =$ number of gallons of pure anti-freeze to be used.
$$1.00x + 0.30(10) = 0.50(x + 10)$$
$$100[1.00x + 0.30(10)] = 100[0.50(x + 10)]$$
$$100(1.00x) + 100[0.30(10)] = 100[0.50(x + 10)]$$
$$100x + 30(10) = 50(x + 10)$$
$$100x + 300 = 50x + 500$$
$$50x + 300 = 500$$
$$50x = 200$$
$$x = 4$$

Thus, 4 gallons of pure anti-freeze should be added to the radiator.

CHECK: 10 gallons of a 30% anti-freeze solution contains 3 gallons of anti-freeze. If we add 4 gallons of anti-freeze, we get 14 gallons in the solution, 7 of which are anti-freeze. This is 50%.

27. Let $x =$ number of pounds of candy that sells for $3.75/lb.
$$3.75x + 5.00(35) = 4.25(x + 35)$$
$$100[3.75x + 5.00(35)] = 100[4.25(x + 35)]$$
$$100(3.75x) + 100[5.00(35)] = 100[4.25(x + 35)]$$
$$375x + 500(35) = 425(x + 35)$$
$$375x + 17500 = 425x + 14875$$
$$17500 = 50x + 14875$$
$$2625 = 50x$$
$$52.5 = x$$

Thus, 52.5 pounds of candy that sells for $3.75/lb should be added to the mixture.

CHECK: 52.5 lbs. at $3.75/lb = 196.87\frac{1}{2}$.

35 lbs. at $5/lb = $175. Total = 371.87\frac{1}{2}$.

87.5 lbs at $4.25/lb = 371.87\frac{1}{2}$.

29. Let t = number of hours that John and Susan must travel before they meet.
$$4t + 8t = 9$$
$$12t = 9$$
$$t = \frac{9}{12} = \frac{3}{4}$$

Since each must travel $\frac{3}{4}$ of an hour or 45 minutes, they will meet at 8:45 a.m.

CHECK: John walks $4\left(\frac{3}{4}\right) = 3$ miles and Susan jogs $8\left(\frac{3}{4}\right) = 6$ miles. $3 + 6 = 9$.

31. Let t = number of hours that John must travel before they meet.
Then $t - \frac{1}{4}$ = number of hours that Susan must travel before they meet.
$$4t + 8\left(t - \frac{1}{4}\right) = 9$$
$$4t + \overset{2}{\cancel{8}} \cdot \frac{1}{\cancel{4}} = 9$$
$$12t - 2 = 9$$
$$12t = 11$$
$$t = \frac{11}{12}$$

Since John must travel $\frac{11}{12}$ hours or 55 minutes before he meets Susan, they will meet at 8:40 a.m.

CHECK: John walks $4\left(\frac{11}{12}\right) = \frac{11}{3} = 3\frac{2}{3}$ miles and Susan jogs $8\left(\frac{11}{12} - \frac{1}{4}\right) = 8\left(\frac{8}{12}\right) = \frac{16}{3} = 5\frac{1}{3}$ miles. $3\frac{2}{3} + 5\frac{1}{3} = 9$.

33. Let t = number of hours that David spends jogging.
Then $2 - t$ = number of hours that David spends walking.
$$5(2 - t) + 9t = 16$$
$$10 - 5t + 9t = 16$$
$$10 + 4t = 16$$
$$4t = 6$$
$$t = \frac{6}{4} = \frac{3}{2}$$

Thus, David jogs for $1\frac{1}{2}$ hours.

CHECK: David jogs $9\left(\frac{3}{2}\right) = \frac{27}{2} = 13\frac{1}{2}$ miles and walks $5\left(\frac{1}{2}\right) = \frac{5}{2} = 2\frac{1}{2}$ miles. $13\frac{1}{2} + 2\frac{1}{2} = 16$.

35. Let r = rate at which additional money is invested.
$$0.072(3200) + r(2800) = 390$$
$$230.4 + 2800r = 390$$
$$2800r = 159.6$$
$$r = \frac{159.6}{2800}$$
$$r = 0.057$$

Sal must invest the additional $2,800 at a 5.7% rate.

CHECK: An investment of $3,200 at 7.2% pays $230.40 investment. An investment of $2,800 at 5.7% pays $159.60 interest. $230.40 + $159.60 = $390.

37. Let p = percent solution for additional hydrochloric acid.
$$0.20(30) + p(60) = 0.40(90)$$
$$6 + 60p = 36$$
$$60p = 30$$
$$p = \frac{30}{60}$$
$$p = 0.50$$

The chemist must use a 50% hydrochloric acid solution.

CHECK: 30 ml of a 20% solution contains 6 ml of acid. 60 ml of a 50% solution contains 30 ml of acid. 90 ml of a 40% solution contains 36 ml of acid and $6 + 30 = 36$.

39. Let w = width of the rectangle.
Then $2w - 1$ = length of the rectangle.
$$w = \frac{1}{7}[2(w) + 2(2w - 1)] + 1$$
$$w = \frac{1}{7}(2w + 4w - 2) + 1$$
$$7w = 7\left[\frac{1}{7}(6w - 2) + 1\right]$$
$$7w = 6w - 2 + 7$$
$$7w = 6w + 5$$
$$w = 5$$

Then $2w - 1 = 2(5) - 1 = 9$.
So the rectangle has a width of 5 and a length of 9.

CHECK: 9 is 1 less than twice 5. The perimeter of a 5×9 rectangle is $2(5) + 2(9) = 28$, and 5 is 1 more than $\frac{1}{7}(28) = 4$.

41. $C = \frac{5}{9}(F - 32)$, where C is the temperature in degrees Celsius and F is the temperature in degrees Fahrenheit.
$$25 < C < 40$$
$$25 < \frac{5}{9}(F - 32) < 40$$
$$9(25) < 9\left[\frac{5}{9}(F - 32)\right] < 9(40)$$
$$225 < 5(F - 32) < 360$$
$$45 < F - 32 < 72$$
$$77 < F < 104$$

Thus, the corresponding temperature range in degrees Fahrenheit is between 77°F and 104°F.

43. Let x = original price of a skirt.
Then $0.80x$ = sale price of the skirt.
$$12.60 < 0.80x < 20.76$$
$$100(12.60) < 100(0.80x) < 100(20.76)$$
$$1260 < 80x < 2076$$
$$15.75 < x < 25.95$$

Thus, the original range of prices on the skirts was between $15.75 and $25.95.

Chapter 4 Review Exercises

1. $\dfrac{-18}{42} = \dfrac{-3 \cdot \cancel{6}}{7 \cdot \cancel{6}} = \dfrac{-3}{7} = -\dfrac{3}{7}$

3. $\dfrac{15x^6}{6x^2} = \dfrac{5x^4 \cdot \cancel{3x^2}}{2 \cdot \cancel{3x^2}} = \dfrac{5x^4}{2}$

5. $\dfrac{-10x^3 y^5}{4xy^{10}} = \dfrac{-5x^2 \cdot \cancel{2xy^5}}{2y^5 \cdot \cancel{2xy^5}} = \dfrac{-5x^2}{2y^5}$
$= -\dfrac{5x^2}{2y^5}$

7. $\dfrac{3t - 7t - t}{-2t^2 - 3t^2} = \dfrac{-5t}{-5t^2} = \dfrac{1(\cancel{-5t})}{t(\cancel{-5t})} = \dfrac{1}{t}$

9. $\dfrac{a}{4} \cdot \dfrac{a}{4} = \dfrac{a^2}{16}$

11. $\dfrac{7a}{6} - \dfrac{5a}{6} = \dfrac{7a - 5a}{6} = \dfrac{\overset{1}{\cancel{2}}a}{\underset{3}{\cancel{6}}} = \dfrac{a}{3}$

13. $\dfrac{4x - 3}{6x} - \dfrac{x - 1}{6x} = \dfrac{4x - 3 - (x - 1)}{6x}$
$= \dfrac{4x - 3 - x + 1}{6} = \dfrac{3x - 2}{6x}$

15. $\dfrac{2y^2 - 3y}{4} - \dfrac{y^2 - 3y}{4} + \dfrac{y^2}{4} = \dfrac{2y^2 - 3y - (y^2 - 3y) + y^2}{4}$
$= \dfrac{2y^2 - 3y - y^2 + 3y + y^2}{4}$
$= \dfrac{\overset{1}{\cancel{2}}y^2}{\underset{2}{\cancel{4}}} = \dfrac{y^2}{2}$

17. $\left(\dfrac{\cancel{x}}{\underset{2}{\cancel{4}}} \cdot \dfrac{\overset{3}{\cancel{6}}}{\cancel{2}y^2} \right) \div (2xy) = \dfrac{3x}{2y^2} \div \dfrac{2xy}{1}$
$= \dfrac{3\cancel{x}}{2y^2} \cdot \dfrac{1}{2\cancel{x}y} = \dfrac{3}{4y^3}$

19. $\dfrac{a}{2} \cdot \dfrac{a}{4} = \dfrac{a^2}{8}$

21. $\dfrac{x^2}{2} - \dfrac{x^2}{6} + \dfrac{x^2}{3} = \dfrac{x^2(3)}{2(3)} - \dfrac{x^2}{6} + \dfrac{x^2(2)}{3(2)}$
$= \dfrac{3x^2}{6} - \dfrac{x^2}{6} + \dfrac{2x^2}{6} = \dfrac{3x^2 - x^2 + 2x^2}{6}$
$= \dfrac{\overset{2}{\cancel{4}}x^2}{\underset{3}{\cancel{6}}} = \dfrac{2x^2}{3}$

23. $\dfrac{4}{x^2} + \dfrac{3}{2x} = \dfrac{4(2)}{x^2(2)} + \dfrac{3(x)}{2x(x)}$
$= \dfrac{8}{2x^2} + \dfrac{3x}{2x^2}$
$= \dfrac{8 + 3x}{2x^2}$

25. $\dfrac{3}{4a^2 b} - \dfrac{5}{6ab} + \dfrac{7}{8b^3} = \dfrac{3(6b^2)}{4a^2 b(6b^2)} - \dfrac{5(4ab^2)}{6ab(4ab^2)} + \dfrac{7(3a^2)}{8b^3(3a^2)}$
$= \dfrac{18b^2}{24a^2 b^3} - \dfrac{20ab^2}{24a^2 b^3} + \dfrac{21a^2}{24a^2 b^3}$
$= \dfrac{18b^2 - 20ab^2 + 21a^2}{24a^2 b^3}$

27. $$\frac{x}{6} - \frac{1}{4} = \frac{7}{12}$$
$$12\left(\frac{x}{6} - \frac{1}{4}\right) = 12\left(\frac{7}{12}\right)$$
$$\frac{\cancel{12}^2}{1} \cdot \frac{x}{\cancel{6}_1} - \frac{\cancel{12}^3}{1} \cdot \frac{1}{\cancel{4}_1} = \frac{\cancel{12}}{1} \cdot \frac{7}{\cancel{12}}$$
$$2x - 3 = 7$$
$$2x = 10$$
$$x = 5$$

29. $$\frac{t+1}{2} + \frac{t+2}{3} < \frac{t+7}{6}$$
$$6\left(\frac{t+1}{2} + \frac{t+2}{3}\right) < 6\left(\frac{t+7}{6}\right)$$
$$\frac{\cancel{6}^3}{1} \cdot \frac{t+1}{\cancel{2}_1} + \frac{\cancel{6}^2}{1} \cdot \frac{t+2}{\cancel{3}_1} < \frac{\cancel{6}}{1} \cdot \frac{t+7}{\cancel{6}}$$
$$3(t+1) + 2(t+2) < t + 7$$
$$3t + 3 + 2t + 4 < t + 7$$
$$5t + 7 < t + 7$$
$$4t < 0$$
$$t < 0$$

31. $$\frac{y+3}{5} - \frac{y-2}{3} = 1$$
$$15\left(\frac{y+3}{5} - \frac{y-2}{3}\right) = 15(1)$$
$$\frac{\cancel{15}^3}{1} \cdot \frac{y+3}{\cancel{5}_1} - \frac{\cancel{15}^5}{1} \cdot \frac{y-2}{\cancel{3}_1} = 15$$
$$3(y+3) - 5(y-2) = 15$$
$$3y + 9 - 5y + 10 = 15$$
$$-2y + 19 = 15$$
$$-2y = -4$$
$$y = 2$$

33. $$\frac{x}{3} = \frac{x+1}{6}$$
$$6 \cdot \frac{x}{3} = 6 \cdot \frac{x+1}{6}$$
$$\frac{\cancel{6}^2}{1} \cdot \frac{x}{\cancel{3}_1} = \frac{\cancel{6}}{1} \cdot \frac{x+1}{\cancel{6}}$$
$$2x = x + 1$$
$$x = 1$$

35. $$2x + 0.2(x+6) = 10$$
$$10[2x + 0.2(x+6)] = 10 \cdot 10$$
$$10(2x) + 10[0.2(x+6)] = 100$$
$$20x + 2(x+6) = 100$$
$$20x + 2x + 12 = 100$$
$$22x + 12 = 100$$
$$22x = 88$$
$$x = 4$$

37. Let x = the number of ounces in 1 kilogram (1000 grams).
$$\frac{1000}{x} = \frac{28.4}{1}$$
$$\cancel{x} \cdot \frac{1000}{\cancel{x}} = x \cdot \frac{28.4}{1}$$
$$1000 = 28.4x$$
$$\frac{1000}{28.4} = x$$
$35.21 = x$ (to the nearest hundredth).
Thus, there are 35.21 ounces in 1 kilogram.

39. Let $x =$ amount invested at 6%.
    Then $2x =$ amount invested at 7%
    and $7000 - 3x =$ amount invested at 8%.

$$0.06x + 0.07(2x) + 0.08(7000 - 3x) \geq 500$$
$$100[0.06x + 0.07(2x) + 0.08(7000 - 3x)] \geq 100(500)$$
$$100(0.06x) + 100[0.07(2x)] + 100[.08(7000 - 3x)] \geq 50000$$
$$6x + 7(2x) + 8(7000 - 3x) \geq 50000$$
$$6x + 14x + 56000 - 24x \geq 50000$$
$$-4x + 56000 \geq 50000$$
$$-4x \geq -6000$$
$$x \leq 1500$$

Thus, the most that can be invested at 6% is $1500.

41. Let $x =$ Bill's present speed.
$5x = 3(x + 20)$
$5x = 3x + 60$
$2x = 60$
$x = 30$

Thus, Bill's present speed is 30 mph.

CHECK: After 5 hours, Bill covered $(30)(5) = 150$ miles. In 3 hours, he would cover $(50)(3) = 150$ miles at the faster rate as well.

Chapter 4 Practice Test

1. (a) $\dfrac{-10}{24} = \dfrac{\cancel{2}(-5)}{\cancel{2}(12)} = \dfrac{-5}{12} = -\dfrac{5}{12}$

   (b) $\dfrac{x^{\cancel{10}8}}{\cancel{x^2}_1} = \dfrac{x^8}{1} = x^8$

   (c) $\dfrac{\cancel{-6a^6}^{2a^3}}{\cancel{-3a^3}_1} = \dfrac{2a^3}{1} = 2a^3$

   (d) $\dfrac{\cancel{25r^2t^5}^{5t^2}}{\cancel{-15r^4t}_{-3r^2}} = \dfrac{5t^2}{-3r^2} = -\dfrac{5t^2}{3r^2}$

3. (a)
$$\dfrac{x}{3} + \dfrac{x}{5} = 8$$
$$15\left(\dfrac{x}{3} + \dfrac{x}{5}\right) = 15 \cdot 8$$
$$\dfrac{\cancel{15}^5}{1} \cdot \dfrac{x}{\cancel{3}_1} + \dfrac{\cancel{15}^3}{1} \cdot \dfrac{x}{\cancel{5}_1} = 120$$
$$5x + 3x = 120$$
$$8x = 120$$
$$x = 15$$

   (b)
$$\dfrac{x-5}{2} + \dfrac{x}{5} \geq 8$$
$$10\left(\dfrac{x-5}{2} + \dfrac{x}{5}\right) \geq 10 \cdot 8$$
$$\dfrac{\cancel{10}^5}{1} \cdot \dfrac{x-5}{\cancel{2}_1} + \dfrac{\cancel{10}^2}{1} \cdot \dfrac{x}{\cancel{5}_1} \geq 80$$
$$5(x - 5) + 2x \geq 80$$
$$5x - 25 + 2x \geq 80$$
$$7x - 25 \geq 80$$
$$ +25 +25$$
$$\overline{7x \geq 105}$$
$$\dfrac{7x}{7} \geq \dfrac{105}{7}$$
$$x \geq 15$$

(c)
$$\frac{a+3}{5} - \frac{a-2}{4} = 1$$
$$20\left(\frac{a+3}{5} - \frac{a-2}{4}\right) = 20 \cdot 1$$
$$\frac{\cancel{20}^4}{1} \cdot \frac{a+3}{\cancel{5}_1} - \frac{\cancel{20}^5}{1} \cdot \frac{a-2}{\cancel{4}_1} = 20$$
$$4(a+3) - 5(a-2) = 20$$
$$4a + 12 - 5a + 10 = 20$$
$$-a + 22 = 20$$
$$\underline{-22 \quad -22}$$
$$-a = -2$$
$$\frac{-a}{-1} = \frac{-2}{-1}$$
$$a = 2$$

(d)
$$0.03t + 0.5t = 10.6$$
$$100(0.03t + 0.5t) = 100(10.6)$$
$$100(0.03t) + 100(0.5t) = 100(10.6)$$
$$3t + 50t = 1060$$
$$53t = 1060$$
$$\frac{\cancel{53}t}{\cancel{53}} = \frac{1060}{53}$$
$$t = 20$$

5. Let x = the number.
$$x + \frac{2}{3}x = 2x - 5$$
$$3\left(x + \frac{2}{3}x\right) = 3(2x - 5)$$
$$3x + \frac{\cancel{3}}{1} \cdot \frac{2}{\cancel{3}}x = 3(2x) - 3(5)$$
$$3x + 2x = 6x - 15$$
$$5x = 6x - 15$$
$$\underline{-6x \quad -6x}$$
$$-x = -15$$
$$\frac{-x}{-1} = \frac{-15}{-1}$$
$$x = 15$$

CHECK: $\dfrac{2}{3}$ of $15 = \dfrac{2}{\cancel{3}_1} \cdot \dfrac{\cancel{15}^5}{1} = \dfrac{10}{1} = 10.$

Then $15 + 10 = 25$, which is 5 less than twice 15.

7. Let x = amount invested at 8%.
Then $7000 - x$ = amount invested at 13%.

$$0.08x + 0.13(7000 - x) = 750$$
$$100[0.08x + 0.13(7000 - x)] = 100(750)$$
$$100(0.08x) + 100[0.13(7000 - x)] = 100(750)$$
$$8x + 13(7000 - x) = 75000$$
$$8x + 91000 - 13x = 75000$$
$$-5x + 91000 = 75000$$
$$\underline{-91000 \quad -91000}$$
$$-5x = -16000$$
$$\frac{\cancel{-5}x}{\cancel{-5}} = \frac{-16000}{-5}$$
$$x = 3200$$

Then $7000 - x = 7000 - 3200 = 3800$.
Thus, \$3200 is invested at 8% and \$3800 is invested at 13%.

CHECK: $3200 + 3800 = 7000$.
The interest on \$3200 at 8% is
\$256 ($= 0.08 \cdot 3200$).
The interest on \$3800 at 13% is
\$494 ($= 0.13 \cdot 3800$).
Then \$256 + \$494 = \$750,
as required.

9. Let x = number of hours that the first person drives.
Then $x - 4$ = number of hours that the second person drives.

$$48x + 55(x - 4) = 604$$
$$48x + 55x - 220 = 604$$
$$103x - 220 = 604$$
$$ +220 +220$$
$$\overline{103x = 824}$$
$$\frac{\cancel{103}x}{\cancel{103}} = \frac{824}{103}$$
$$x = 8$$

Thus, they will be 604 kilometers apart after the first person drives for 8 hours, or at 7:00 p.m.

CHECK: In 8 hours, the first person covers $48(8) = 384$ km.
In $8 - 4 = 4$ hours, the second person covers $55(4) = 220$ km.
Together they cover $384 + 220 = 604$ km.

Chapter 5
Graphing Straight Lines

Exercises 5.1

1-19. (odd)

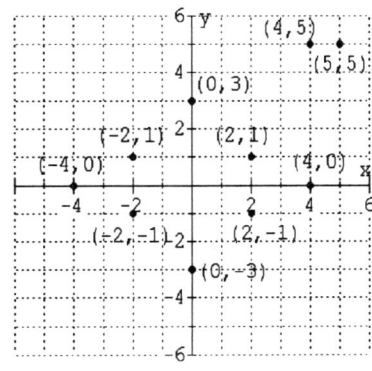

21. Quadrant I
23. Quadrant III
25. on y-axis
27. Quadrant II
29. on both x-axis and y-axis (origin)
31. Quadrant IV
33. $(1, 7)$ satisfies the equation $2y - 4x = 10$, since $2(7) - 4(1) = 14 - 4 = 10$.
35. $(0, -1)$ satisfies the equation $y = 4x - 1$, since $4(0) - 1 = 0 - 1 = -1$.
37. All three points satisfy the equation $6x - 4y - 8 = 0$, since
$6(-2) - 4(-5) - 8 = -12 + 20 - 8 = 0$;
$6(6) - 4(7) - 8 = 36 - 28 - 8 = 0$;
$6(-10) - 4(-17) - 8 = -60 + 68 - 8 = 0$.
39. $(-6, -10)$ satisfies the equation $\frac{2}{3}x - \frac{1}{2}y = 1$, since $\frac{2}{3}(-6) - \frac{1}{2}(-10) = -4 + 5 = 1$.
41. $(1, -6)$ satisfies the equation $y = x^2 - 3x - 4$, since $(1)^2 - 3(1) - 4 = 1 - 3 - 4 = -6$.
43. The last two ordered pairs satisfy the equation $y = \frac{1}{x+2}$, since $\frac{1}{1+2} = \frac{1}{3}$ and $\frac{1}{-1+2} = 1$.
45. $4y - 3x = 7$

$x = -1$:
$4y - 3(-1) = 7$
$4y + 3 = 7$
$4y = 4$
$y = 1$

$y = -5$:
$4(-5) - 3x = 7$
$-20 - 3x = 7$
$-3x = 27$
$x = -9$

$y = 0$:
$4(0) - 3x = 7$
$0 - 3x = 7$
$-3x = 7$
$x = -\frac{7}{3}$

$x = 0$:
$4y - 3(0) = 7$
$4y - 0 = 7$
$4y = 7$
$y = \frac{7}{4}$

x	y
-1	1
-9	-5
$-\frac{7}{3}$	0
0	$\frac{7}{4}$

47. $y = \frac{2}{3}x + 1$

$x = 8:$
$y = \frac{2}{3}(8) + 1$
$y = \frac{19}{3}$

$y = 8:$
$8 = \frac{2}{3}x + 1$
$7 = \frac{2}{3}x$
$\frac{21}{2} = x$

$y = -4:$
$-4 = \frac{2}{3}x + 1$
$-5 = \frac{2}{3}x$
$-\frac{15}{2} = x$

$x = -4:$
$y = \frac{2}{3}(-4) + 1$
$y = -\frac{5}{3}$

x	y
8	$\frac{19}{3}$
$\frac{21}{2}$	8
$-\frac{15}{2}$	-4
-4	$-\frac{5}{3}$

49. $\{(x, y) \mid y = x + 2 \text{ and } x = -3, 0, 4\}$

x	y	(x, y)
-3	-1	$(-3, -1)$
0	2	$(0, 2)$
4	6	$(4, 6)$

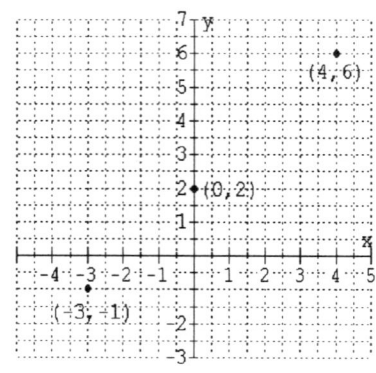

50. $\{(x, y) \mid x = y + 2 \text{ and } x = -3, 0, 4\}$

x	y	(x, y)
-3	-5	$(-3, -5)$
0	-2	$(0, -2)$
4	2	$(4, 2)$

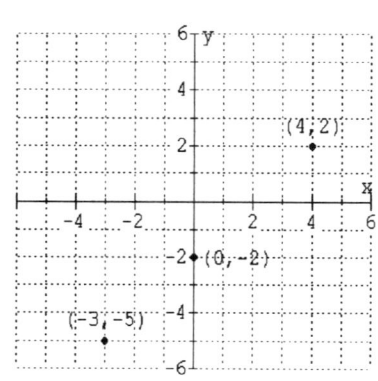

51. $(2, 3), (0, 6), (6, -3), (4, 0)$

52. $(-2, -16), (0, -8), (3, 4), (2, 0)$

53. An ordered pair (x, y) is a pair of numbers in which the order of appearance matters. For instance, $(3, 4)$ and $(4, 3)$ are different ordered pairs, even though both involve the same two numbers.

54. The notation $\{x, y\}$ stands for the set containing the two elements x and y. Here, order does not matter. That is, $\{3, 4\}$ and $\{4, 3\}$ are equal sets. When we write (x, y), we mean that x comes first and y second.

55.

x	y	(x,y)
-1	1	$(-1,1)$
0	0	$(0,0)$
1	1	$(1,1)$
2	4	$(2,4)$

The graph of $y = x^2$ is not a straight line.

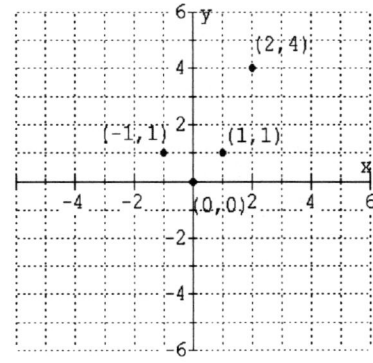

56.

x	y	(x,y)
-1	-1	$(-1,-1)$
0	0	$(0,0)$
1	1	$(1,1)$
2	8	$(2,8)$

The graph of $y = x^3$ is not a straight line.

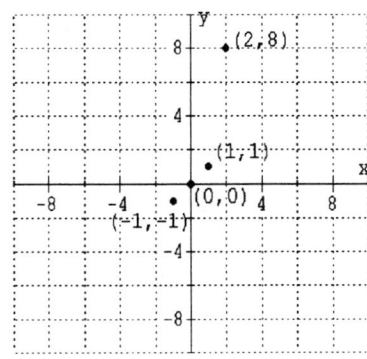

57. Let $x =$ number of HW problems Marina had.
$$\frac{1}{3}x + 8 + \frac{2}{5}x = x$$
$$15\left(\frac{1}{3}x + 8 + \frac{2}{5}x\right) = 15x$$
$$\overset{5}{\cancel{15}}\left(\frac{1}{\cancel{3}}x\right) + 15(8) + \overset{3}{\cancel{15}}\left(\frac{2}{\cancel{5}}x\right) = 15x$$
$$5x + 120 + 6x = 15x$$
$$11x + 120 = 15x$$
$$120 = 4x$$
$$30 = x$$

Thus, Marina had 30 homework problems.

Exercises 5.2

1. $3x - 7y = 21$

$y = 15:$
$3x - 7(15) = 21$
$3x - 105 = 21$
$3x = 126$
$x = 42$
$(42, 15)$

$x = 14:$
$3(14) - 7y = 21$
$42 - 7y = 21$
$-7y = -21$
$y = 3$
$(14, 3)$

$x = -2:$
$3(-2) - 7y = 21$
$-6 - 7y = 21$
$-7y = 27$
$y = -\frac{27}{7}$
$\left(-2, -\frac{27}{7}\right)$

3. $2y + 9x = 36$

$x = 6:$
$2y + 9(6) = 36$
$2y + 54 = 36$
$2y = -18$
$y = -9$
$(6, -9)$

$x = 0:$
$2y + 9(0) = 36$
$2y + 0 = 36$
$2y = 36$
$y = 18$
$(0, 18)$

$y = 0:$
$2(0) + 9x = 36$
$0 + 9x = 36$
$9x = 36$
$x = 4$
$(4, 0)$

5. $3x - 5y = 10$

$x = -\dfrac{2}{3}:$
$3\left(-\dfrac{2}{3}\right) - 5y = 10$
$-2 - 5y = 10$
$-5y = 12$
$y = -\dfrac{12}{5}$
$\left(-\dfrac{2}{3}, -\dfrac{12}{5}\right)$

$y = -\dfrac{4}{5}:$
$3x - 5\left(-\dfrac{4}{5}\right) = 10$
$3x + 4 = 10$
$3x = 6$
$x = 2$
$\left(2, -\dfrac{4}{5}\right)$

$x = 5:$
$3(5) - 5y = 10$
$15 - 5y = 10$
$-5y = -5$
$y = 1$
$(5, 1)$

7. $y = -\dfrac{1}{2}x + 5$

$x = -6:$
$y = -\dfrac{1}{2}(-6) + 5$
$y = 8$
$(-6, 8)$

$y = 4:$
$4 = -\dfrac{1}{2}x + 5$
$-1 = -\dfrac{1}{2}x$
$2 = x$
$(2, 4)$

$x = 3:$
$y = -\dfrac{1}{2}(3) + 5$
$y = \dfrac{7}{2}$
$\left(3, \dfrac{7}{2}\right)$

9. To find the x-intercept, we set $y = 0$ and solve for x.
$x - y = 7$
$x - 0 = 7$
$x = 7$
Therefore, the x-intercept is 7.

To find the y-intercept, we set $x = 0$ and solve for y.
$x - y = 7$
$0 - y = 7$
$-y = 7$
$y = -7$
Therefore, the y-intercept is -7.

11. To find the x-intercept, we set $y = 0$ and solve for x.
$y - x = -4$
$0 - x = -4$
$-x = -4$
$x = 4$
Therefore, the x-intercept is 4.

To find the y-intercept, we set $x = 0$ and solve for y.
$y - x = -4$
$y - 0 = -4$
$y = -4$
Therefore, the y-intercept is -4.

13. To find the x-intercept, we set $y = 0$ and solve for x.
$2x + 3y = 12$
$2x + 3(0) = 12$
$2x + 0 = 12$
$2x = 12$
$x = 6$
Therefore, the x-intercept is 6.

To find the y-intercept, we set $x = 0$ and solve for y.
$2x + 3y = 12$
$2(0) + 3y = 12$
$0 + 3y = 12$
$3y = 12$
$y = 4$
Therefore, the y-intercept is 4.

15. To find the x-intercept, we set $y = 0$ and solve for x.
$y = 4x$
$0 = 4x$
$0 = x$
Therefore, the x-intercept is 0.

To find the y-intercept, we set $x = 0$ and solve for y.
$y = 4x$
$y = 4(0)$
$y = 0$
Therefore, the y-intercept is 0.

17. To find the x-intercept, we set $y = 0$ and solve for x.
$2x = 5y$
$2x = 5(0)$
$2x = 0$
$x = 0$
Therefore, the x-intercept is 0.

To find the y-intercept, we set $x = 0$ and solve for y.
$2x = 5y$
$2(0) = 5y$
$0 = 5y$
$0 = y$
Therefore, the y-intercept is 0.

19. To find the x-intercept, we set $y = 0$ and solve for x. Since the equation $x = 5$ does not contain y, its solution is just $x = 5$. So the x-intercept is 5. To find the y-intercept, we set $x = 0$ and solve for y. But this gives $0 = 5$, which is a contradiction. So the equation $x = 5$ has no y-intercept.

21. To find the x-intercept, we set $y = 0$ and solve for x.
$x - 4 = 3y$
$x - 4 = 3(0)$
$x - 4 = 0$
$x = 4$
Therefore, the x-intercept is 4.

To find the y-intercept, we set $x = 0$ and solve for y.
$x - 4 = 3y$
$0 - 4 = 3y$
$-4 = 3y$
$-\dfrac{4}{3} = y$
Therefore, the y-intercept is $-\dfrac{4}{3}$.

23. $x + y = -5$

x-intercept:
$x + (0) = -5$
$x + 0 = -5$
$x = -5$
Plot $(-5, 0)$

y-intercept:
$(0) + y = -5$
$0 + y = -5$
$y = -5$
Plot $(0, -5)$

check point: choose $x = 1$
$(1) + y = -5$
$1 + y = -5$
$y = -6$
Plot $(1, -6)$

The graph of $x + y = -5$

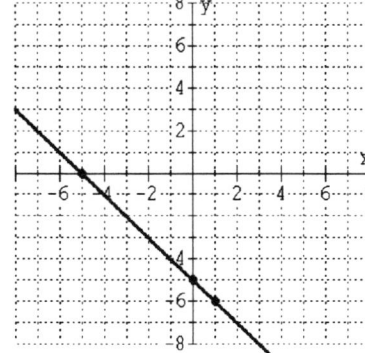

25. $x - y = 8$
    x-intercept: y-intercept:
    $x - 0 = 8$ $0 - y = 8$
    $x = 8$ $-y = 8$
     $y = -8$
    Plot $(8, 0)$ Plot $(0, -8)$

    check point: choose $x = 4$
    $4 - y = 8$
    $ -y = 4$
    $y = -4$
    Plot $(4, -4)$

The graph of $x - y = 8$

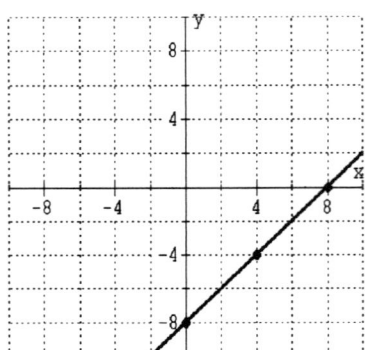

27. $3x - 4y = 12$
    x-intercept: y-intercept:
    $3x - 4(0) = 12$ $3(0) - 4y = 12$
    $3x - 0 = 12$ $0 - 4y = 12$
    $3x = 12$ $-4y = 12$
    $x = 4$ $y = -3$
    Plot $(4, 0)$ Plot $(0, -3)$

    check point: choose $x = 8$
    $3(8) - 4y = 12$
    $24 - 4y = 12$
    $-4y = -12$
    $y = 3$
    Plot $(8, 3)$

The graph of $3x - 4y = 12$

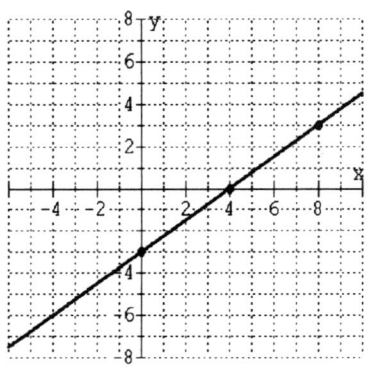

29. $y = 2x - 10$
    x-intercept: y-intercept:
    $0 = 2x - 10$ $y = 2(0) - 10$
    $10 = 2x$ $y = -10$
    $5 = x$
    Plot $(5, 0)$ Plot $(0, -10)$

    check point: choose $x = 3$
    $y = 2(3) - 10$
    $y = -4$
    Plot $(3, -4)$

The graph of $y = 2x - 10$

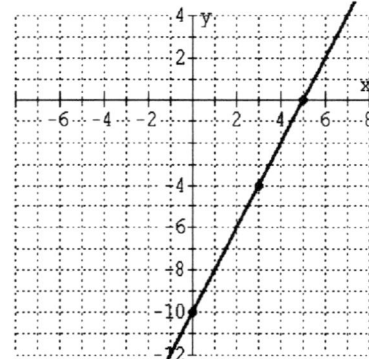

31. $y = -4x$
   x-intercept:
   $(0) = -4x$
   $0 = -4x$
   $0 = x$
   Plot $(0, 0)$

   second point: choose $x = 1$
   $y = -4(1)$
   $y = -4$
   Plot $(1, -4)$

   check point: choose $x = -1$
   $y = -4(-1)$
   $y = 4$
   Plot $(-1, 4)$

The graph of $y = -4x$

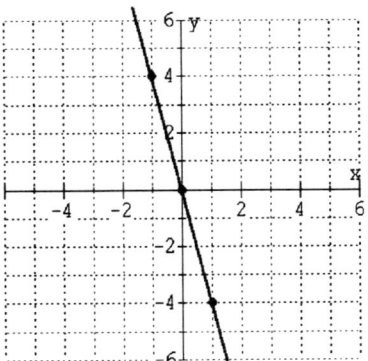

33. $y + 7 = x - 5$
   x-intercept: y-intercept:
   $0 + 7 = x - 5$ $y + 7 = 0 - 5$
   $7 = x - 5$ $y + 7 = -5$
   $12 = x$ $y = -12$
   Plot $(12, 0)$ Plot $(0, -12)$

   check point: choose $x = 5$
   $y + 7 = 5 - 5$
   $y + 7 = 0$
   $y = -7$
   Plot $(5, -7)$

The graph of $y + 7 = x - 5$

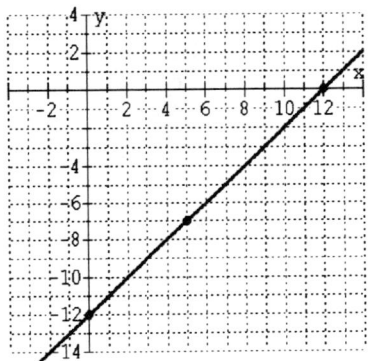

35. $y = 5$

Its graph is a line parallel to the x-axis and 5 units above it.

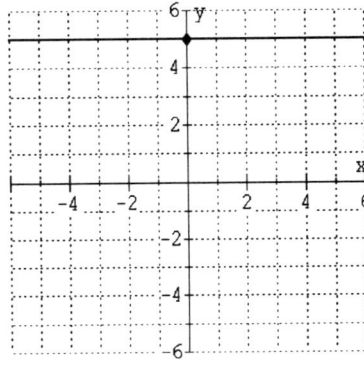

37. $x = -4$

Its graph is a line parallel to the y-axis and 4 units to the left of it.

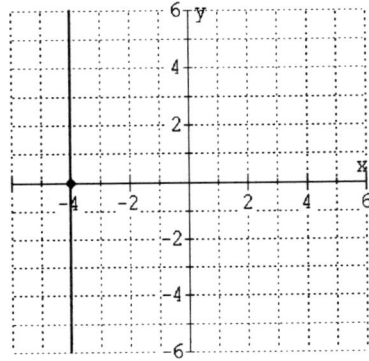

39. $y = \dfrac{x-3}{2}$

x-intercept:
$0 = \dfrac{x-3}{2}$
$0 = x - 3$
$3 = x$

Plot $(3, 0)$

y-intercept:
$y = \dfrac{0-3}{2}$
$y = -\dfrac{3}{2}$

Plot $\left(0, -\dfrac{3}{2}\right)$

check point: choose $x = 1$
$y = \dfrac{1-3}{2}$
$y = -1$
Plot $(1, -1)$

The graph of $y = \dfrac{x-3}{2}$

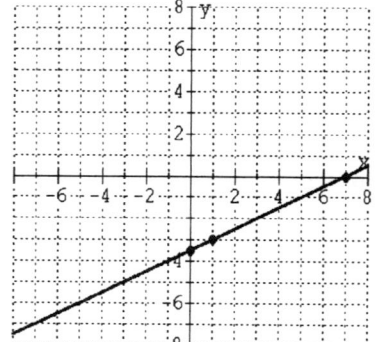

41. $2(x - 3) = 4(y + 2)$

x-intercept:
$2(x - 3) = 4[(0) + 2]$
$2(x - 3) = 4(0 + 2)$
$2x - 6 = 8$
$2x = 14$
$x = 7$

Plot $(7, 0)$

y-intercept:
$2[(0) - 3] = 4(y + 2)$
$2(0 - 3) = 4(y + 2)$
$-6 = 4y + 8$
$-14 = 4y$
$-\dfrac{7}{2} = y$

Plot $\left(0, -\dfrac{7}{2}\right)$

check point: choose $x = 1$
$2[(1) - 3] = 4(y + 2)$
$2(1 - 3) = 4(y + 2)$
$2(-2) = 4(y + 2)$
$-4 = 4y + 8$
$-12 = 4y$
$-3 = y$
Plot $(1, -3)$

The graph of $2(x - 3) = 4(y + 2)$

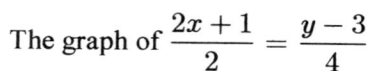

43. $\dfrac{2x+1}{2} = \dfrac{y-3}{4}$

$4\left(\dfrac{2x+1}{2}\right) = 4\left(\dfrac{y-3}{4}\right)$
$2(2x + 1) = y - 3$
$4x + 2 = y - 3$

x-intercept:
$4x + 2 = 0 - 3$
$4x + 2 = -3$
$4x = -5$
$x = -\dfrac{5}{4}$

Plot $\left(-\dfrac{5}{4}, 0\right)$

y-intercept:
$4(0) + 2 = y - 3$
$2 = y - 3$
$5 = y$

Plot $(0, 5)$

check point: choose $x = -1$
$4(-1) + 2 = y - 3$
$-2 = y - 3$
$1 = y$

Plot $(-1, 1)$

The graph of $\dfrac{2x+1}{2} = \dfrac{y-3}{4}$

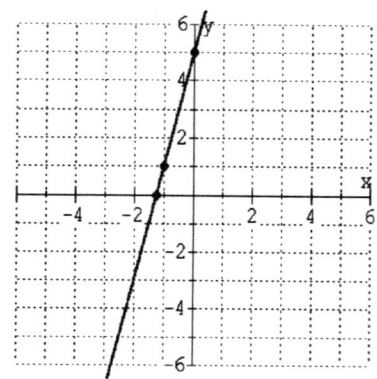

45. $d = 3t + 4$

    t-intercept: d-intercept:

    $0 = 3t + 4$ $d = 3(0) + 4$
    $-4 = 3t$ $d = 0 + 4$
    $-\dfrac{4}{3} = t$ $d = 4$

    Plot $\left(-\dfrac{4}{3}, 0\right)$ Plot $(0, 4)$

    check point: choose $t = -1$
    $d = 3(-1) + 4$
    $d = -3 + 4$
    $d = 1$
    Plot $(-1, 1)$

The graph of $d = 3t + 4$

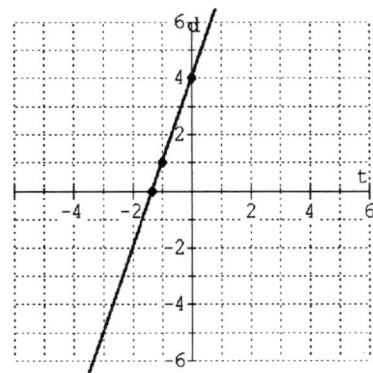

47. $h = -6t + 30$

    t-intercept: h-intercept:

    $0 = -6t + 30$ $h = -6(0) + 30$
    $6t = 30$ $h = 0 + 30$
    $t = 5$ $h = 30$

    Plot $(5, 0)$ Plot $(0, 30)$

    check point: choose $t = 3$
    $h = -6(3) + 30$
    $h = -18 + 30$
    $h = 12$
    Plot $(3, 12)$

The graph of $h = -6t + 30$

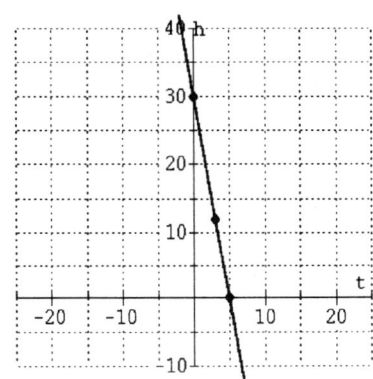

49. $u - 4v = 8$

    u-intercept: v-intercept:

    $u - 4(0) = 8$ $(0) - 4v = 8$
    $u - 0 = 8$ $-4v = 8$
    $u = 8$ $v = -2$

    Plot $(8, 0)$ Plot $(0, -2)$

    check point: choose $v = -1$
    $u - 4(-1) = 8$
    $u + 4 = 8$
    $u = 4$
    Plot $(4, -1)$

The graph of $u - 4v = 8$

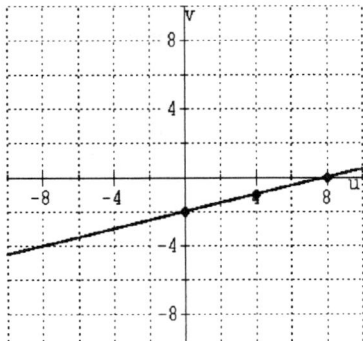

51. (a) $125v + 165t = 9110$

    (b) v-intercept:
    $$125v + 165(0) = 9110$$
    $$125v + 0 = 9110$$
    $$125v = 9110$$
    $$v \approx 72.88$$

    t-intercept:
    $$125(0) + 165t = 9110$$
    $$0 + 165t = 9110$$
    $$165t = 9110$$
    $$t \approx 55.21$$

    (c) $$125(28) + 165t = 9110$$
    $$3500 + 165t = 9110$$
    $$165t = 5610$$
    $$t = 34$$
    The store orders 34 TVs.

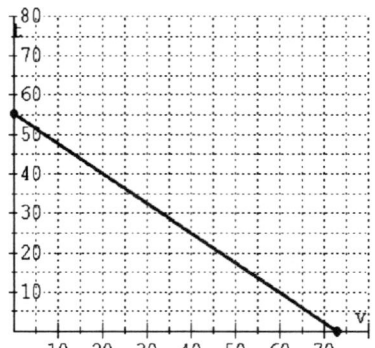

53. (a) $12a + 8c = 360$

    (b) a-intercept:
    $$12a + 8(0) = 360$$
    $$12a + 0 = 360$$
    $$12a = 360$$
    $$a = 30$$

    c-intercept:
    $$12(0) + 8c = 360$$
    $$0 + 8c = 360$$
    $$8c = 360$$
    $$c = 45$$

    (c) $$12(25) + 8c = 360$$
    $$300 + 8c = 360$$
    $$8c = 60$$
    $$c = 7.5$$
    Mary checked 7.5 credit reports.

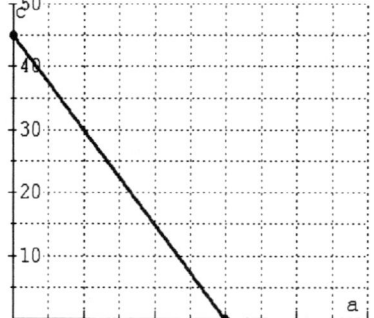

55.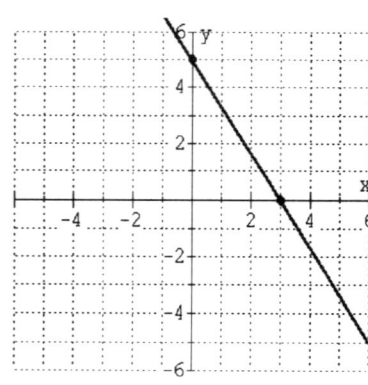

56. Some points on this line are $(-4, -4)$, $(-3, -3)$, $(0, 0)$, $(2, 2)$, and $(5, 5)$. Since the y-coordinate of any such point is equal to its x-coordinate, an equation of this line would be $y = x$.

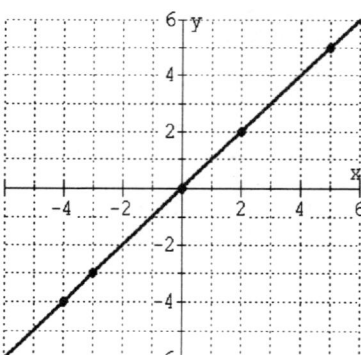

57. Some points on this line are $(-4, 4)$, $(-3, 3)$, $(0, 0)$, $(2, -2)$, and $(5, -5)$. Since the y-coordinate of any such point is equal to the negative of its x-coordinate, an equation of this line would be $y = -x$.

The graph of $y = -x$

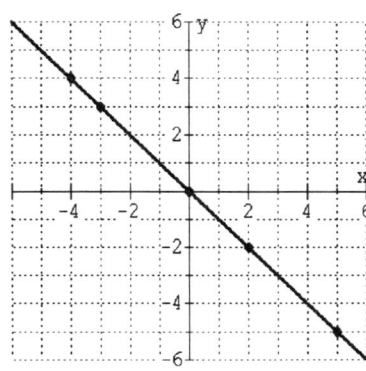

58. The x-coordinate of a point where a graph intersects the x-axis is called an x-intercept of the graph. To find an x-intercept, set $y = 0$ and solve for x. The y-coordinate of a point where a graph intersects the y-axis is called a y-intercept of the graph. To find a y-intercept, set $x = 0$ and solve for y.

59. $c = 3n + 2$

    (a) n-intercept:
    $$\begin{aligned} 0 &= 3n + 2 \\ -2 &= 3n \\ -\frac{2}{3} &= n \end{aligned}$$
    Plot $\left(-\frac{2}{3}, 0\right)$

    c-intercept:
    $$\begin{aligned} c &= 3(0) + 2 \\ c &= 0 + 2 \\ c &= 2 \end{aligned}$$
    Plot $(0, 2)$

    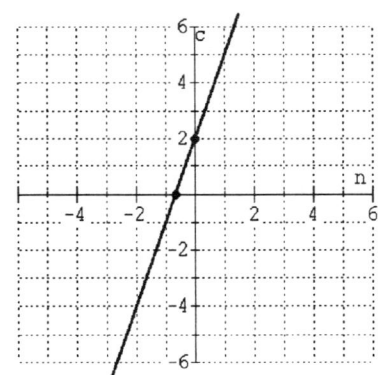

    (b) As n increases, c increases.

    (c) n-intercept:
    Plot $\left(0, -\frac{2}{3}\right)$

    c-intercept:
    Plot $(2, 0)$

    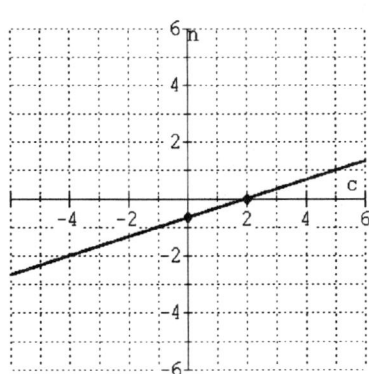

    (d) As c increases, n increases.

    (e) The appearance is different, but the relationship between the variables is unchanged.

61. $(2x^2)(5xy)(-y^2) + x^3y^3 = -10x^3y^3 + x^3y^3$
$= -9x^3y^3$

63. Let r = interest rate.
$$0.04(1300) + 800r = 100$$
$$52 + 800r = 100$$
$$800r = 48$$
$$r = 0.06 = 6\%$$
The $800 needs to be invested at 6%.

Exercises 5.3

1. $m = \dfrac{y_2 - y_1}{x_2 - x_1} = \dfrac{9 - 5}{6 - 3} = \dfrac{4}{3}$

3. $m = \dfrac{y_2 - y_1}{x_2 - x_1} = \dfrac{6 - 3}{2 - 1} = \dfrac{3}{1} = 3$

5. $m = \dfrac{y_2 - y_1}{x_2 - x_1} = \dfrac{4 - (-2)}{-1 - 3} = \dfrac{6}{-4} = -\dfrac{3}{2}$

7. $m = \dfrac{y_2 - y_1}{x_2 - x_1} = \dfrac{-4 - (-2)}{-3 - (-1)} = \dfrac{-2}{-2} = 1$

9. $m = \dfrac{y_2 - y_1}{x_2 - x_1} = \dfrac{2 - 0}{0 - 3} = -\dfrac{2}{3}$

11. $m = \dfrac{y_2 - y_1}{x_2 - x_1} = \dfrac{7 - 7}{-3 - 4} = \dfrac{0}{-7} = 0$

13. m is undefined. (If we tried to use the formula, we would have obtained $m = \dfrac{9 - 6}{2 - 2} = \dfrac{3}{0}$, and division by zero is undefined.)

15. $m = \dfrac{y_2 - y_1}{x_2 - x_1} = \dfrac{\frac{1}{2} - 2}{1 - \frac{3}{4}} = \dfrac{\frac{1}{2} - \frac{4}{2}}{\frac{4}{4} - \frac{3}{4}}$
$= \dfrac{-\frac{3}{2}}{\frac{1}{4}} = -\dfrac{3}{2} \cdot \dfrac{4}{1} = -6$

17. $m = \dfrac{y_2 - y_1}{x_2 - x_1} = \dfrac{\frac{1}{4} - \frac{1}{5}}{\frac{3}{2} - (-\frac{1}{3})} = \dfrac{\frac{5}{20} - \frac{4}{20}}{\frac{9}{6} + \frac{2}{6}}$
$= \dfrac{\frac{1}{20}}{\frac{11}{6}} = \dfrac{1}{20} \cdot \dfrac{6}{11} = \dfrac{3}{110}$

19. $m = \dfrac{y_2 - y_1}{x_2 - x_1} = \dfrac{3 - (-1)}{-1 - 3} = \dfrac{4}{-4} = -1$

21. $m = \dfrac{y_2 - y_1}{x_2 - x_1} = \dfrac{-5 - (-5)}{2 - (-3)} = \dfrac{0}{5} = 0$

23. $m = \dfrac{y_2 - y_1}{x_2 - x_1} = \dfrac{0 - a}{a - 0} = \dfrac{-a}{a} = -1$

25. $m = \dfrac{y_2 - y_1}{x_2 - x_1} = \dfrac{2b - b}{2a - a} = \dfrac{b}{a}$

27. $m = \dfrac{y_2 - y_1}{x_2 - x_1} = \dfrac{4.72 - 2.65}{1.3 - 0.8} = \dfrac{2.07}{0.5}$
$= 4.14$

29. $m = \dfrac{y_2 - y_1}{x_2 - x_1} = \dfrac{4.9 - (-1.05)}{-2.16 - 3.7}$
$= \dfrac{4.9 + 1.05}{-2.16 - 3.7} = \dfrac{5.95}{-5.86} \approx -1.02$

31. $m = \dfrac{y_2 - y_1}{x_2 - x_1} = \dfrac{8.77 - 8.77}{-1.4 - 9.62} = \dfrac{0}{-11.02} = 0$

33.

35.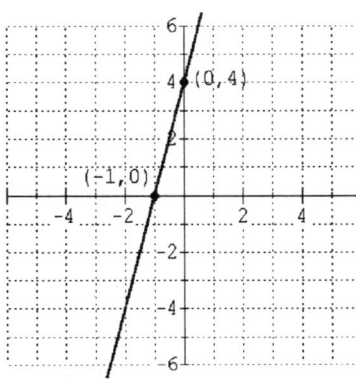

37.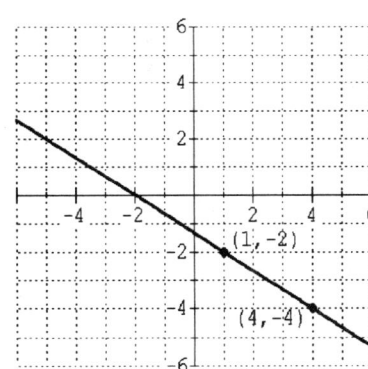

39.

41.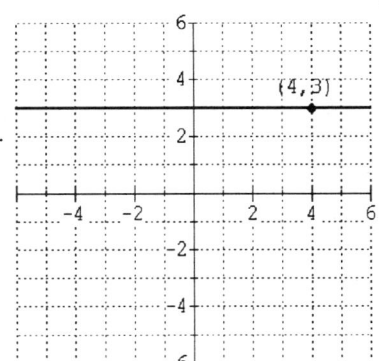

43. A slope of 0 means that the line is horizontal.

45. No slope means that the line is vertical.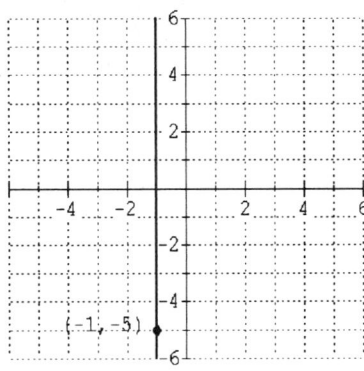

47. $\dfrac{-5}{4}$ 49. 5 51. $\dfrac{7}{3}$ 53. 0

55. positive 57. zero 59. m_3, m_2, m_1

61. (a) 40 calories

(b) Consider the points $(2, 10)$ and $(4, 20)$: slope $= \dfrac{c_2 - c_1}{m_2 - m_1} = \dfrac{20 - 10}{4 - 2} = \dfrac{10}{2} = 5.$

(c) 5 calories per minute are burned.

63. (a) 16 gallons (b) 180 miles (c) $\dfrac{360 \text{ miles}}{16 \text{ gal}} = 22.5 \text{ mpg}$ $\dfrac{180 \text{ miles}}{8 \text{ gal}} = 22.5 \text{ mpg}$

(d) Consider the points $(360, 16)$ and $(180, 8)$: slope $= \dfrac{g_2 - g_1}{m_2 - m_1} = \dfrac{16 - 8}{360 - 180} = \dfrac{8}{180} = \dfrac{2}{45}.$

(e) For every 2 gallons of gas you can go 45 miles.

65. (a) $A = 15d + 80$

(b)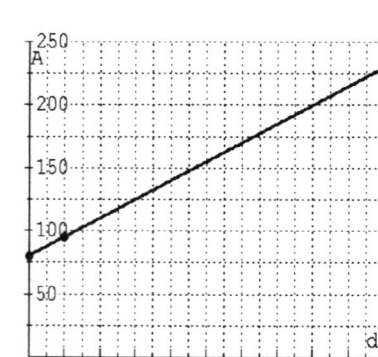

(c) Consider the points $(0, 80)$ and $(1, 95)$:

slope $= \dfrac{A_2 - A_1}{d_2 - d_1} = \dfrac{95 - 80}{1 - 0} = 15.$

67. First, compute the slope of the line through $(-1, -2)$ and $(2, 0)$: $m = \dfrac{0 - (-2)}{2 - (-1)} = \dfrac{2}{3}.$

Next, compute the slope of the line through $(2, 0)$ and $(5, 2)$: $m = \dfrac{2 - 0}{5 - 2} = \dfrac{2}{3}.$

Since these two lines have equal slopes, they are either parallel or actually one line. But the two lines in question cannot be parallel, since $(2, 0)$ is a point on each and parallel lines have no point in common. We can then conclude that the three points $(-1, -2)$, $(2, 0)$, and $(5, 2)$ all lie on a straight line. (Points with this property are called <u>collinear</u>.)

69. LCD $= 12$

$$\dfrac{x}{3} - \dfrac{x-1}{4} + \dfrac{x+1}{2} = \dfrac{(4)x}{(4)3} - \dfrac{(3)(x-1)}{(3)4} + \dfrac{(6)(x+1)}{(6)2}$$

$$= \dfrac{4x}{12} - \dfrac{3(x-1)}{12} + \dfrac{6(x+1)}{12}$$

$$= \dfrac{4x - 3(x-1) + 6(x+1)}{12}$$

$$= \dfrac{4x - 3x + 3 + 6x + 6}{12}$$

$$= \dfrac{7x + 9}{12}$$

Exercises 5.4

1. $y - y_1 = m(x - x_1)$
   $(x_1, y_1) = (1, 5), m = 3$
   $y - 5 = 3(x - 1)$ or
   $\quad y = 3x + 2$

3. $y - y_1 = m(x - x_1)$
   $(x_1, y_1) = (-1, 4), m = \dfrac{1}{2}$
   $y - 4 = \dfrac{1}{2}[x - (-1)]$
   $y - 4 = \dfrac{1}{2}(x + 1)$ or
   $\quad y = \dfrac{1}{2}x + \dfrac{9}{2}$

5. $y - y_1 = m(x - x_1)$
   $(x_1, y_1) = (-3, -5), m = -\dfrac{2}{3}$
   $y - (-5) = -\dfrac{2}{3}[x - (-3)]$
   $y + 5 = -\dfrac{2}{3}(x + 3)$ or
   $\quad y = -\dfrac{2}{3}x - 7$

7. $y = mx + b$
   $m = 5, b = 6$
   $y = 5x + 6$

9. $y = mx + b$
   $m = \dfrac{1}{4}, b = -2$
   $y = \dfrac{1}{4}x + (-2)$ or
   $y = \dfrac{1}{4}x - 2$

11. $y - y_1 = m(x - x_1)$
    $(x_1, y_1) = (-2, 0), m = -\dfrac{3}{4}$
    $y - 0 = -\dfrac{3}{4}[x - (-2)]$
    $y = -\dfrac{3}{4}(x + 2)$ or
    $y = -\dfrac{3}{4}x - \dfrac{3}{2}$

13. $y - y_1 = m(x - x_1)$
    $(x_1, y_1) = (4, -2), m = 1$
    $y - (-2) = 1(x - 4)$
    $y + 2 = x - 4$
    $\quad y = x - 6$

15. $y - y_1 = m(x - x_1)$
    $(x_1, y_1) = (5, 6), m = 0$
    $y - 6 = 0(x - 5)$
    $y - 6 = 0$ or
    $\quad y = 6$

17. $m = \dfrac{y_2 - y_1}{x_2 - x_1} = \dfrac{9 - 3}{5 - 2} = \dfrac{6}{3} = 2$
    $(x_1, y_1) = (2, 3)$
    $y - y_1 = m(x - x_1)$
    $y - 3 = 2(x - 2)$
    $\quad y = 2x - 1$

19. $m = \dfrac{y_2 - y_1}{x_2 - x_1} = \dfrac{-2 - 4}{2 - (-1)} = \dfrac{-6}{3} = -2$
    $(x_1, y_1) = (-1, 4)$
    $y - y_1 = m(x - x_1)$
    $y - 4 = -2[x - (-1)]$
    $y - 4 = -2(x + 1)$ or $y = -2x + 2$

21. $m = \dfrac{y_2 - y_1}{x_2 - x_1} = \dfrac{5 - 2}{0 - 5} = -\dfrac{3}{5}$

 $b = 5$

 $y = mx + b$

 $y = -\dfrac{3}{5}x + 5$

23. $m = \dfrac{y_2 - y_1}{x_2 - x_1} = \dfrac{-1 - 0}{0 - (-1)} = \dfrac{-1}{1} = -1$

 $b = -1$

 $y = mx + b$

 $y = (-1)x + (-1)$ or

 $y = -x - 1$

25. Since the line is horizontal, $m = 0$.

 $(x_1, y_1) = (2, 3)$

 $y - y_1 = m(x - x_1)$

 $y - 3 = 0(x - 2)$

 $y - 3 = 0$

 $y = 3$

27. Since the line is vertical, it has no slope. Therefore, we cannot use either the point-slope form or the slope-intercept form of an equation of a line. We know that all points on a vertical line have the same x-coordinate. Since our line passes through $(4, -3)$, its equation is $x = 4$.

29. An x-intercept of 5 means $(5, 0)$ is on the line. A y-intercept of 2 means $(0, 2)$ is on the line. Then $m = \dfrac{y_2 - y_1}{x_2 - x_1} = \dfrac{2 - 0}{0 - 5} = -\dfrac{2}{5}$ and $b = 2$, so that an equation (in slope-intercept form) is

 $y = -\dfrac{2}{5}x + 2$.

31. An x-intercept of -3 means $(-3, 0)$ is on the line. A y-intercept of 4 means $(0, 4)$ is on the line. Then $m = \dfrac{y_2 - y_1}{x_2 - x_1} = \dfrac{4 - 0}{0 - (-3)} = \dfrac{4}{3}$ and $b = 4$, so that an equation (in slope-intercept form) is

 $y = \dfrac{4}{3}x + 4$.

33. $m = \dfrac{y_2 - y_1}{x_2 - x_1} = \dfrac{82 - 75}{3 - 2} = \dfrac{7}{1} = 7$

 $y - y_1 = m(x - x_1)$

 $y - 75 = 7(x - 2)$

 $y - 75 = 7x - 14$

 $y = 7x + 61$

 When $x = 5$,

 $y = 7(5) + 61 = 35 + 61 = 96$.

 The grade would be 96 if a student studies for 5 hours.

35. $m = \dfrac{y_2 - y_1}{x_2 - x_1} = \dfrac{5 - 8}{8 - 6} = \dfrac{-3}{2} = -\dfrac{3}{2}$

 $y - y_1 = m(x - x_1)$

 $y - 8 = -\dfrac{3}{2}(x - 6)$

 $y - 8 = -\dfrac{3}{2}x + 9$

 $y = -\dfrac{3}{2}x + 17$

 When $x = 4$,

 $y = -\dfrac{3}{2}(4) + 17 = -6 + 17 = 11$.

 The waiting time would be 11 minutes if 4 clerks are working.

37. $y = 5x + 7$

 $y = mx + b$. So $m = 5$.

39. $y = -3x - 1$

 $y = mx + b$. So $m = -3$.

41. $x + y = 7$
    $-x -x$
    $\overline{y = -x + 7}$
    $y = mx + b$. So $m = -1$.

43. $2y + 3x = 6$
    $-3x -3x$
    $\overline{2y = -3x + 6}$
    $\dfrac{2y}{2} = \dfrac{-3x + 6}{2}$
    $y = \dfrac{-3x}{2} + \dfrac{6}{2}$
    $y = -\dfrac{3}{2}x + 3$
    $y = mx + b$. So $m = -\dfrac{3}{2}$.

45. $2x - 5y + 7 = 0$
    $-2x -2x$
    $\overline{-5y + 7 = -2x}$
    $-7 -7$
    $\overline{-5y = -2x - 7}$
    $\dfrac{-5y}{-5} = \dfrac{-2x - 7}{-5}$
    $y = \dfrac{-2x}{-5} + \dfrac{-7}{-5}$
    $y = \dfrac{2}{5}x + \dfrac{7}{5}$
    $y = mx + b$. So $m = \dfrac{2}{5}$.

47. $y = mx + b$
    $m = \dfrac{2 - 0}{3 - 0} = \dfrac{2}{3}$
    $b = 0$
    So $y = \dfrac{2}{3}x$.

49. $y = mx + b$
    $m = \dfrac{3 - 0}{0 - 5} = -\dfrac{3}{5}$
    $b = 3$
    So $y = -\dfrac{3}{5}x + 3$.

51. $y - y_1 = m(x - x_1)$
    $m = \dfrac{3 - (-1)}{6 - (-6)} = \dfrac{4}{12} = \dfrac{1}{3}$
    $y - 3 = \dfrac{1}{3}(x - 6)$
    $y - 3 = \dfrac{1}{3}x - 2$
    $y = \dfrac{1}{3}x + 1$

53. The line is horizontal, so $m = 0$.
    $b = -5$
    $y = mx + b$
    $y = 0x - 5$
    $y = -5$

55. By comparison with $y = mx + b$, the line whose equation is $y = 3x - 8$ has slope $= 3$. Since parallel lines have equal slopes, the line in question also has a slope of 3.

57. The line whose equation is $y = -x + 4$ has slope $= -1$. So the line in question also has a slope of -1, and it passes through $(-3, 0)$.

$y - y_1 = m(x - x_1)$
$y - 0 = -1[x - (-3)]$
$y = -1(x + 3)$
$y = -x - 3$

59. (a) $m = \dfrac{c_2 - c_1}{f_2 - f_1} = \dfrac{80 - 73}{100 - 80} = \dfrac{7}{20}$

$c - c_1 = m(f - f_1)$

$c - 73 = \dfrac{7}{20}(f - 80)$

$c - 73 = \dfrac{7}{20}f - 28$

$c = \dfrac{7}{20}f + 45$

(b) When $f = 90$, $c = \dfrac{7}{20}(90) + 45$

$= 31.5 + 45 = 76.5$

The cost of selling 90 franks is $76.50.

(c) When $c = 90.50$, $90.50 = \dfrac{7}{20}f + 45$

$45.50 = \dfrac{7}{20}f$

$\dfrac{20}{7}(45.50) = f$

$130 = f$

130 franks can be sold for $90.50.

(d) When $f = 0$,

$c = \dfrac{7}{20}(0) + 45 = 0 + 45 = 45.$

So the vendor's fixed costs are $45.

61. (a) $m = \dfrac{N_2 - N_1}{s_2 - s_1} = \dfrac{82 - 80}{18 - 15} = \dfrac{2}{3}$

$N - N_1 = m(s - s_1)$

$N - 80 = \dfrac{2}{3}(s - 15)$

$N - 80 = \dfrac{2}{3}s - 10$

$N = \dfrac{2}{3}s + 70$

(b) When $s = 20$,

$N = \dfrac{2}{3}(20) + 70 = 13.33 + 70 = 83.33.$

The jogger's heart rate will be 83.33 beats per minute.

(c) When $s = 0$,

$N = \dfrac{2}{3}(0) + 70 = 0 + 70 = 70.$

The jogger's heart rate at rest is 70 beats per minute.

63. (a) $m = \dfrac{c_2 - c_1}{n_2 - n_1} = \dfrac{3.20 - 1.70}{2 - 1}$

$= \dfrac{1.50}{1} = 1.50$

$c - c_1 = m(n - n_1)$

$c - 1.70 = 1.50(n - 1)$

$c - 1.70 = 1.50n - 1.50$

$c = 1.50n + 0.20$

(b) When $n = 3.5$,

$c = 1.50(3.5) + 0.20$

$= 5.25 + 0.20 = 5.45$

The cost of 3.5 oz of coffee would be $5.45.

(c) When $c = 6.50$, $6.50 = 1.50n + 0.20$

$6.30 = 1.50n$

$4.2 = n$

The package contains 4.2 oz of coffee.

(d)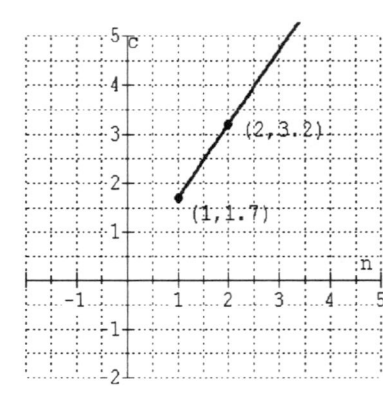

65. (a) $m = \dfrac{s_2 - s_1}{n_2 - n_1} = \dfrac{115 - 62}{-2 - 6} = \dfrac{53}{-8} = -6.625$

$s - s_1 = m(n - n_1)$

$s - 62 = -6.625(n - 6)$

$s - 62 = -6.625n + 39.75$

$s = -6.625n + 101.75$

(b) When $n = 4.5$,

$s = -6.625(4.5) + 101.75$

$\approx -29.81 + 101.75$

≈ 71.94

On a 4.5° incline, the maximum speed would be 71.94 mph.

(c) When $n = -2.8$,

$s = -6.625(-2.8) + 101.75$

$\approx 18.55 + 101.75$

≈ 120.30

On a 2.8° decline, the maximum speed would be 120.30 mph.

(d)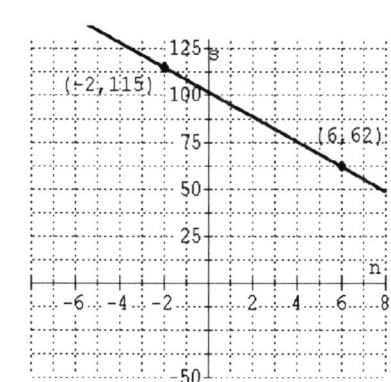

(e) When $n = 0$,

$s = -6.625(0) + 101.75$

$= 0 + 101.75$

$= 101.75$

The s-intercept of this graph is 101.75. This means that the maximum speed on level ground is 101.75 mph.

(f) When $s = 0$,

$0 = -6.625n + 101.75$

$6.625n = 101.75$

$n \approx 15.36$

The n-intercept of this graph is 15.36. This means that the car cannot make it up an incline which is 15.36° or more.

67. (a) $m = \dfrac{w_2 - w_1}{t_2 - t_1} = \dfrac{0.7 - 1.5}{33 - 24}$

$= \dfrac{-0.8}{9} \approx -0.09$

$w - w_1 = m(t - t_1)$

$w - 1.5 = -0.09(t - 24)$

$w - 1.5 = -0.09t + 2.16$

$w = -0.09t + 3.66$

(b) When $t = 28$, $w = -0.09(28) + 3.66$

$\approx -2.52 + 3.66$

≈ 1.14

At 28° C, the width of a gap in this roadway would be 1.14 cm.

(c) When $w = 0$, $0 = -0.09t + 3.66$

$0.09t = 3.66$

$t \approx 40.67$

The gap would close completely when the temperature is 40.67° C.

(d) It is unlikely to occur, since 40.67° C is more than 105° F.

69. The point-slope form is the most flexible since it accommodates *any point* and the slope, whereas the slope-intercept form is most useful for determining the slope of a line given its equation.

70. There are many possible answers to this question. The following answers all depend on the "best fit" line we choose in part (a).
    (a) $n = 0.85h - 14.5$
    (b) If $h = 150$, we get $n = 133$ phone calls.
    (c) The predicted value from the equation (133 calls) agrees very well with the actual number of calls (130).
    (d) If $h = 225$ we get $n = 176.75$ which we would round off to 177 calls.

71. (a)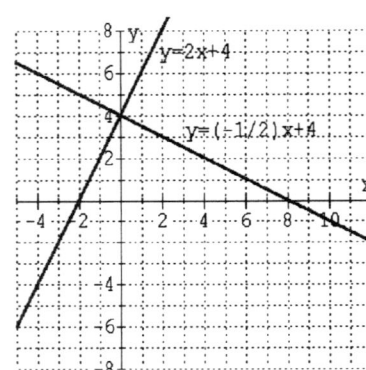
    (b) From the graph, it appears that these lines are perpendicular.
    (c) The line whose equation is $y = 2x + 4$ has slope $= 2$. The line whose equation is $y = -\frac{1}{2}x + 4$ has slope $= -\frac{1}{2}$. The product of these slopes is equal to -1. Put another way, the slopes are negative reciprocals of one another.
    (d) The line whose equation is $y = \frac{2}{5}x + 7$ has slope $= \frac{2}{5}$. Therefore, any line that is perpendicular to this line must have slope $= -\frac{5}{2}$.

73. Let $r =$ the alcohol percentage of the added solution.
$$0.60(40) + r(30) = 0.45(70)$$
$$100[0.60(40) + r(30)] = 100[0.45(70)]$$
$$100[0.60(40)] + 100[r(30)] = 100[0.45(70)]$$
$$60(40) + 3000r = 45(70)$$
$$2400 + 3000r = 3150$$
$$3000r = 750$$
$$r = 0.25$$
The added solution is 25% alcohol.

Chapter 5 Review Exercises

1, 3, 5.

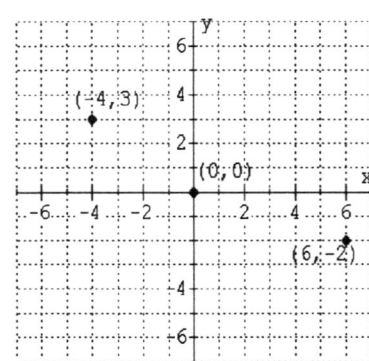

7. $2x + 4y = 14$
    $2(3) + 4y = 14$
    $6 + 4y = 14$
    $4y = 8$
    $y = 2$
    Therefore, the ordered pair is $(3, 2)$.

9. $x + 2y = 4$
    $x + 2(-3) = 4$
    $x - 6 = 4$
    $x = 10$
    Therefore, the ordered pair is $(10, -3)$.

11. $x - y = 0$
    $0 - y = 0$
    $-y = 0$
    $y = 0$
    Therefore, the ordered pair is $(0, 0)$.

13. $x - y = 8$

    x-intercept:
    $x - 0 = 8$
    $x = 8$
    Plot $(8, 0)$

    y-intercept:
    $0 - y = 8$
    $-y = 8$
    $y = -8$
    Plot $(0, -8)$

    check point:
    choose $x = 4$
    $(4) - y = 8$
    $4 - y = 8$
    $-y = 4$
    $y = -4$
    Plot $(4, -4)$

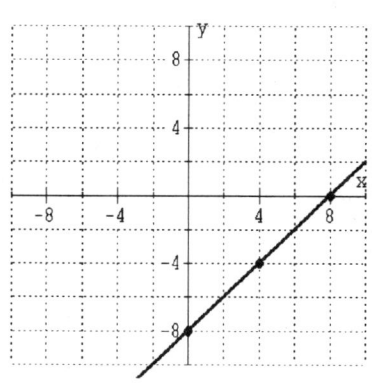

15. $3x + 7y = 21$

    x-intercept:
    $3x + 7(0) = 21$
    $3x + 0 = 21$
    $3x = 21$
    $x = 7$
    Plot $(7, 0)$

    y-intercept:
    $3(0) + 7y = 21$
    $0 + 7y = 21$
    $7y = 21$
    $y = 3$
    Plot $(0, 3)$

    check point:
    choose $y = 6$
    $3x + 7(6) = 21$
    $3x + 42 = 21$
    $3x = -21$
    $x = -7$
    Plot $(-7, 6)$

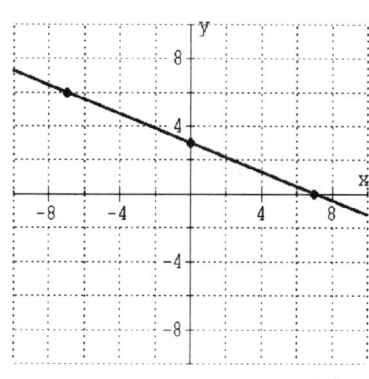

17. $y = 3x + 2$

    x-intercept:
    $0 = 3x + 2$
    $-2 = 3x$
    $-\frac{2}{3} = x$
    Plot $(-\frac{2}{3}, 0)$

    y-intercept:
    $y = 3(0) + 2$
    $y = 0 + 2$
    $y = 2$
    Plot $(0, 2)$

    check point:
    choose $x = 1$
    $y = 3(1) + 2$
    $y = 3 + 2$
    $y = 5$
    Plot $(1, 5)$

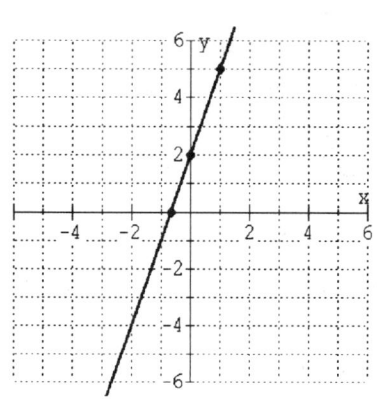

19. $x - 2y = 4$

    x-intercept:
    $x - 2(0) = 4$
    $x - 0 = 4$
    $x = 4$
    Plot $(4, 0)$

    y-intercept:
    $0 - 2y = 4$
    $-2y = 4$
    $y = -2$
    Plot $(0, -2)$

    check point:
    choose $y = -1$
    $x - 2(-1) = 4$
    $x + 2 = 4$
    $x = 2$
    Plot $(2, -1)$

    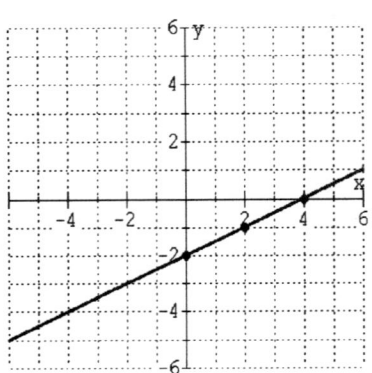

21. $x = 4$
    Its graph is a line parallel to the y-axis and 4 units to the right of it.

    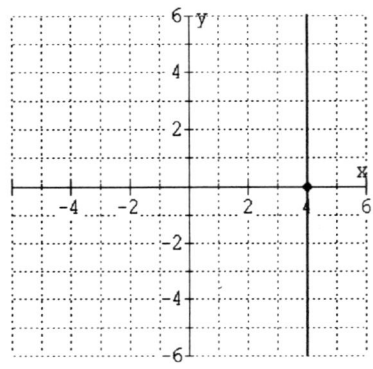

23. $3x + 2y = 12$

    x-intercept:
    $3x + 2(0) = 12$
    $3x + 0 = 12$
    $3x = 12$
    $x = 4$
    Plot $(4, 0)$

    y-intercept:
    $3(0) + 2y = 12$
    $0 + 2y = 12$
    $2y = 12$
    $y = 6$
    Plot $(0, 6)$

    check point:
    choose $x = 2$
    $3(2) + 2y = 12$
    $6 + 2y = 12$
    $2y = 6$
    $y = 3$
    Plot $(2, 3)$

    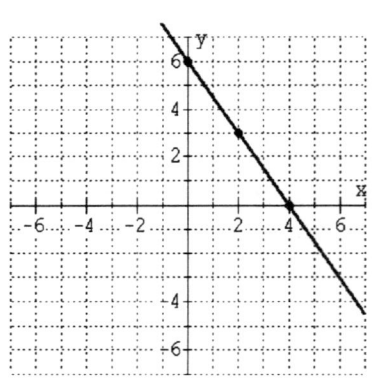

25. $3y = 6$
    This equation is equivalent to $y = 2$ (divide both sides by 3), whose graph is a line parallel to the x-axis and 2 units above it.

    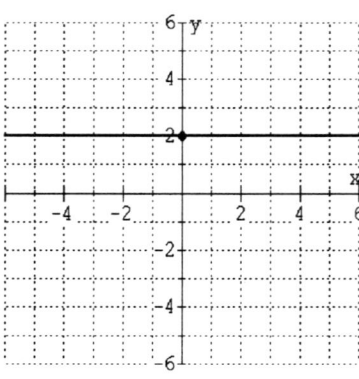

27. $m = \dfrac{y_2 - y_1}{x_2 - x_1} = \dfrac{7 - 5}{1 - 2} = \dfrac{2}{-1} = -2$

29. $m = \dfrac{y_2 - y_1}{x_2 - x_1} = \dfrac{2 - (-2)}{3 - (-3)}$
$= \dfrac{2 + 2}{3 + 3} = \dfrac{4}{6} = \dfrac{2}{3}$

31. $m = \dfrac{y_2 - y_1}{x_2 - x_1} = \dfrac{8 - 8}{-1 - 1} = \dfrac{0}{-2} = 0$

33. $(x_1, y_1) = (3, 2)$, $m = 4$
$y - y_1 = m(x - x_1)$
$y - 2 = 4(x - 3)$ or $y = 4x - 10$

35. $m = \dfrac{y_2 - y_1}{x_2 - x_1} = \dfrac{0 - (-6)}{1 - 2} = \dfrac{0 + 6}{1 - 2}$
$= \dfrac{6}{-1} = -6$
$(x_1, y_1) = (2, -6)$
$y - y_1 = m(x - x_1)$
$y - (-6) = -6(x - 2)$
$y + 6 = -6(x - 2)$ or $y = -6x + 6$

37. All points on a horizontal line have the same y-coordinate. Since the point $(3, 8)$ is on our line, an equation for this horizontal line must be $y = 8$.

39. $-\dfrac{3}{4}$

41. The line whose equation is $y = 5x - 1$ has slope $= 5$. So the line in question has slope $= 5$, and it passes through $(-3, 5)$.
$y - y_1 = m(x - x_1)$
$y - 5 = 5[x - (-3)]$
$y - 5 = 5(x + 3)$
$y - 5 = 5x + 15$
$y = 5x + 20$

43. $y = mx + b$
$m = \dfrac{4 - 0}{3 - 0} = \dfrac{4}{3}$
$b = 0$
So $y = \dfrac{4}{3}x$

45. Let $h =$ number of overtime hours.
$d =$ number of defective items.
$m = \dfrac{d_2 - d_1}{h_2 - h_1} = \dfrac{17 - 12}{10 - 8} = \dfrac{5}{2}$
$d - d_1 = m(h - h_1)$
$d - 12 = \dfrac{5}{2}(h - 8)$
$d - 12 = \dfrac{5}{2}h - 20$
$d = \dfrac{5}{2}h - 8$

When $h = 20$, $d = \dfrac{5}{2}(20) - 8 = 50 - 8 = 42$.
If the worker puts in 20 overtime hours, you would expect to find 42 defective items.

Chapter 5 Practice Test

1. $$\frac{2y - x}{5} = x - y$$
$$\frac{2(-4) - (-3)}{5} \stackrel{?}{=} (-3) - (-4)$$
$$\frac{-5}{5} \stackrel{?}{=} 1$$
$$-1 \neq 1$$

   So the point $(-3, -4)$ does not satisfy the given equation.

3. (a) $3x - 5y = 15$

x-intercept:	y-intercept:	check point:
$3x - 5(0) = 15$	$3(0) - 5y = 15$	choose $x = -5$
$3x - 0 = 15$	$0 - 5y = 15$	$3(-5) - 5y = 15$
$3x = 15$	$-5y = 15$	$-15 - 5y = 15$
$x = 5$	$y = -3$	$-5y = 30$
Plot $(5, 0)$	Plot $(0, -3)$	$y = -6$
		Plot $(-5, -6)$

   (b) $y = 3x - 7$

x-intercept:	y-intercept:	check point:
$0 = 3x - 7$	$y = 3(0) - 7$	choose $x = 2$
$7 = 3x$	$y = 0 - 7$	$y = 3(2) - 7$
$\frac{7}{3} = x$	$y = -7$	$y = 6 - 7$
		$y = -1$
Plot $\left(\frac{7}{3}, 0\right)$	Plot $(0, -7)$	Plot $(2, -1)$

   (c) $y + 2x = 0$

x-intercept:	second point:	check point:
$0 + 2x = 0$	choose $x = 1$	choose $x = -1$
$2x = 0$	$y + 2(1) = 0$	$y + 2(-1) = 0$
$x = 0$	$y + 2 = 0$	$y - 2 = 0$
Plot $(0, 0)$	$y = -2$	$y = 2$
(This implies that $y = 0$ is the y-intercept.)	Plot $(1, -2)$	Plot $(-1, 2)$

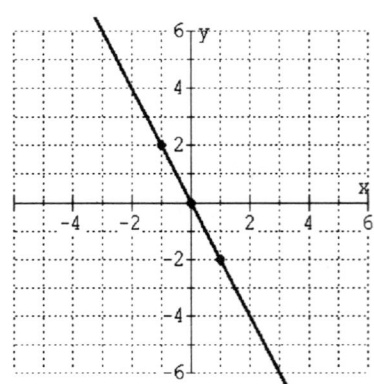

(d) $x = 4$
This is an equation of a line parallel to the y-axis and 4 units to the right of it.

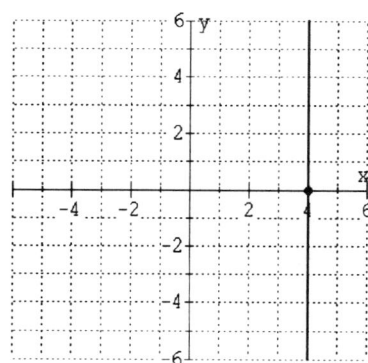

(e) $y = -3$
This is an equation of a line parallel to the x-axis and 3 units below it.

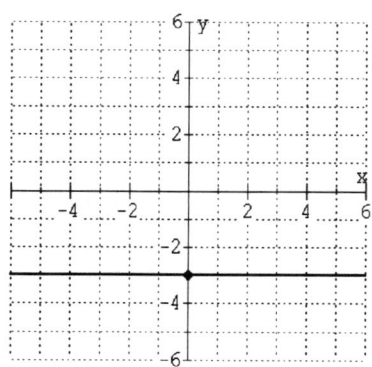

5. $m = \dfrac{4}{3}, (x_1, y_1) = (4, -1)$
$y - y_1 = m(x - x_1)$
$y - (-1) = \dfrac{4}{3}(x - 4)$
$y + 1 = \dfrac{4}{3}(x - 4)$ or
$y = \dfrac{4}{3}x - \dfrac{19}{3}$

7. $y = 5$ is the horizontal line passing through $(3, 5)$.
$x = 3$ is the vertical line passing through $(3, 5)$.

9. The line whose equation is $y = -3x + 4$ has slope $= -3$. So the line in equation has slope $= -3$ and it passes through $(-4, 0)$.
$y - y_1 = m(x - x_1)$
$y - 0 = -3[x - (-4)]$
$y - 0 = -3(x + 4)$
$y = -3x - 12$

11. Let $C =$ number of phone calls.
$A =$ amount of charity pledged.
$m = \dfrac{A_2 - A_1}{C_2 - C_1} = \dfrac{115 - 80}{21 - 15} = \dfrac{35}{6}$
$A - A_1 = m(C - C_1)$
$A - 80 = \dfrac{35}{6}(C - 15)$
$A - 80 = \dfrac{35}{6}C - \dfrac{175}{2}$
$A = \dfrac{35}{6}C - \dfrac{15}{2}$

When $C = 30$, $A = \dfrac{35}{6}(30) - \dfrac{15}{2}$
$= 175 - 7.50$
$= 167.50.$

If they make 30 phone calls, the amount of charity pledged will be $167.50.

Chapter 6
Interpreting Graphs and Systems of Equations

Exercises 6.1

1. (a) B   (b) B   (c) A   (d) B

3. (a) 2,000 calls   (b) 4:00 a.m.   (c) 10,000 calls
    (d) From 4:00 p.m. to 4:00 a.m.   (e) 10:00 a.m. to 4:00 p.m.
    (f) 10:00 a.m. to 4:00 p.m.

5. (a) This graph denies the belief that temperature drops steadily as altitude increases. Notice that the temperature increases as the altitude increases from 12 to 50 miles and again when the altitude is greater than 80 miles.
    (b) The temperature decreases when the altitude is between 0 and 12 miles and also when it is between 50 and 80 miles.
    (c) For the first 5 miles that the balloon rises, the temperature will increase; for the last 5 miles, it will decrease.

7. (a) Immediately after the students memorized the list of 20 words, the group remembered all twenty.
    (b) after 4 hours
    (c) during the first two hours
    (d) 16 words are forgotten during the first 4 hours; 2 more words are forgotten during the next 4 hours.

9. (a) 150 deer   (b) between 1996 and 1998   (c) 800 deer

11. (a) The purchase price of Machine A is $10,000; the purchase price of Machine B is $7500.
    (b) It takes Machine A 8 years and it takes Machine B 10 years.
    (c) The machines have the same value 5 years after they were purchased.
    (d) The value of Machine A is greater than the value of Machine B for the first five years. The opposite is true during the next five years.

13. (a) Thursday; $45   (b) Tuesday
    (c) Thursday; $70 − $45 = $25   (d) $70 − $50 = $20

15. Let m = number of minutes that elapse from the beginning of the experiment.
    T = temperature of the metal bar (in °C).

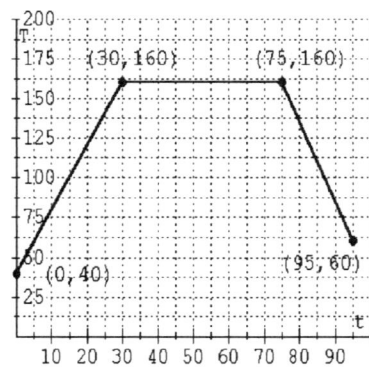

17. $\dfrac{6x^2}{25yz^3} \div 15xyz = \dfrac{6x^2}{25yz^3} \div \dfrac{15xyz}{1}$

$= \dfrac{\cancel{6}x^2}{25yz^3} \cdot \dfrac{1}{\cancel{15}xyz}$ (with 2 and 5)

$= \dfrac{2x}{125y^2z^4}$

19. Let x = number of pairs of dress slacks.
Then $36 - x$ = number of pairs of casual slacks.
$$40x + 25(36 - x) = 1185$$
$$40x + 900 - 25x = 1185$$
$$15x + 900 = 1185$$
$$15x = 285$$
$$x = 19$$
Then $36 - x = 36 - 19 = 17$.
So 19 pairs of dress slacks and 17 pairs of casual slacks were bought.

Exercises 6.2

1. (a)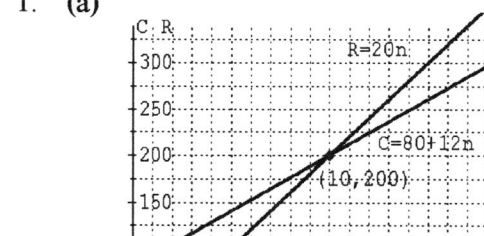

    (b) $(10, 200)$

    (c) In order to break even, the potter must sell 10 pieces.

3. $x + y = 4$
   Set $y = 0$ and solve for x to get an x-intercept of 4.
   Set $x = 0$ and solve for y to get a y-intercept of 4.

   $x - y = 2$
   Set $y = 0$ and solve for x to get an x-intercept of 2.
   Set $x = 0$ and solve for y to get a y-intercept of -2.

   The lines cross at the point $(3, 1)$. So the system
   $\begin{cases} x + y = 4 \\ x - y = 2 \end{cases}$ is satisfied by the point $(3, 1)$.

   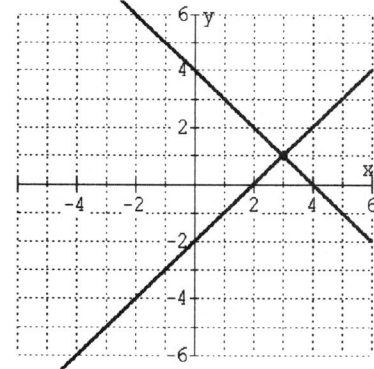

114

5. $3x + y = 3$
   Set $y = 0$ and solve for x to get an x-intercept of 1.
   Set $x = 0$ and solve for y to get a y-intercept of 3.

   $x - y = 1$
   Set $y = 0$ and solve for x to get an x-intercept of 1.
   Set $x = 0$ and solve for y to get a y-intercept of -1.

   The lines cross at the point $(1, 0)$. So the system
   $\begin{cases} 3x + y = 3 \\ x - y = 1 \end{cases}$ is satisfied by the point $(1, 0)$.

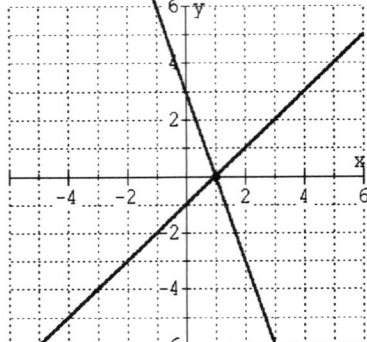

7. $3x + y = 6$
   Set $y = 0$ and solve for x to get an x-intercept of 2.
   Set $x = 0$ and solve for y to get a y-intercept of 6.

   $6x + 2y = 12$
   Set $y = 0$ and solve for x to get an x-intercept of 2.
   Set $x = 0$ and solve for y to get a y-intercept of 6.

   Since these lines have the same x-intercept and the same y-intercept, the lines coincide. Therefore, there are infinitely many solutions to the system.
   $\begin{cases} 3x + y = 6 \\ 6x + 2y = 12 \end{cases}$: $\{(x, y) | 3x + y = 6\}$.

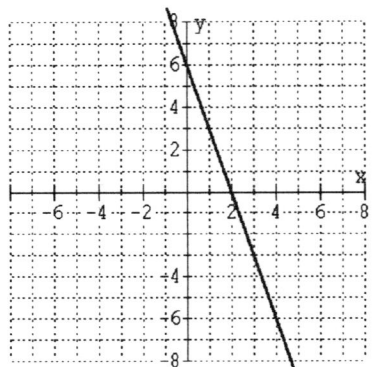

9. $x + y = 0$
   Set $y = 0$ and solve for x to get an x-intercept of 0.
   This is also a y-intercept.

   $x - y = 0$
   Set $y = 0$ and solve for x to get an x-intercept of 0.
   This is also a y-intercept.

   The lines cross at the point $(0, 0)$. So the system
   $\begin{cases} x + y = 0 \\ x - y = 0 \end{cases}$ is satisfied by the point $(0, 0)$.

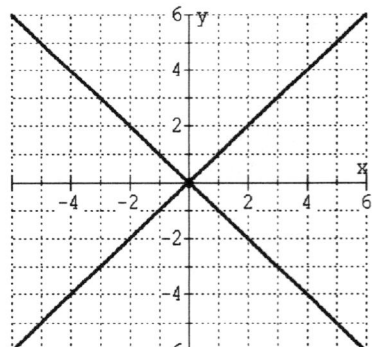

11. $3x - 2y = 6$
    Set $y = 0$ and solve for x to get an x-intercept of 2.
    Set $x = 0$ and solve for y to get a y-intercept of -3.

    $x + y = 2$
    Set $y = 0$ and solve for x to get an x-intercept of 2.
    Set $x = 0$ and solve for y to get a y-intercept of 2.

    The lines cross at the point $(2, 0)$. So the system
    $\begin{cases} 3x - 2y = 6 \\ x + y = 2 \end{cases}$ is satisfied by the point $(2, 0)$.

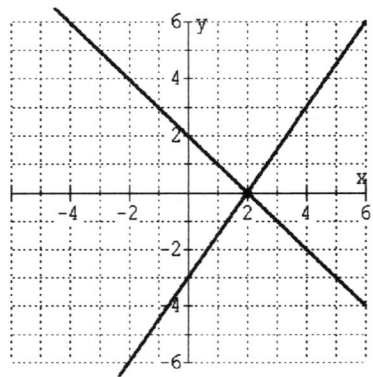

13. $3x - y = 9$
    Set $y = 0$ and solve for x to get an x-intercept of 3.
    Set $x = 0$ and solve for y to get a y-intercept of -9.

    $6x - 2y = 18$
    Set $y = 0$ and solve for x to get an x-intercept of 3.
    Set $x = 0$ and solve for y to get a y-intercept of -9.

    Since these lines have the same x- and y-intercepts, the lines coincide. The system has infinitely many solutions.
    $\begin{cases} 3x - y = 9 \\ 6x - 2y = 18 \end{cases}$: $\{(x, y) | 3x - y = 9\}$.

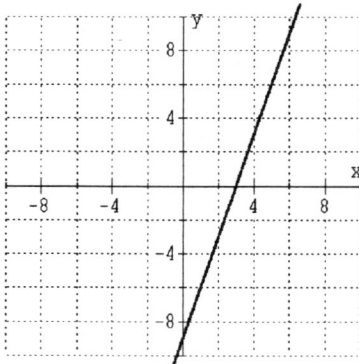

15. $5x - 3y = 15$
    Set $y = 0$ and solve for x to get an x-intercept of 3.
    Set $x = 0$ and solve for y to get a y-intercept of -5.

    $2x - y = 4$
    Set $y = 0$ and solve for x to get an x-intercept of 2.
    Set $x = 0$ and solve for y to get a y-intercept of -4.

    The lines cross at the point $(-3, -10)$. So the system
    $\begin{cases} 5x - 3y = 15 \\ 2x - y = 4 \end{cases}$ is satisfied by the point $(-3, -10)$.

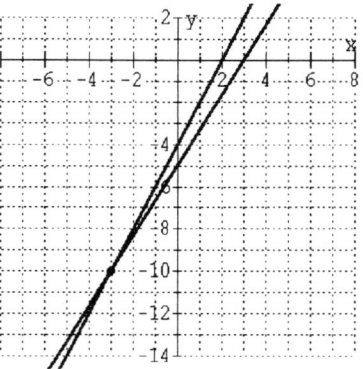

17. $y = 2x + 6$
    Set $y = 0$ and solve for x to get an x-intercept of -3.
    Set $x = 0$ and solve for y to get a y-intercept of 6.

    $y = x + 1$
    Set $y = 0$ and solve for x to get an x-intercept of -1.
    Set $x = 0$ and solve for y to get a y-intercept of 1.

    The lines cross at the point $(-5, -4)$. So the system
    $\begin{cases} y = 2x + 6 \\ y = x + 1 \end{cases}$ is satisfied by the point $(-5, -4)$.

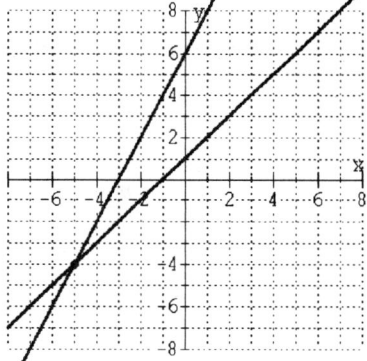

19. $y = x - 3 \quad m = 1$
    Set $x = 0$ and solve for y to get a y-intercept of -3.

    $y = x + 4 \quad m = 1$
    Set $x = 0$ and solve for y to get a y-intercept of 4.

    These lines are parallel since their slopes are the same and their y-intercepts are different. Therefore, the system
    $\begin{cases} y = x - 3 \\ y = x + 4 \end{cases}$ has no solution.

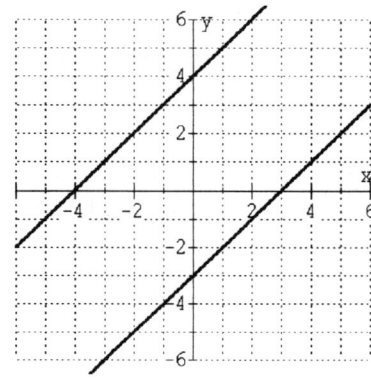

21. $2x = y - 8$
    Set $y = 0$ and solve for x to get an x-intercept of -4.
    Set $x = 0$ and solve for y to get a y-intercept of 8.

    $3x = y + 6$
    Set $y = 0$ and solve for x to get an x-intercept of 2.
    Set $x = 0$ and solve for y to get a y-intercept of -6.

    The lines cross at the point $(14, 36)$. So the system
    $\begin{cases} 2x = y - 8 \\ 3x = y + 6 \end{cases}$ is satisfied by the point $(14, 36)$.

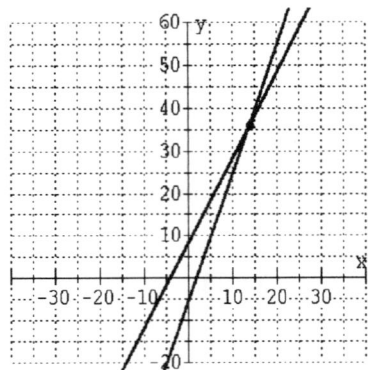

23. $3x + y = 3$
    Set $y = 0$ and solve for x to get an x-intercept of 1.
    Set $x = 0$ and solve for y to get a y-intercept of 3.

    $y = x + 5$
    Set $y = 0$ and solve for x to get an x-intercept of -5.
    Set $x = 0$ and solve for y to get a y-intercept of 5.

    The lines cross at the point $\left(-\dfrac{1}{2}, \dfrac{9}{2}\right)$. So the system
    $\begin{cases} 3x + y = 3 \\ y = x + 5 \end{cases}$ is satisfied by the point $\left(-\dfrac{1}{2}, \dfrac{9}{2}\right)$.

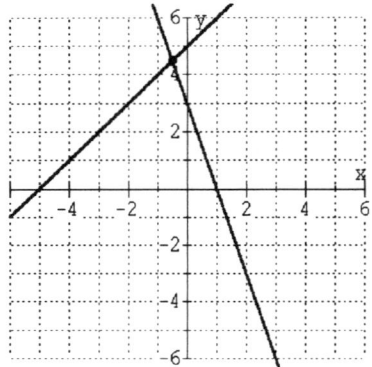

25. $5x - 4y = 20$
    Set $y = 0$ and solve for x to get an x-intercept of 4.
    Set $x = 0$ and solve for y to get a y-intercept of -5.

    $y = x - 3$
    Set $y = 0$ and solve for x to get an x-intercept of 3.
    Set $x = 0$ and solve for y to get a y-intercept of -3.

    The lines cross at the point $(8, 5)$. So the system
    $\begin{cases} 5x - 4y = 20 \\ y = x - 3 \end{cases}$ is satisfied by the point $(8, 5)$.

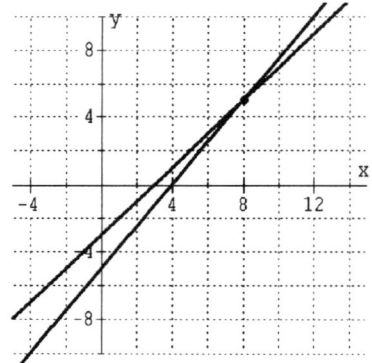

27. $2x + y = 10 \quad m = -2$
    Set $y = 0$ and solve for x to get an x-intercept of 5.
    Set $x = 0$ and solve for y to get a y-intercept of 10.

    $2y = 20 - 4x \quad m = -2$
    Set $y = 0$ and solve for x to get an x-intercept of 5.
    Set $x = 0$ and solve for y to get a y-intercept of 10.

    Since these lines have the same x- and y-intercepts, the lines coincide. The system has infinitely many solutions.
    $\begin{cases} 2x + y = 10 \\ 2y = 20 - 4x \end{cases} : \{(x, y) | 2x + y = 10\}$.

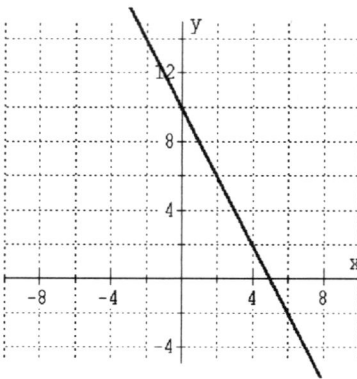

29. $x + y = 8$

   Set $y = 0$ and solve for x to get an x-intercept of 8.
   Set $x = 0$ and solve for y to get a y-intercept of 8.

   $y = x$
   Set $y = 0$ and solve for x to get an x-intercept of 0.
   This is also a y-intercept.

   The lines cross at the point $(4, 4)$. So the system
   $\begin{cases} x + y = 8 \\ y = x \end{cases}$ is satisfied by the point $(4, 4)$.

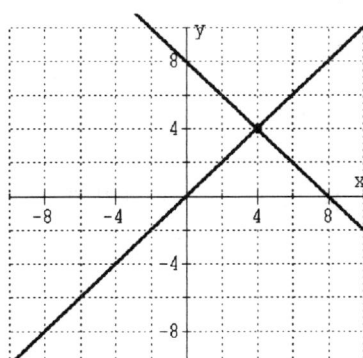

31. $x + y = 6$

   Set $y = 0$ and solve for x to get an x-intercept of 6.
   Set $x = 0$ and solve for y to get a y-intercept of 6.

   $y = -2$
   This is a horizontal line two units below the x-axis.

   The lines cross at the point $(8, -2)$. So the system
   $\begin{cases} x + y = 6 \\ y = -2 \end{cases}$ is satisfied by the point $(8, -2)$.

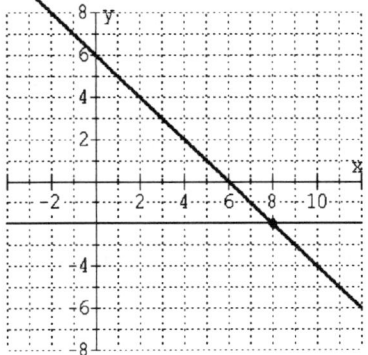

33. $2x - y = 2$

   Set $y = 0$ and solve for x to get an x-intercept of 1.
   Set $x = 0$ and solve for y to get a y-intercept of -2.

   $x = -4$
   This is a vertical line four units left of the y-axis.

   The lines cross at the point $(-4, -10)$. So the system
   $\begin{cases} 2x - y = 2 \\ x = -4 \end{cases}$ is satisfied by the point $(-4, -10)$.

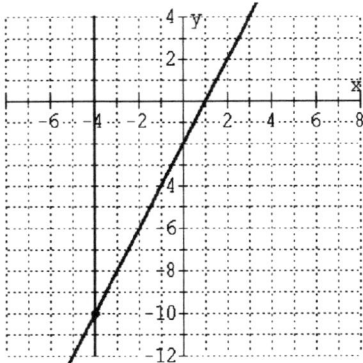

35. $y = 4$
   This is a horizontal line four units above the x-axis.

   $x = -1$
   This is a vertical line one unit to the left of the y-axis.

   The lines cross at the point $(-1, 4)$. So the system
   $\begin{cases} y = 4 \\ x = -1 \end{cases}$ is satisfied by the point $(-1, 4)$.

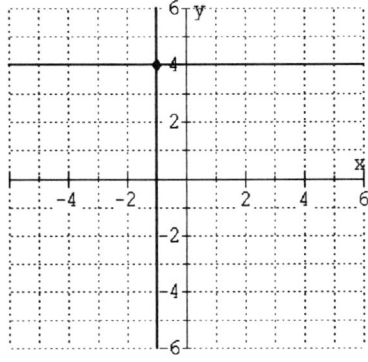

37. A solution to a system of equations is an ordered pair that satisfies both equations simultaneously.

38. It is impossible for a system of two linear equations to have exactly two solutions. It this were possible, then we would be able to draw two straight lines that intersect exactly twice. But two straight lines either do not intersect at all (when they are parallel), intersect in one point, or intersect in infinitely many points (when they coincide).

39. 1-b, 2-e, 3-a, 4-c, 5-d, 6-f

41. $9 - 5(x+4) > 9$
$9 - 5x - 20 > 9$
$-5x - 11 > 9$
$-5x > 20$
$x < -4$

43. $1 \leq 7 - 2x < 13$
$-6 \leq -2x < 6$
$3 \geq x > -3$
or
$-3 < x \leq 3$

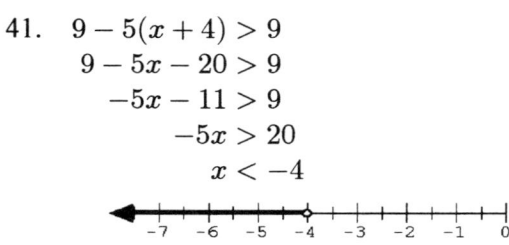

45. Let $8x =$ number of members who voted in favor.
Then $7x =$ number of members who voted against.
$8x + 7x = 435$
$15x = 435$
$x = 29$

When $x = 29$, $8x = 8(29) = 232$. So 232 members were in favor of the bill.

Exercises 6.3

1. $\begin{cases} x + y = 1 \\ x - y = 3 \end{cases}$ Add.
$\quad 2x \quad = 4$
$\quad x \quad = 2$

$x + y = 1$
$2 + y = 1$
$y = -1$
Solution: $(2, -1)$

3. $\begin{cases} 2x + y = 5 \\ x - y = 4 \end{cases}$ Add.
$\quad 3x \quad = 9$
$\quad x \quad = 3$

$2x + y = 5$
$2(3) + y = 5$
$y = -1$
Solution: $(3, -1)$

5. $\begin{cases} 7x + 2y - 15 = 0 \\ 3x - 2y + 5 = 0 \end{cases}$ Add.
$10x \quad - 10 = 0$
$10x \quad = 10$
$\quad x = 1$

$3x - 2y + 5 = 0$
$3(1) - 2y + 5 = 0$
$-2y + 8 = 0$
$-2y = -8$
$y = 4$
Solution: $(1, 4)$

7. $\begin{cases} 2x + y = 15 \\ x - 2y = 0 \end{cases}$ $\xrightarrow[\text{multiply by } -2]{\text{as is}}$

$2x + y = 15$
$-2x + 4y = 0$ Add.
$\overline{\quad 5y = 15}$
$\quad y = 3$

$2x + y = 15$
$2x + (3) = 15$
$2x + 3 = 15$
$2x = 12$
$x = 6$
Solution: $(6, 3)$

9. $\begin{cases} 3x + 2y = -11 \\ x + 3y = 1 \end{cases}$ $\xrightarrow{\text{as is}}$ $\xrightarrow{\text{multiply by } -3}$

$\begin{aligned} 3x + 2y &= -11 \\ -3x - 9y &= -3 \quad \text{Add.} \\ \hline -7y &= -14 \\ y &= 2 \end{aligned}$

$\begin{aligned} x + 3y &= 1 \\ x + 3(2) &= 1 \\ x + 6 &= 1 \\ x &= -5 \end{aligned}$

Solution: $(-5, 2)$

11. $\begin{cases} 4s - 5t = 4 \\ 2s + 10t = 7 \end{cases}$ $\xrightarrow{\text{multiply by 2}}$ $\xrightarrow{\text{as is}}$

$\begin{aligned} 8s - 10t &= 8 \\ 2s + 10t &= 7 \quad \text{Add.} \\ \hline 10s &= 15 \\ s &= \dfrac{3}{2} \end{aligned}$

$\begin{aligned} 2s + 10t &= 7 \\ 2\left(\dfrac{3}{2}\right) + 10t &= 7 \\ 10t &= 4 \\ t &= \dfrac{2}{5} \end{aligned}$

Solution: $\left(\dfrac{3}{2}, \dfrac{2}{5}\right)$

13. $\begin{cases} 12c - 20d = 19 \\ 18c - 12d = 15 \end{cases}$ $\xrightarrow{\text{multiply by 3}}$ $\xrightarrow{\text{multiply by } -2}$

$\begin{aligned} 36c - 60d &= 57 \\ -36c + 24d &= -30 \quad \text{Add.} \\ \hline -36d &= 27 \\ d &= -\dfrac{3}{4} \end{aligned}$

$\begin{aligned} 12c - 20d &= 19 \\ 12c - 20\left(-\dfrac{3}{4}\right) &= 19 \\ 12c + 15 &= 19 \\ 12c &= 4 \\ c &= \dfrac{1}{3} \end{aligned}$

Solution: $\left(\dfrac{1}{3}, -\dfrac{3}{4}\right)$

15. $\begin{cases} x + 2y = 9 \\ y = 3x + 1 \end{cases}$

Substitute the value of y given in the second equation into the first.

$\begin{aligned} x + 2(3x + 1) &= 9 \\ x + 6x + 2 &= 9 \\ 7x + 2 &= 9 \\ 7x &= 7 \\ x &= 1 \end{aligned}$

When $x = 1$,
$y = 3(1) + 1 = 3 + 1 = 4$
Solution: $(1, 4)$

17. $\begin{cases} 2x - 3y = 8 \\ x = 2y + 4 \end{cases}$

Substitute the value of x given in the second equation into the first.

$\begin{aligned} 2(2y + 4) - 3y &= 8 \\ 4y + 8 - 3y &= 8 \\ y + 8 &= 8 \\ y &= 0 \end{aligned}$

When $y = 0$,
$x = 2(0) + 4 = 4$
Solution: $(4, 0)$

19. $\begin{cases} 5x + 4y = 6 \\ x - y = 3 \end{cases}$
Isolate x in the second equation, then substitute into the first.
$x = y + 3$
$5(y + 3) + 4y = 6$
$5y + 15 + 4y = 6$
$9y + 15 = 6$
$9y = -9$
$y = -1$
When $y = -1$,
$x = -1 + 3 = 2$
Solution: $(2, -1)$

21. $\begin{cases} 2r - 5s = 9 \\ s = 1 - r \end{cases}$
Substitute the value of s given in the second equation into the first.
$2r - 5(1 - r) = 9$
$2r - 5 + 5r = 9$
$7r - 5 = 9$
$7r = 14$
$r = 2$
When $r = 2$,
$s = 1 - 2 = -1$
Solution: $(2, -1)$

23. $\begin{cases} 7x + 2y = 9 \\ 2x + 3y = 5 \end{cases}$
Isolate x in the second equation, then substitute into the first.
$2x = -3y + 5$
$x = -\frac{3}{2}y + \frac{5}{2}$
$7\left(-\frac{3}{2}y + \frac{5}{2}\right) + 2y = 9$
$-\frac{21}{2}y + \frac{35}{2} + 2y = 9$
$-21y + 35 + 4y = 18$
$-17y + 35 = 18$
$-17y = -17$
$y = 1$
When $y = 1$,
$x = -\frac{3}{2}(1) + \frac{5}{2} = 1$
Solution: $(1, 1)$

25. $\begin{cases} 6u - w = 2 \\ 2u - 3w = 2 \end{cases}$
Isolate w in the first equation, then substitute into the second.
$6u - 2 = w$
$2u - 3(6u - 2) = 2$
$2u - 18u + 6 = 2$
$-16u + 6 = 2$
$-16u = -4$
$u = \frac{1}{4}$
When $u = \frac{1}{4}$,
$w = 6\left(\frac{1}{4}\right) - 2 = -\frac{1}{2}$
Solution: $\left(\frac{1}{4}, -\frac{1}{2}\right)$

27. $\begin{cases} 4w - 3t = 8 \\ 6w - t = 5 \end{cases}$
Isolate t in the second equation, then substitute into the first.
$6w - 5 = t$
$4w - 3(6w - 5) = 8$
$4w - 18w + 15 = 8$
$-14w + 15 = 8$
$-14w = -7$
$w = \frac{1}{2}$
When $w = \frac{1}{2}$,
$t = 6\left(\frac{1}{2}\right) - 5 = -2$
Solution: $\left(-2, \frac{1}{2}\right)$

29. $\begin{cases} r + 2t = 10 \\ 3r + t = -15 \end{cases}$ $\xrightarrow{\text{as is}}$ $\xrightarrow{\text{multiply by } -2}$ $\begin{array}{r} r + 2t = 10 \\ -6r - 2t = 30 \\ \hline -5r = 40 \\ r = -8 \end{array}$ Add. $\begin{array}{r} r + 2t = 10 \\ (-8) + 2t = 10 \\ -8 + 2t = 10 \\ 2t = 18 \\ t = 9 \end{array}$

Solution: $(-8, 9)$

31. $\begin{cases} 6x + y = 6 \\ 4x + 1 = y \end{cases}$ Rewrite the second equation in the system as $4x - y = -1$. $\begin{array}{r} 6x + y = 6 \\ 4x - y = -1 \\ \hline 10x = 5 \\ x = \dfrac{1}{2} \end{array}$ Add. $\begin{array}{r} 6x + y = 6 \\ 6\left(\dfrac{1}{2}\right) + y = 6 \\ 3 + y = 6 \\ y = 3 \end{array}$

Solution: $\left(\dfrac{1}{2}, 3\right)$

33. $\begin{cases} 8a + 6b = -3 \\ 12a + 9b = -5 \end{cases}$ $\xrightarrow{\text{multiply by } 3}$ $\xrightarrow{\text{multiply by } -2}$ $\begin{array}{r} 24a + 18b = -9 \\ -24a - 18b = 10 \\ \hline 0 = 1 \end{array}$ Add.

Contradiction
No solution

35. $\begin{cases} 11a - 2b = 30 \\ 3a + 3b = -6 \end{cases}$ $\xrightarrow{\text{multiply by } 3}$ $\xrightarrow{\text{multiply by } 2}$ $\begin{array}{r} 33a - 6b = 90 \\ 6a + 6b = -12 \\ \hline 39a = 78 \\ a = 2 \end{array}$ Add. $\begin{array}{r} 11a - 2b = 30 \\ 11(2) - 2b = 30 \\ 22 - 2b = 30 \\ -2b = 8 \\ b = -4 \end{array}$

Solution: $(2, -4)$

37. $\begin{cases} 2x + 3 = 4y \\ 6x = 9 - 12y \end{cases}$ $\begin{array}{l} 2x - 4y = -3 \\ 6x + 12y = 9 \end{array}$ $\xrightarrow{\text{multiply by } -3}$ $\xrightarrow{\text{as is}}$ $\begin{array}{r} -6x + 12y = 9 \\ 6x + 12y = 9 \\ \hline 24y = 18 \\ y = \dfrac{3}{4} \end{array}$ Add. $\begin{array}{r} 6x = 9 - 12y \\ 6x = 9 - 12\left(\dfrac{3}{4}\right) \\ 6x = 0 \\ x = 0 \end{array}$
Rewrite each equation.

Solution: $\left(0, \dfrac{3}{4}\right)$

39. $\begin{cases} 5x + 2y = 4y + 9 \\ y = x - 3 \end{cases}$ $\begin{array}{l} 5x - 2y = 9 \\ -x + y = -3 \end{array}$ $\xrightarrow{\text{as is}}$ $\xrightarrow{\text{multiply by } 2}$ $\begin{array}{r} 5x - 2y = 9 \\ -2x + 2y = -6 \\ \hline 3x = 3 \\ x = 1 \end{array}$ Add. $\begin{array}{r} y = x - 3 \\ y = (1) - 3 \\ y = 1 - 3 \\ y = -2 \end{array}$
Rewrite each equation.

Solution: $(1, -2)$

41. $\begin{cases} a - b = 1 \\ \dfrac{a}{3} + \dfrac{b}{5} = 1 \end{cases}$ $\xrightarrow{\text{as is}}$ $a - b = 1$ $\xrightarrow{\text{multiply by 3}}$ $3a - 3b = 3$ $\qquad a - b = 1$

$\xrightarrow{\text{multiply by 15}}$ $5a + 3b = 15$ $\xrightarrow{\text{as is}}$ $5a + 3b = 15$ Add. $\dfrac{9}{4} - b = 1$

$\overline{\;8a \; = 18}$

$a = \dfrac{9}{4}$ $\qquad -b = -\dfrac{5}{4}$

$b = \dfrac{5}{4}$

Solution: $\left(\dfrac{9}{4}, \dfrac{5}{4}\right)$

43. $\begin{cases} \dfrac{w}{2} - \dfrac{t}{5} = 11 \\ \dfrac{w}{8} - \dfrac{t}{9} = 0 \end{cases}$ $\xrightarrow{\text{multiply by 10}}$ $5w - 2t = 110$ $\xrightarrow{\text{multiply by } -4}$ $-20w + 8t = -440$ $\qquad \dfrac{w}{8} - \dfrac{t}{9} = 0$

$\xrightarrow{\text{multiply by 72}}$ $9w - 8t = 0$ $\xrightarrow{\text{as is}}$ $9w - 8t = 0$ $\qquad \dfrac{40}{8} - \dfrac{t}{9} = 0$

Add. $\overline{-11w \; = -440}$

$w = 40$ $\qquad 5 - \dfrac{t}{9} = 0$

Solution: $(45, 40)$ $\qquad 5 = \dfrac{t}{9}$

$t = 45$

45. $\begin{cases} \dfrac{2r}{5} - \dfrac{3t}{8} = 14 \\ \dfrac{4r}{5} + \dfrac{3t}{4} = 4 \end{cases}$ $\xrightarrow{\text{multiply by 40}}$ $16r - 15t = 560$ $\qquad \dfrac{4r}{5} + \dfrac{3t}{4} = 4$

$\xrightarrow{\text{multiply by 20}}$ $16r + 15t = 80$ Add. $\dfrac{4(20)}{5} + \dfrac{3t}{4} = 4$

$\overline{32r \; = 640}$

$r = 20$ $\qquad 16 + \dfrac{3t}{4} = 4$

$\dfrac{3t}{4} = -12$

$t = -16$

Solution: $(20, -16)$

47. $\begin{cases} 5x - 3y = 20 \\ 7x + 2y = 28 \end{cases}$ $\xrightarrow{\text{multiply by 2}}$ $10x - 6y = 40$ $\qquad 7x + 2y = 28$

$\xrightarrow{\text{multiply by 3}}$ $21x + 6y = 84$ Add. $7(4) + 2y = 28$

$\overline{31x \; = 124}$

$x = 4$ $\qquad 28 + 2y = 28$

$2y = 0$

$y = 0$

Solution: $(4, 0)$

49. $\begin{cases} 7x + 4y = 5 \\ 4x + 3y = 0 \end{cases}$ $\xrightarrow{\text{multiply by 3}}$ $21x + 12y = 15$ $\qquad 4x + 3y = 0$

$\xrightarrow{\text{multiply by } -4}$ $-16x - 12y = 0$ Add. $4(3) + 3y = 0$

$\overline{5x \; = 15}$

$x = 3$ $\qquad 12 + 3y = 0$

$3y = -12$

$y = -4$

Solution: $(3, -4)$

51. $\begin{cases} \dfrac{x}{2} + \dfrac{y}{3} = 1 \\ \dfrac{x}{4} - y = 11 \end{cases}$ $\xrightarrow{\text{multiply by } 6}$ $3x + 2y = 6$ $\xrightarrow{\text{multiply by } 2}$ $6x + 4y = 12$ $\qquad \dfrac{x}{2} + \dfrac{y}{3} = 1$

$\xrightarrow{\text{multiply by } 4}$ $x - 4y = 44$ $\xrightarrow{\text{as is}}$ $\underline{x - 4y = 44}$ Add. $\dfrac{(8)}{2} + \dfrac{y}{3} = 1$

$\hspace{10em} 7x \hspace{3em} = 56$ $\qquad 4 + \dfrac{y}{3} = 1$

$\hspace{12em} x = 8$ $\qquad \dfrac{y}{3} = -3$

$\hspace{10em}$ Solution: $(8, -9)$ $\qquad y = -9$

53. $\begin{cases} \dfrac{x+y}{2} = 4 \\ 3x = 5 - 3y \end{cases}$ $\xrightarrow{\text{multiply by } 2}$ $x + y = 8$ $\xrightarrow{\text{multiply by } -3}$ $-3x - 3y = -24$

$\xrightarrow{\text{add } 3y}$ $3x + 3y = 5$ $\xrightarrow{\text{as is}}$ $\underline{3x + 3y = 5}$ Add.

$\hspace{18em} 0 = -19$

$\hspace{16em}$ Contradiction

$\hspace{16em}$ No solution

55. $\begin{cases} 0.4x + 0.2y = 8 \\ 0.7x - 0.3y = 1 \end{cases}$ $\xrightarrow{\text{multiply by } 10}$ $4x + 2y = 80$ $\xrightarrow{\text{multiply by } 3}$ $12x + 6y = 240$

$\xrightarrow{\text{multiply by } 10}$ $7x - 3y = 10$ $\xrightarrow{\text{multiply by } 2}$ $\underline{14x - 6y = 20}$ Add.

$\hspace{18em} 26x \hspace{2em} = 260$

$\hspace{20em} x = 10$

$\hspace{12em} 0.4x + 0.2y = 8 \qquad 0.2y = 4$

$\hspace{12em} 0.4(10) + 0.2y = 8 \qquad 2y = 40$

$\hspace{14em} 4 + 0.2y = 8 \qquad y = 20$

$\hspace{14em}$ Solution: $(10, 20)$

57. $\begin{cases} 5.2x + 3y = 14 \\ 0.3x - 2y = 9.5 \end{cases}$ $\xrightarrow{\text{multiply by } 20}$ $104x + 60y = 280$ $\qquad 5.2x + 3y = 14$

$\xrightarrow{\text{multiply by } 30}$ $\underline{9x - 60y = 285}$ Add. $\qquad 5.2(5) + 3y = 14$

$\hspace{10em} 113x \hspace{2em} = 565$ $\qquad 26 + 3y = 14$

$\hspace{12em} x \hspace{2em} = 5$ $\qquad 3y = -12$

$\hspace{20em} y = -4$

$\hspace{14em}$ Solution: $(5, -4)$

59. $\begin{cases} 3.6x + 2.9y = 23.71 \\ 1.7x + 4.5y = 14.64 \end{cases}$ $\xrightarrow{\text{multiply by } -1.7}$ $-6.12x - 4.93y = -40.307$

$\xrightarrow{\text{multiply by } 3.6}$ $\underline{6.12x + 16.2y = 52.704}$ Add.

$\hspace{16em} 11.27y = 12.397$

$\hspace{18em} y = 1.1$

$\hspace{12em} 1.7x + 4.5y = 14.64$

$\hspace{12em} 1.7x + 4.5(1.1) = 14.64$

$\hspace{14em} 1.7x + 4.95 = 14.64$

$\hspace{18em} 1.7x = 9.69$

$\hspace{20em} x = 5.7$

$\hspace{14em}$ Solution: $(5.7, 1.1)$

61. $\begin{cases} 11.2x - 2.6y = 22.84 \\ 6.7x + 15.3y = 3.55 \end{cases}$ $\xrightarrow[\text{multiply by 2.6}]{\text{multiply by 15.3}}$ $\begin{array}{r} 171.36x - 39.78y = 349.452 \\ 17.42x + 39.78y = 9.23 \quad \text{Add.} \\ \hline 188.78x = 358.682 \\ x = 1.9 \end{array}$

$$6.7x + 15.3y = 3.55$$
$$6.7(1.9) + 15.3y = 3.55$$
$$12.73 + 15.3y = 3.55$$
$$15.3y = -9.18$$
$$y = -0.6$$

Solution: $(1.9, -0.6)$

63. The student forgot to multiply the right hand side of the second equation by 2.

64. The student mistakenly multiplied 0 times 2 in the second equation and got a product of 2 rather than 0.

Exercises 6.4

1. Let $x =$ first number.
   Let $y =$ second number.

   $\begin{cases} x + y = 130 \\ x - y = 28 \end{cases}$ Add. $\quad\quad x + y = 130$
   $\phantom{\begin{cases}} 2x = 158$ $\quad\quad\quad\quad\quad 79 + y = 130$
   $\phantom{\begin{cases}} x = 79$ $\quad\quad\quad\quad\quad\quad\quad y = 51$

   Thus, the numbers are 79 and 51.
   CHECK: $79 + 51 \stackrel{\checkmark}{=} 130$ and $79 - 51 \stackrel{\checkmark}{=} 28$.

3. Let $n =$ number of nickels that Sam has.
   Let $q =$ number of quarters that Sam has.

   $\begin{cases} n + q = 80 \\ 5n + 25q = 1360 \end{cases}$ $\xrightarrow[\text{as is}]{\text{multiply by } -5}$ $\begin{array}{r} -5n - 5q = -400 \\ 5n + 25q = 1360 \quad \text{Add.} \\ \hline 20q = 960 \\ q = 48 \end{array}$ $\quad\quad n + q = 80$
   $\quad\quad\quad\quad\quad\quad\quad\quad\quad\quad\quad\quad\quad\quad\quad\quad\quad\quad\quad\quad\quad\quad\quad\quad\quad\quad\quad n + 48 = 80$
   $\quad\quad\quad\quad\quad\quad\quad\quad\quad\quad\quad\quad\quad\quad\quad\quad\quad\quad\quad\quad\quad\quad\quad\quad\quad\quad\quad n = 32$

   Thus, Sam has 32 nickels and 48 quarters.
   CHECK: $32 + 48 \stackrel{\checkmark}{=} 80$ and $5(32) + 25(48) = 160 + 1200 \stackrel{\checkmark}{=} 1360$.

5. Let $x =$ speed of the slower car.
   Let $y =$ speed of the faster car.

   $\begin{cases} y = x + 15 \\ 5x + 5y = 275 \end{cases}$ $\xrightarrow[\text{as is}]{\text{subtract } x \text{ from both sides}}$ $\begin{array}{l} -x + y = 15 \\ 5x + 5y = 275 \end{array}$ $\xrightarrow[\text{as is}]{\text{multiply by } 5}$ $\begin{array}{r} -5x + 5y = 75 \\ 5x + 5y = 275 \quad \text{Add.} \\ \hline 10y = 350 \\ y = 35 \end{array}$

   $35 = x + 15$
   $20 = x \quad\quad$ The speed of the slower car is 20 kph and the speed of the faster car is 35 kph.

CHECK: The difference of the speeds is $35 - 20 = 15$ kph. The slower car travels $5(20) = 100$ km and the faster car travels $5(35) = 175$ kph. The distance between the cars after 5 hours is $100 + 175 = 275$ km.

7. Let $x =$ amount of money that Carmen invested at 6%.
   Let $y =$ amount of money that Carmen invested at 7%.
   $$\begin{cases} x + y = 1700 \\ 0.06x + 0.07y = 110 \end{cases} \xrightarrow{\text{as is}} \begin{array}{c} x + y = 1700 \\ \xrightarrow{\text{multiply by 100}} 6x + 7y = 11000 \end{array} \xrightarrow{\text{multiply by } -6} \begin{array}{r} -6x - 6y = -10200 \\ 6x + 7y = 11000 \\ \hline \text{Add.:} \quad y = 800 \end{array}$$

   $x + y = 1700$
   $x + 800 = 1700$
   $x = 900$ Thus, Carmen invested \$900 at 6% and \$800 at 7%.

   CHECK: $\$900 + \$800 \stackrel{\checkmark}{=} \$1700$ and $0.06(\$900) + 0.07(\$800) = \$54 + \$56 \stackrel{\checkmark}{=} \$110$

9. Let $x =$ price of a cassette.
   Let $y =$ price of a CD.
   $$\begin{cases} 4x + 6y = 107.66 \\ 5x + 3y = 76.30 \end{cases} \xrightarrow{\text{as is}} \begin{array}{r} 4x + 6y = 107.66 \\ \xrightarrow{\text{multiply by } -2} -10x - 6y = -152.60 \\ \hline -6x \quad\quad = -44.94 \\ x = 7.49 \end{array} \begin{array}{r} \text{Add.} \end{array} \begin{array}{r} 4x + 6y = 107.66 \\ 4(7.49) + 6y = 107.66 \\ 29.96 + 6y = 107.66 \\ 6y = 77.70 \\ y = 12.95 \end{array}$$

   Thus, a cassette costs \$7.49 and a CD costs \$12.95.

   CHECK: $4(\$7.49) + 6(\$12.95) = \$29.96 + \$77.70 \stackrel{\checkmark}{=} \$107.66$ and
   $5(\$7.49) + 3(\$12.95) = \$37.45 + \$38.85 \stackrel{\checkmark}{=} \$76.30$.

11. Let $L =$ length of the rectangle.
    Let $W =$ width of the rectangle.
    $$\begin{cases} L = 2W \\ 2L + 2W = 28 \end{cases}$$ Substitute the value of L from the first equation into the second to get
    $2(2W) + 2W = 28$
    $4W + 2W = 28$
    $6W = 28$
    $W = \dfrac{28}{6} = \dfrac{14}{3}$ So $L = 2W = 2\left(\dfrac{14}{3}\right) = \dfrac{28}{3}$

    Thus, the width of the rectangle is $\dfrac{14}{3}$ inches and the length is $\dfrac{28}{3}$ inches.

    CHECK: $\dfrac{28}{3}$ is twice as large as $\dfrac{14}{3}$.

    The perimeter of the rectangle is $2\left(\dfrac{28}{3}\right) + 2\left(\dfrac{14}{3}\right) = \dfrac{56}{3} + \dfrac{28}{3} = \dfrac{84}{3} = 28$, as required.

13. Let $x =$ price of regular selection.
    Let $y =$ price of special selection.
    $$\begin{cases} 2x + 3y = 56.90 \\ 3x + 4y = 80.85 \end{cases} \xrightarrow{\text{multiply by } 3}_{\text{multiply by } -2}$$

    $6x + 9y = 170.7$
    $\underline{-6x - 8y = -161.7}$ Add.
    $y = 9$

    $2x + 3y = 56.90$
    $2x + 3(9) = 56.90$
    $2x + 27 = 56.90$
    $2x = 29.90$
    $x = 14.95$

    Thus, the regular selection is \$14.95 and the specially discounted selection is \$9.
    CHECK: $2(\$14.95) + 3(\$9) = \$29.90 + \$27 \stackrel{\checkmark}{=} \$56.90$ and
    $3(\$14.95) + 4(\$9) = \$44.85 + \$36 \stackrel{\checkmark}{=} \$80.85$.

15. Let $x =$ larger number.
    Let $y =$ smaller number.
    $$\begin{cases} \dfrac{x}{y} = \dfrac{6}{5} \\ x - y = 8 \end{cases}$$
    Solve the second equation for x, obtaining $x = y + 8$. Then substitute this result into the first equation.

    $\dfrac{y+8}{y} = \dfrac{6}{5}$
    $\dfrac{5\cancel{y}}{1} \cdot \dfrac{y+8}{\cancel{y}} = \dfrac{\cancel{5}y}{1} \cdot \dfrac{6}{\cancel{5}}$
    $5(y+8) = 6y$
    $5y + 40 = 6y$
    $40 = y$

    Then $x = y + 8 = 40 + 8 = 48$. Thus, the numbers are 48 and 40.
    CHECK: $\dfrac{48}{40} = \dfrac{6(8)}{5(8)} \stackrel{\checkmark}{=} \dfrac{6}{5}$ and $48 - 40 \stackrel{\checkmark}{=} 8$.

17. Let $x =$ speed of the first plane.
    Let $y =$ speed of the second plane.
    $$\begin{cases} x = 2y \\ 4x + 4y = 1800 \end{cases}$$
    Substitute the value of x given, in the first equation into the second.

    $4(2y) + 4y = 1800$
    $8y + 4y = 1800$
    $12y = 1800$
    $y = 150$

    Then $x = 2y = 2(150) = 300$. Thus, one plane is flying at 150 mph while the other is flying at 300 mph.
    CHECK: $2(150) \stackrel{\checkmark}{=} 300$ and $4(150) + 4(300) = 600 + 1200 \stackrel{\checkmark}{=} 1800$ miles.

19. Let $x =$ cost of a receiver.
    Let $y =$ cost of a turntable.
    $$\begin{cases} 8x + 4y = 2060 \\ 5x + 6y = 1690 \end{cases} \xrightarrow{\text{multiply by } 3}_{\text{multiply by } -2}$$

    $24x + 12y = 6180$
    $\underline{-10x - 12y = -3380}$ Add.
    $14x = 2800$
    $x = 200$

    $8x + 4y = 2060$
    $8(200) + 4y = 2060$
    $1600 + 4y = 2060$
    $4y = 460$
    $y = 115$

    Thus, a receiver costs \$200 and a turntable costs \$115.
    CHECK: $8(\$200) + 4(\$115) = \$1600 + \$460 \stackrel{\checkmark}{=} \$2060$ and
    $5(\$200) + 6(\$115) = \$1000 + \$690 \stackrel{\checkmark}{=} \$1690$.

21. Let $x =$ cost of plain donut.
    Let $y =$ cost of cream-filled donut.
    $$\begin{cases} 10x + 5y = 3.70 \\ 5x + 10y = 4.10 \end{cases} \xrightarrow[\text{as is}]{\text{multiply by } -2} \begin{array}{r} -20x - 10y = -7.40 \\ 5x + 10y = 4.10 \\ \hline -15x = -3.30 \\ x = 0.22 \end{array} \quad \begin{array}{l} 5x + 10y = 4.10 \\ 5(0.22) + 10y = 4.10 \\ 1.10 + 10y = 4.10 \\ 10y = 3 \\ y = 0.30 \end{array}$$

    Thus, a plain donut costs $0.22.
    CHECK: $10(\$0.22) + 5(\$0.30) = \$2.20 + \$1.50 \stackrel{\checkmark}{=} \$3.70$ and
    $5(\$0.22) + 10(\$0.30) = \$1.10 + \$3.00 \stackrel{\checkmark}{=} \$4.10$.

23. Let $x =$ number of $7 books bought.
    Let $y =$ number of $9 books bought.
    $$\begin{cases} x + y = 35 \\ 7x + 9y = 271 \end{cases} \xrightarrow[\text{as is}]{\text{multiply by } -7} \begin{array}{r} -7x - 7y = -245 \\ 7x + 9y = 271 \\ \hline 2y = 26 \\ y = 13 \end{array} \quad \begin{array}{l} x + y = 35 \\ x + 13 = 35 \\ x = 22 \end{array}$$

    Thus, the bookstore bought 22 books at $7 each and 13 books at $9 each.
    CHECK: $22 + 13 \stackrel{\checkmark}{=} 35$ and $(\$7)(22) + (\$9)(13) = \$154 + \$117 \stackrel{\checkmark}{=} \$271$.

25. Let $x =$ speed of the first car.
    Let $y =$ speed of the second car.
    $$\begin{cases} 4x + 4y = 480 \\ x = y + 40 \end{cases}$$
    Substitute the value of x given, in the second equation into the first.
    $$\begin{array}{l} 4(y + 40) + 4y = 480 \\ 4y + 160 + 4y = 480 \\ 8y + 160 = 480 \\ 8y = 320 \\ y = 40 \end{array}$$

    Then $x = y + 40 = 40 + 40 = 80$. Thus, one car is traveling at 80 kph while the other is traveling at 40 kph.
    CHECK: $40 + 40 \stackrel{\checkmark}{=} 80$ and $4(40) + 4(80) = 160 + 320 \stackrel{\checkmark}{=} 480$ km.

27. Let $p =$ speed (in mph) of the plane in still air.
    Let $w =$ speed (in mph) of the wind.
    $$\begin{cases} p + w = 150 \\ p - w = 90 \end{cases} \text{Add.} \quad \begin{array}{r} 2p = 240 \\ p = 120 \end{array} \quad \begin{array}{l} p + w = 150 \\ 120 + w = 150 \\ w = 30 \end{array}$$

    Thus, the speed of the plane in still air is 120 mph and the speed of the wind is 30 mph.
    CHECK: With the tailwind, the speed of the plane is increased by the speed of the wind, giving $120 + 30 \stackrel{\checkmark}{=} 150$ mph. With the headwind, the speed of the plane is decreased by the speed of the wind, giving $120 - 30 \stackrel{\checkmark}{=} 90$ mph.

29. Let x = number of pounds of $3.35 blend.
Let y = number of pounds of $2.75 blend.

$$\begin{cases} x + y = 60 \\ 3.35x + 2.75y = 3.10(60) \end{cases}$$

Solve the first equation for x, obtaining $x = 60 - y$. Then substitute this result into the second equation.

$3.35(60 - y) + 2.75y = 186$
$201 - 3.35y + 2.75y = 186$
$201 - 0.6y = 186$
$-0.6y = -15$
$y = 25$

Then $x = 60 - y = 60 - 25 = 35$. Thus, 35 pounds of the $3.35 blend should be mixed with 25 pounds of the $2.75 blend.

CHECK: $35 + 25 \stackrel{\checkmark}{=} 60$ pounds and
$$\frac{35(\$3.35) + 25(\$2.75)}{60 \text{ lb}} = \frac{\$117.25 + \$68.75}{60 \text{ lb}} = \frac{\$186}{60 \text{ lb}} \stackrel{\checkmark}{=} \$3.10 \text{ per pound.}$$

31. Let d = number of desktop setups.
Let C = total cost of system.

$$\begin{cases} C = 100,000 + 800d \\ C = 16,000 + 1200d \end{cases}$$

Substitute the value of C from the first equation into the second.

$100,000 + 800d = 16,000 + 1200d$
$100,000 = 16,000 + 400d$
$84,000 = 400d$
$210 = d$

The two systems will cost the same when there are 210 desktops setups.

CHECK: $100,000 + 800(210) = 100,000 + 168,000 \stackrel{\checkmark}{=} \$268,000$.
$16,000 + 1200(210) = 16,000 + 252,000 \stackrel{\checkmark}{=} \$268,000$.

33. Let c = cost of a computer.
Let p = cost of a printer.

$$\begin{cases} 10c + 10p = 10,000 \\ 12c + 2p = 10,000 \end{cases}$$

$\xrightarrow{\text{as is}}$ $10c + 10p = 10,000$
$\xrightarrow{\text{multiply by } -5}$ $-60c - 10p = -50,000$
Add. $\overline{-50c \quad\quad = -40,000}$
$c = 800$

$10c + 10p = 10,000$
$10(800) + 10p = 10,000$
$8,000 + 10p = 10,000$
$10p = 2000$
$p = 200$

Thus, a computer costs $800 and a printer costs $200.

CHECK: $10(\$800) + 10(\$200) = \$8,000 + \$2,000 \stackrel{\checkmark}{=} \$10,000$ and
$12(\$800) + 2(\$200) = \$9,600 + \$400 \stackrel{\checkmark}{=} \$10,000$.

35. Let w = width.
Let l = length.

$$\begin{cases} 2w + 2l = 46 \\ 2w = 1 + l \end{cases}$$

Solve the second equation for l, obtaining $l = 2w - 1$. Then substitute this result into the first equation.

$2w + 2(2w - 1) = 46$
$2w + 4w - 2 = 46$
$6w - 2 = 46$
$6w = 48$
$w = 8$

Then $l = 2w - 1 = 2(8) - 1 = 15$. Thus, the width is 8 cm and the length is 15 cm.
CHECK: $2(8) + 2(15) = 16 + 30 \stackrel{\checkmark}{=} 46$ cm and $2(8) = 16$ cm which is $1 + 15 \stackrel{\checkmark}{=} 16$ cm.

37. Let r = number of red marbles.
    Let b = number of blue marbles.
    $$\begin{cases} r+b = 36 \\ 2r = b-6 \end{cases} \xrightarrow[\text{subtract } b]{\text{as is}} \begin{array}{r} r+b = 36 \\ 2r-b = -6 \\ \hline \text{Add. } 3r = 30 \\ r = 10 \end{array} \quad \begin{array}{l} r+b = 36 \\ 10+b = 36 \\ b = 26 \end{array}$$

    Thus, there are 10 red marbles and 26 blue marbles.
    CHECK: $10 + 26 \stackrel{\checkmark}{=} 36$ marbles and $2(10) = 20$ red marbles which is 6 less than 26 blue marbles.

Chapter 6 Review Exercises

1. $x + y = 4$
   Set $y = 0$ and solve for x to get an x-intercept of 4.
   Set $x = 0$ and solve for y to get a y-intercept of 4.

   $x - y = 0$
   Here, both the x and y intercepts are 0. To find a second point on this line, choose $y = 1$ and find $x = 1$. This gives $(1, 1)$.

   The lines cross at the point $(2, 2)$. So the system
   $\begin{cases} x + y = 4 \\ x - y = 0 \end{cases}$ is satisfied by the point $(2, 2)$.

   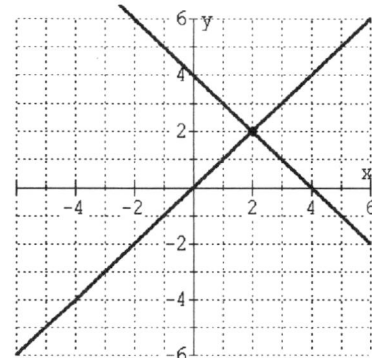

3. $x - 2y = 8$
   Set $y = 0$ and solve for x to get an x-intercept of 8.
   Set $x = 0$ and solve for y to get a y-intercept of -4.

   $y = x - 5$
   Set $y = 0$ and solve for x to get an x-intercept of 5.
   Set $x = 0$ and solve for y to get a y-intercept of -5.

   The lines cross at the point $(2, -3)$. So the system
   $\begin{cases} x - 2y = 8 \\ y = x - 5 \end{cases}$ is satisfied by the point $(2, -3)$.

   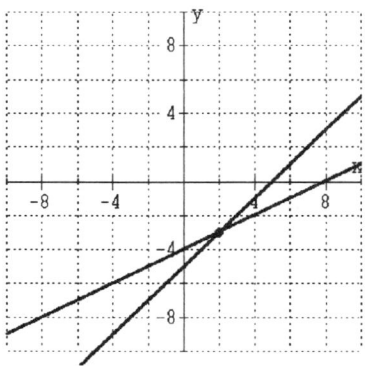

5. $y = x$

Here, both the x and y intercepts are 0. To find a second point on this line, choose $x = 1$ and find $y = 1$. This gives $(1, 1)$.

$3x - 2y = 6$

Set $y = 0$ and solve for x to get an x-intercept of 2.
Set $x = 0$ and solve for y to get a y-intercept of -3.

The lines cross at the point $(6, 6)$. So the system
$\begin{cases} y = x \\ 3x - 2y = 6 \end{cases}$ is satisfied by the point $(6, 6)$.

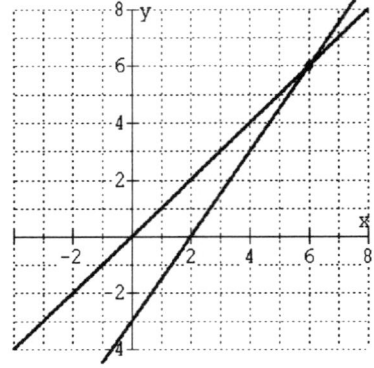

7. $\begin{cases} x + y = 4 \\ x - y = 6 \end{cases}$ Add. $\begin{aligned} x + y &= 4 \\ 5 + y &= 4 \\ y &= -1 \end{aligned}$

   $\begin{aligned} 2x &= 10 \\ x &= 5 \end{aligned}$

   Solution: $(5, -1)$

9. $\begin{cases} y = 2x - 3 \\ x = 3y - 2 \end{cases}$ Substitute the value of x given in the second equation into the first.

   $y = 2(3y - 2) - 3$
   $y = 6y - 4 - 3$
   $y = 6y - 7$
   $-5y = -7$
   $y = \dfrac{7}{5}$

   When $y = \dfrac{7}{5}$,
   $x = 3\left(\dfrac{7}{5}\right) - 2 = \dfrac{11}{5}$

   Solution: $\left(\dfrac{11}{5}, \dfrac{7}{5}\right)$

11. $\begin{cases} 2x - y = 10 \\ x + 3y = -16 \end{cases}$ $\xrightarrow[\text{as is}]{\text{multiply by 3}}$ $\begin{aligned} 6x - 3y &= 30 \\ x + 3y &= -16 \quad \text{Add} \\ \hline 7x &= 14 \\ x &= 2 \end{aligned}$ $\begin{aligned} 2x - y &= 10 \\ 2(2) - y &= 10 \\ 4 - y &= 10 \\ -y &= 6 \\ y &= -6 \end{aligned}$

   Solution: $(2, -6)$

13. $\begin{cases} x - 5y = 1 \\ 3x - 2y = 3 \end{cases}$ $\xrightarrow[\text{as is}]{\text{multiply by } -3}$ $\begin{aligned} -3x + 15y &= -3 \\ 3x - 2y &= 3 \quad \text{Add} \\ \hline 13y &= 0 \\ y &= 0 \end{aligned}$ $\begin{aligned} x - 5y &= 1 \\ x - 5(0) &= 1 \\ x - 0 &= 1 \\ x &= 1 \end{aligned}$

   Solution: $(1, 0)$

15. $\begin{cases} 4x - 3y = 10 \\ 9x + 2y = 5 \end{cases}$ $\xrightarrow[\text{multiply by 3}]{\text{multiply by 2}}$ $\begin{aligned} 8x - 6y &= 20 \\ 27x + 6y &= 15 \quad \text{Add} \\ \hline 35x &= 35 \\ x &= 1 \end{aligned}$ $\begin{aligned} 4x - 3y &= 10 \\ 4(1) - 3y &= 10 \\ 4 - 3y &= 10 \\ -3y &= 6 \\ y &= -2 \end{aligned}$

   Solution: $(1, -2)$

17. $\begin{cases} \dfrac{x}{2} + y = 5 \\ 2y = 8 - x \end{cases}$ $\xrightarrow{\text{multiply by 2}}$ $\xrightarrow{\text{add } x \text{ to both sides}}$ $\begin{array}{l} x + 2y = 10 \\ x + 2y = 8 \text{ Subtract} \\ \hline 0 = 2 \end{array}$

This is a contradiction.

Therefore, the system of equations has no solution.

19. $\begin{cases} x + y - 8 = 2x - 4 \\ 2(y - x) = 8 \end{cases}$ $\xrightarrow{\text{subtract } 2x \text{ from both sides}}_{\text{divide both sides by 2}}$ $\begin{array}{l} y - x - 8 = -4 \\ y - x = 4 \end{array}$ $\xrightarrow{\text{add 8 to both sides}}_{\text{as is}}$ $\begin{array}{l} y - x = 4 \\ y - x = 4 \\ \hline \text{Subtract:} 0 = 0 \end{array}$

This is an identity.

Therefore, the system of equations has infinitely many solutions: $\{(x, y) \mid y - x = 4\}$.

21. Let x = number of gallons pure water in the mixture.
    Let y = number of gallons of 30% alcohol solution in the mixture.
    $\begin{cases} x + y = 30 \\ 0x + 0.30y = 0.25(30) \end{cases}$ $\xrightarrow{\text{as is}}_{\text{multiply by 100}}$ $\begin{array}{l} x + y = 30 \\ 30y = 750 \end{array}$

    From the second equation, find $y = 25$. Then $30 = x + y = x + 25$, so $x = 5$. Thus, 5 gallons of water should be added to 25 gallons of a 30% alcohol solution to produce 30 gallons of a 25% alcohol solution.

    CHECK: $0.30(25) = 7.5$ gallons of pure alcohol and $0.25(30) = 7.5$ gallons of pure alcohol.

23. Let x = walking speed (in kph).
    Let y = jogging speed (in kph).
    $\begin{cases} x + y = 17 \\ 2x + \dfrac{1}{2}y = 16 \end{cases}$ $\xrightarrow{\text{as is}}_{\text{multiply by } -2}$ $\begin{array}{l} x + y = 17 \\ -4x - y = -32 \text{ Add} \\ \hline -3x = -15 \\ x = 5 \end{array}$ $\begin{array}{l} x + y = 17 \\ 5 + y = 17 \\ y = 12 \end{array}$

    Thus, their walking speed is 5 kph and their jogging speed is 12 kph.

    CHECK: $1(5) + 1(12) \stackrel{\checkmark}{=} 17$ and $2(5) + \dfrac{1}{2}(12) = 10 + 6 \stackrel{\checkmark}{=} 16$.

25. (a) 7 (b) decreasing
    (c) 0; occurs when $w = 9$ (d) 9; occurs when $w = 7$
    (e) the value of R when $w = 5$.

Chapter 6 Practice Test

1. (a) $32 per share
   (b) between 10:00 a.m. and 11:00 a.m. and between 12:00 noon and 1:00 p.m.
   (c) $30 per share
   (d) Yes. This occurs between 11:00 a.m. and 12:00 noon.
   (e) rising

3. $\begin{cases} 3x - 4 = y - 1 \\ 9 + 3y = x \end{cases}$ Substitute the value of x given in the second equation into the first equation.

$3(9 + 3y) - 4 = y - 1$ When $y = -3$,   Solution: $(0, -3)$

$27 + 9y - 4 = y - 1$ $x = 9 + 3(-3) = 9 - 9 = 0$

$9y + 23 = y - 1$

$8y + 23 = -1$

$8y = -24$

$y = -3$

5. $\begin{cases} \dfrac{3x}{2} - y = 6 \\ x - \dfrac{2y}{3} = 5 \end{cases}$ $\xrightarrow{\text{multiply by 2}}$ $3x - 2y = 12$

                  $\xrightarrow{\text{multiply by 3}}$ $3x - 2y = 15$ Subtract

                                    $0 = -3$

This is a contradiction.

Therefore, the system of equations has no solution.

7. $2x - y = -8$

Set $y = 0$ and solve for x to get an x-intercept of -4.
Set $x = 0$ and solve for y to get a y-intercept of 8.

$x + 2y = 6$

Set $y = 0$ and solve for x to get an x-intercept of 6.
Set $x = 0$ and solve for y to get a y-intercept of 3.

The lines cross at the point $(-2, 4)$. So the system
$\begin{cases} 2x - y = -8 \\ x + 2y = 6 \end{cases}$ is satisfied by the point $(-2, 4)$.

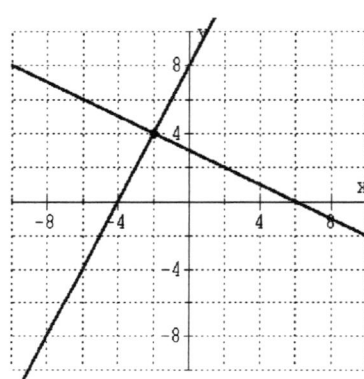

9. Let $x =$ amount invested at 4%.
Let $y =$ amount invested at 5%.

$\begin{cases} x + y = 5550 \\ 0.04x + 0.05y = 259 \end{cases}$ $\xrightarrow[\text{multiply by 100}]{\text{as is}}$ $\begin{aligned} x + y &= 5550 \\ 4x + 5y &= 25900 \end{aligned}$ $\xrightarrow[\text{as is}]{\text{multiply by } -4}$ $\begin{aligned} -4x - 4y &= -22200 \\ 4x + 5y &= 25900 \end{aligned}$

                                                                                                     Add: $y = 3700$

$x + y = 5550$

$x + 3700 = 5550$

$x = 1850$

Thus, $1850 is invested at 4% and $3700 is invested at 5%.

CHECK: $\$1850 + \$3700 \stackrel{\checkmark}{=} \$5500$

$0.04(\$1850) + 0.05(\$3700) = \$74 + \$185 \stackrel{\checkmark}{=} \$259.$

Chapters 4-6
Cumulative Review

1. $\dfrac{-24}{42} = \dfrac{(-4)(6)}{(7)(6)} = \dfrac{-4}{7} = -\dfrac{4}{7}$

3. $\dfrac{36s^8t^9}{20s^8t^8} = \dfrac{9t}{5s}$

5. $\dfrac{6t}{25} \cdot \dfrac{10}{t} = \dfrac{12}{5}$

7. $\dfrac{6x}{25} + \dfrac{10}{x} = \dfrac{6x(x)}{25(x)} + \dfrac{10(25)}{x(25)}$
$= \dfrac{6x^2}{25x} + \dfrac{250}{25x}$
$= \dfrac{6x^2 + 250}{25x}$

9. $\dfrac{3t-5}{6t^2} + \dfrac{9t+5}{6t^2} = \dfrac{3t-5+9t+5}{6t^2}$
$= \dfrac{12t}{6t^2} = \dfrac{2}{t}$

11. $\dfrac{12x^3y^2}{35z^2} \div \dfrac{20xy}{14z} = \dfrac{12x^3y^2}{35z^2} \cdot \dfrac{14z}{20xy}$
$= \dfrac{6x^2y}{25z}$

13. $\dfrac{5}{3x} - \dfrac{7}{2x} = \dfrac{5(2)}{3x(2)} - \dfrac{7(3)}{2x(3)}$
$= \dfrac{10}{6x} - \dfrac{21}{6x} = \dfrac{10-21}{6x}$
$= \dfrac{-11}{6x} = -\dfrac{11}{6x}$

15. $\dfrac{5}{6x^2y} - \dfrac{9}{10xy^3} = \dfrac{5(5y^2)}{6x^2y(5y^2)} - \dfrac{9(3x)}{10xy^3(3x)}$
$= \dfrac{25y^2}{30x^2y^3} - \dfrac{27x}{30x^2y^3}$
$= \dfrac{25y^2 - 27x}{30x^2y^3}$

17. $\left(8 \cdot \dfrac{4}{x}\right) \div \dfrac{16}{x^2} = \dfrac{32}{x} \div \dfrac{16}{x^2}$
$= \dfrac{32}{x} \cdot \dfrac{x^2}{16}$
$= \dfrac{2x}{1} = 2x$

19. $2 + \dfrac{3}{x} - \dfrac{1}{x^2} = \dfrac{2}{1} + \dfrac{3}{x} - \dfrac{1}{x^2}$
$= \dfrac{2(x^2)}{1(x^2)} + \dfrac{3(x)}{x(x)} - \dfrac{1}{x^2}$
$= \dfrac{2x^2}{x^2} + \dfrac{3x}{x^2} - \dfrac{1}{x^2}$
$= \dfrac{2x^2 + 3x - 1}{x^2}$

21. $\dfrac{x}{3} - \dfrac{x}{4} = \dfrac{x-4}{6}$ LCD = 12
$12\left(\dfrac{x}{3} - \dfrac{x}{4}\right) = 12\left(\dfrac{x-4}{6}\right)$
$\dfrac{12}{1} \cdot \dfrac{x}{3} - \dfrac{12}{1} \cdot \dfrac{x}{4} = \dfrac{12}{1} \cdot \dfrac{x-4}{6}$
$4x - 3x = 2(x-4)$
$x = 2x - 8$
$-x = -8$
$x = 8$

23. $$\frac{a}{5} - \frac{a}{6} = \frac{a}{30}$$

LCD = 30

$$30\left(\frac{a}{5} - \frac{a}{6}\right) = 30\left(\frac{a}{30}\right)$$

$$\frac{\overset{6}{\cancel{30}}}{1} \cdot \frac{a}{\underset{1}{\cancel{5}}} - \frac{\overset{5}{\cancel{30}}}{1} \cdot \frac{a}{\underset{1}{\cancel{6}}} = \frac{\overset{1}{\cancel{30}}}{1} \cdot \frac{a}{\underset{1}{\cancel{30}}}$$

$$6a - 5a = a$$
$$a = a \quad \text{Identity}$$

25. $$\frac{7-2y}{4} - \frac{5-4y}{6} = \frac{8y+5}{9} \quad \text{LCD} = 36$$

$$36\left(\frac{7-2y}{4} - \frac{5-4y}{6}\right) = 36\left(\frac{8y+5}{9}\right)$$

$$\frac{\overset{9}{\cancel{36}}}{1} \cdot \frac{7-2y}{\underset{1}{\cancel{4}}} - \frac{\overset{6}{\cancel{36}}}{1} \cdot \frac{5-4y}{\underset{1}{\cancel{6}}} = \frac{\overset{4}{\cancel{36}}}{1} \cdot \frac{8y+5}{\underset{1}{\cancel{9}}}$$

$$9(7-2y) - 6(5-4y) = 4(8y+5)$$
$$63 - 18y - 30 + 24y = 32y + 20$$
$$6y + 33 = 32y + 20$$
$$33 = 26y + 20$$
$$13 = 26y$$
$$\frac{\overset{1}{\cancel{13}}}{\underset{2}{\cancel{26}}} = y$$
$$\frac{1}{2} = y$$

27. $$0.8x - 0.07(x - 5) = 58.75 \quad \text{LCD} = 100$$
$$100[0.8x - 0.07(x-5)] = 100(58.75)$$
$$100(0.8x) - 100[0.07(x-5)] = 100(58.75)$$
$$80x - 7(x-5) = 5875$$
$$80x - 7x + 35 = 5875$$
$$73x + 35 = 5875$$
$$73x = 5840$$
$$x = 80$$

29. $y = 2x - 6$

x-intercept:	y-intercept:	check point:
$0 = 2x - 6$	$y = 2(0) - 6$	choose $x = 2$
$6 = 2x$	$y = -6$	$y = 2(2) - 6$
$3 = x$		$y = -2$
Plot $(3, 0)$	Plot $(0, -6)$	Plot $(2, -2)$

The graph of $y = 2x - 6$

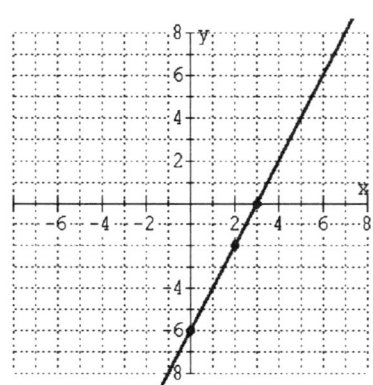

31. $3y - 6x = 12$

    x-intercept:
    $3(0) - 6x = 12$
    $0 - 6x = 12$
    $-6x = 12$
    $x = -2$
    Plot $(-2, 0)$

    y-intercept:
    $3y - 6(0) = 12$
    $3y - 0 = 12$
    $3y = 12$
    $y = 4$
    Plot $(0, 4)$

    check point:
    choose $y = 2$
    $3(2) - 6x = 12$
    $6 - 6x = 12$
    $-6x = 6$
    $x = -1$
    Plot $(-1, 2)$

    The graph of $3y - 6x = 12$
    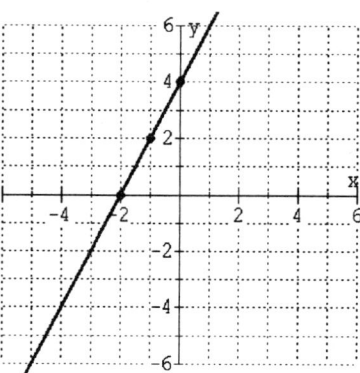

33. $x - 2 = 0$
    Add 2 to both sides of this equation to get the equivalent equation $x = 2$. The graph of the equation is a line parallel to the y-axis and 2 units to the right of it.

    The graph of $x - 2 = 0$
    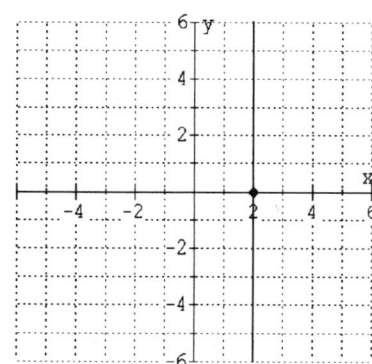

35. $y = 5x$

    x-intercept:
    $0 = 5x$
    $0 = x$
    Plot $(0, 0)$
    (This implies that the y-intercept is 0.)

    check point: choose $x = -1$
    $y = 5(-1)$
    $y = -5$
    Plot $(-1, -5)$

    second point: choose $x = 1$
    $y = 5(1)$
    $y = 5$
    Plot $(1, 5)$

    The graph of $y = 5x$
    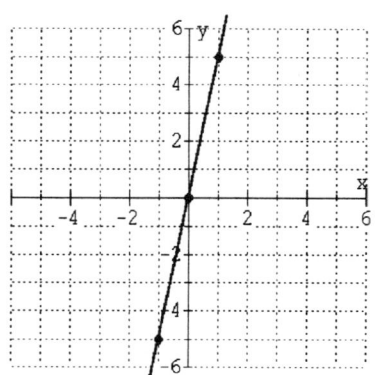

37. $m = \dfrac{y_2 - y_1}{x_2 - x_1} = \dfrac{4 - (-1)}{-3 - 2} = \dfrac{4 + 1}{-3 - 2}$
    $= \dfrac{5}{-5} = -1$

39. $m = \dfrac{y_2 - y_1}{x_2 - x_1} = \dfrac{4 - 4}{-1 - 2} = \dfrac{0}{-3} = 0$

41. $m = 4, (x_1, y_1) = (2, 3)$
    $y - y_1 = m(x - x_1)$
    $y - 3 = 4(x - 2)$ or $y = 4x - 5$

43. $m = -\dfrac{3}{4}, b = 3$
    $y = mx + b$
    $y = -\dfrac{3}{4}x + 3$

45. $m = \dfrac{y_2 - y_1}{x_2 - x_1} = \dfrac{-2 - 5}{2 - (-3)} = \dfrac{-7}{5}$

   $m = -\dfrac{7}{5}$ $(x_1, y_1) = (-3, 5)$

   $y - y_1 = m(x - x_1)$

   $y - 5 = -\dfrac{7}{5}[x - (-3)]$

   $y - 5 = -\dfrac{7}{5}(x + 3)$ or

   $y = -\dfrac{7}{5}x + \dfrac{4}{5}$

47. $2x - y = 7$

   Set $y = 0$ and solve for x to get an x-intercept of $\dfrac{7}{2}$.
   Set $x = 0$ and solve for y to get a y-intercept of -7.

   $x + 2y = 6$
   Set $y = 0$ and solve for x to get an x-intercept of 6.
   Set $x = 0$ and solve for y to get a y-intercept of 3.

   The lines cross at the point $(4, 1)$. So the system
   $\begin{cases} 2x - y = 7 \\ x + 2y = 6 \end{cases}$ is satisfied by the point $(4, 1)$.

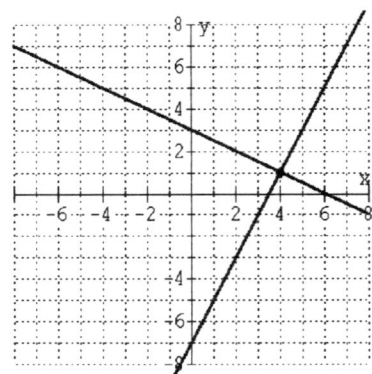

49. $\begin{cases} 2x - y = 7 \\ x + 2y = 6 \end{cases}$ $\xrightarrow{\text{multiply by 2}}$ $\begin{array}{r} 4x - 2y = 14 \\ x + 2y = 6 \\ \hline 5x = 20 \\ x = 4 \end{array}$ Add.

   $2x - y = 7$
   $2(4) - y = 7$
   $8 - y = 7$
   $-y = -1$
   $y = 1$

   Solution: $(4, 1)$

51. $\begin{cases} 4x - 3y = 0 \\ 2x - y = \dfrac{1}{3} \end{cases}$ $\xrightarrow{\text{as is}}$ $\xrightarrow{\text{multiply by } -3}$ $\begin{array}{r} 4x - 3y = 0 \\ -6x + 3y = -1 \\ \hline -2x = -1 \\ x = \dfrac{1}{2} \end{array}$ Add.

   $4x - 3y = 0$
   $4\left(\dfrac{1}{2}\right) - 3y = 0$
   $2 - 3y = 0$
   $2 = 3y$
   $\dfrac{2}{3} = y$

   Solution: $\left(\dfrac{1}{2}, \dfrac{2}{3}\right)$

53. $\begin{cases} y = 5x - 4 \\ x = 3y + 12 \end{cases}$

   Substitute the value of y given in the first equation into the second.

   $x = 3(5x - 4) + 12$
   $x = 15x - 12 + 12$ Then $y = 5x - 4$
   $x = 15x$ $= 5(0) - 4$
   $0 = 14x$ $= -4$
   $0 = x$ Solution: $(0, -4)$

55. $\begin{cases} x + \dfrac{y}{2} = 5 \\ 2x + y = 10 \end{cases}$ $\xrightarrow{\text{multiply by 2}}$ $\xrightarrow{\text{as is}}$ $\begin{array}{r} 2x + y = 10 \\ 2x + y = 10 \\ \hline 0 = 0 \end{array}$ Subtract.

   This is an identity.
   Therefore, the system has infinitely many solutions: $\{(x, y) \mid 2x + y = 10\}$.

57. Let $x =$ number of cheaper tickets sold.
    Then $360 - x =$ number of expensive tickets sold.
    $$6.25x + 8.75(360 - x) = 2850$$
    $$100[6.25x + 8.75(360 - x)] = 100 \cdot 2850$$
    $$100(6.25x) + 100[8.75(360 - x)] = 100 \cdot 2850$$
    $$625x + 875(360 - x) = 285000$$
    $$625x + 315000 - 875x = 285000$$
    $$-250x + 315000 = 285000$$
    $$-250x = -30000$$
    $$x = 120$$
    Then $360 - x = 360 - 120 = 240$. Thus, 120 tickets at \$6.25 each and 240 tickets at \$8.75 were sold.

59. Let $x =$ number of votes that Party A received.
    $$\frac{x}{15700} = \frac{8}{5}$$
    $$15700 \cdot \frac{x}{15700} = 15700 \cdot \frac{8}{5}$$
    $$\cancel{15700} \cdot \frac{x}{\cancel{15700}} = \overset{3140}{\cancel{15700}} \cdot \frac{8}{\cancel{5}}$$
    $$x = 25,120$$
    Party A received 25,120 votes.

61. Let $t =$ number of hours needed for the faster car to overtake the slower one.
    $$80t = 65\left(t + \frac{1}{4}\right)$$
    $$80t = 65t + 65 \cdot \frac{1}{4}$$
    $$15t = \frac{65}{4}$$
    $$t = \frac{\left(\frac{65}{4}\right)}{15} = \frac{65}{4} \cdot \frac{1}{15} = \frac{65}{60} = 1\frac{5}{60}$$
    The faster car overtakes the slower one after 1 hour and 5 minutes.

63. (a) 18
    (b) No. It decreases from 1960 to 1970 and from 1970 to 1980.
    (c) 1900 to 1910; 6 people per square mile.
    (d) 1910 to 1920.

Chapters 4 - 6
Cumulative Practice Test

1. $\dfrac{\overset{4s}{\cancel{12s^2}} \cancel{t^3}}{\underset{1}{\cancel{5d^2}}} \cdot \dfrac{\overset{\overset{1}{\cancel{3d^3}}}{\cancel{15d^5}}}{\underset{\underset{1}{\cancel{3t}}}{\cancel{9st^4}}} = \dfrac{4sd^3}{t}$

3. $\dfrac{11a}{9x} - \dfrac{a}{9x} + \dfrac{5a}{9x} = \dfrac{11a - a + 5a}{9x} = \dfrac{\overset{5}{\cancel{15a}}}{\underset{3}{\cancel{9x}}} = \dfrac{5a}{3x}$

5. $\dfrac{5}{6ab^2} + \dfrac{4}{9b} = \dfrac{5(3)}{6ab^2(3)} + \dfrac{4(2ab)}{9b(2ab)}$
   $= \dfrac{15}{18ab^2} + \dfrac{8ab}{18ab^2}$
   $= \dfrac{15 + 8ab}{18ab^2}$

7. $\dfrac{a}{6} - \dfrac{a}{9} = 18$
   $18\left(\dfrac{a}{6} - \dfrac{a}{9}\right) = 18(18)$
   $\dfrac{\overset{3}{\cancel{18}}}{1} \cdot \dfrac{a}{\underset{1}{\cancel{6}}} - \dfrac{\overset{2}{\cancel{18}}}{1} \cdot \dfrac{a}{\underset{1}{\cancel{9}}} = 324$
   $3a - 2a = 324$
   $a = 324$

9. $x - 3y = 0$

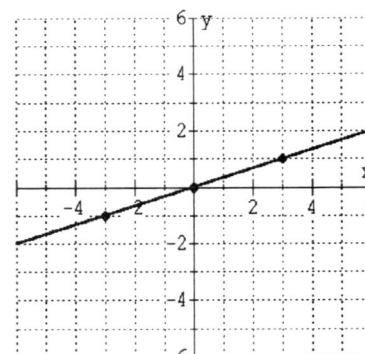

11. $x - 3 = 6$
(This is equivalent to the equation $x = 9$.)

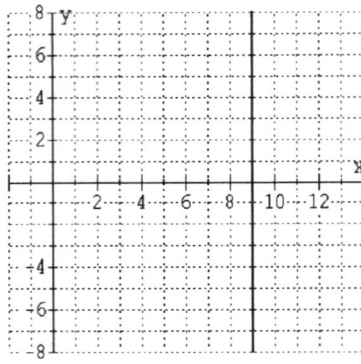

13. $m = \dfrac{y_2 - y_1}{x_2 - x_1} = \dfrac{3 - (-4)}{-1 - 2} = \dfrac{7}{-3}$

$m = -\dfrac{7}{3} \quad (x_1, y_1) = (2, -4)$

$y - y_1 = m(x - x_1)$

$y - (-4) = -\dfrac{7}{3}(x - 2)$

$y + 4 = -\dfrac{7}{3}(x - 2)$ or

$y = -\dfrac{7}{3}x + \dfrac{2}{3}$

15. $\begin{cases} \dfrac{x}{2} - y = 5 \\ -x + 2y = 8 \end{cases} \xrightarrow{\text{multiply by 2}} \begin{array}{l} x - 2y = 10 \\ -x + 2y = 8 \end{array}$

$\xrightarrow{\text{as is}}$ Add. $\quad 0 = 18$

Contradiction
No solution

17. Let $x =$ number of 34¢ stamps Jamie bought.
Let $y =$ number of 50¢ stamps Jamie bought.

$\begin{cases} x + y = 28 \\ 34x + 50y = 1096 \end{cases} \xrightarrow{\text{multiply by } -34} \begin{array}{l} -34x - 34y = -952 \\ 34x + 50y = 1096 \end{array}$ Add

$\begin{array}{l} 16y = 144 \\ y = 9 \end{array}$

$\begin{array}{l} x + y = 28 \\ x + 9 = 28 \\ x = 19 \end{array}$

Thus, Jamie bought nineteen 34¢ stamps and nine 50¢ stamps.
CHECK: $19 + 9 \stackrel{\vee}{=} 28$ and $34(19) + 50(9) = 646 + 450 \stackrel{\vee}{=} 1096$.

19. Let $t =$ amount of time Terry walks (in hours).

$6t + 8\left(t - \dfrac{1}{3}\right) = 9$

$6t + 8t - \dfrac{8}{3} = 9$

$14t - \dfrac{8}{3} = 9$

$14t = \dfrac{35}{3}$

$t = \dfrac{\overset{5}{\cancel{35}}}{3} \cdot \dfrac{1}{\underset{2}{\cancel{14}}} = \dfrac{5}{6}$

Since Terry walks for $\dfrac{5}{6}$ hours
or for 50 minutes, Terry and
Tom will be 9 km apart at 11:50 a.m.
CHECK: From 11:00 a.m. to 11:50 a.m.,
Terry walks $6\left(\dfrac{5}{6}\right) = 5$ km. From
11:20 a.m. to 11:50 a.m., Tom walks
$8\left(\dfrac{1}{2}\right) = 4$ km. $5 \text{ km} + 4 \text{ km} \stackrel{\vee}{=} 9 \text{ km}$

21. Graph C.

Chapter 7
Exponents and Polynomials

Exercises 7.1

1. $x^3 x^2 = x^{3+2} = x^5$

3. $(x^3)^2 = x^{3 \cdot 2} = x^6$

5. $x^3 x x^5 = x^{3+1+5} = x^9$

7. $10^4 10^5 = 10^{4+5} = 10^9$

9. $2^3 3^4$ cannot be simplified
   $(2^3 \cdot 3^4 = 8 \cdot 81 = 648)$

11. $\dfrac{y^3 y^5}{y^2 y^4} = \dfrac{y^{3+5}}{y^{2+4}} = \dfrac{y^8}{y^6} = y^{8-6} = y^2$

13. $\dfrac{9u^9 v^8}{3u^3 v^4} = \dfrac{9}{3} u^{9-3} v^{8-4} = 3u^6 v^4$

15. $\dfrac{(a^3)^5}{(a^4)^2} = \dfrac{a^{3 \cdot 5}}{a^{4 \cdot 2}} = \dfrac{a^{15}}{a^8} = a^{15-8} = a^7$

17. $(-x^2)^4 = (-1)^4 (x^2)^4 = 1 \cdot x^{2 \cdot 4} = x^8$

19. $(x^2 y)^2 = (x^2)^2 y^2 = x^{2 \cdot 2} y^2 = x^4 y^2$

21. $(x^2 y^3)^5 = (x^2)^5 (y^3)^5$
    $= x^{2 \cdot 5} y^{3 \cdot 5} = x^{10} y^{15}$

23. $(2r^3 s^5)^4 = 2^4 (r^3)^4 (s^5)^4$
    $= 16 r^{3 \cdot 4} s^{5 \cdot 4} = 16 r^{12} s^{20}$

25. $(-x^3 y)^3 = (-1)^3 (x^3)^3 y^3$
    $= -1 x^{3 \cdot 3} y^3 = -x^9 y^3$

27. $\left(\dfrac{x^3}{y^2}\right)^4 = \dfrac{(x^3)^4}{(y^2)^4} = \dfrac{x^{3 \cdot 4}}{y^{2 \cdot 4}} = \dfrac{x^{12}}{y^8}$

29. $(2x^3)^4 (3x^2)^2 = 2^4 (x^3)^4 \cdot 3^2 (x^2)^2$
    $= 16 x^{3 \cdot 4} \cdot 9 x^{2 \cdot 2}$
    $= (16 \cdot 9) x^{12} x^4$
    $= 144 x^{12+4} = 144 x^{16}$

31. $\dfrac{(x^4 y^2)^3}{x^5 (y^3)^2} = \dfrac{(x^4)^3 (y^2)^3}{x^5 (y^3)^2} = \dfrac{x^{4 \cdot 3} y^{2 \cdot 3}}{x^5 y^{3 \cdot 2}}$
    $= \dfrac{x^{12} \cancel{y^6}}{x^5 \cancel{y^6}} = \dfrac{x^{12}}{x^5} = x^{12-5} = x^7$

33. $\dfrac{(3x^5 y^4)^2}{9(x^3 y)^3} = \dfrac{3^2 (x^5)^2 (y^4)^2}{9 (x^3)^3 y^3} = \dfrac{9 x^{5 \cdot 2} y^{4 \cdot 2}}{9 x^{3 \cdot 3} y^3}$
    $= \dfrac{\cancel{9} x^{10} y^8}{\cancel{9} x^9 y^3} = \dfrac{x^{10}}{x^9} \cdot \dfrac{y^8}{y^3}$
    $= x^{10-9} y^{8-3} = x y^5$

35. $\left(\dfrac{x^2 y}{4u}\right)^4 = \dfrac{(x^2 y)^4}{(4u)^4} = \dfrac{(x^2)^4 y^4}{4^4 u^4}$
    $= \dfrac{x^{2 \cdot 4} y^4}{256 u^4} = \dfrac{x^8 y^4}{256 u^4}$

37. $\left(\dfrac{2x^3 y^4}{x y^6}\right)^5 = \left(2 \cdot \dfrac{x^3}{x} \cdot \dfrac{y^4}{y^6}\right)^5$
    $= \left(2 x^{3-1} \cdot \dfrac{1}{y^2}\right)^5 = \left(\dfrac{2x^2}{y^2}\right)^5$
    $= \dfrac{(2x^2)^5}{(y^2)^5} = \dfrac{2^5 (x^2)^5}{(y^2)^5}$
    $= \dfrac{32 x^{2 \cdot 5}}{y^{2 \cdot 5}} = \dfrac{32 x^{10}}{y^{10}}$

39. $\left(\dfrac{-3a^2 b^3}{2c}\right)^3 = \dfrac{(-3a^2 b^3)^3}{(2c)^3}$
    $= \dfrac{(-3)^3 (a^2)^3 (b^3)^3}{2^3 c^3}$
    $= \dfrac{-27 a^{2 \cdot 3} b^{3 \cdot 3}}{8 c^3}$
    $= \dfrac{-27 a^6 b^9}{8 c^3} = -\dfrac{27 a^6 b^9}{8 c^3}$

41. $\dfrac{-3^2}{(-3)^2} = \dfrac{\cancel{-9}^{-1}}{\cancel{9}_{1}} = \dfrac{-1}{1} = -1$

43. $\dfrac{-x^2}{(-x)^2} = \dfrac{\cancel{-x^2}^{-1}}{(-1)^2\cancel{x^2}_{1}} = \dfrac{-1}{1} = -1$

45. $\dfrac{-u^3}{(-u)^3} = \dfrac{-u^3}{(-1)^3 u^3} = \dfrac{\cancel{-u^3}^{1}}{\cancel{-u^3}_{1}} = \dfrac{1}{1} = 1$

47. $\dfrac{(-x^2)^4}{-(x^3)^2} = \dfrac{(-1)^4(x^2)^4}{-(x^3)^2} = \dfrac{1 \cdot x^{2 \cdot 4}}{-x^{3 \cdot 2}}$
$= \dfrac{x^8}{-x^6} = -x^{8-6} = -x^2$

49. $\dfrac{-2^4 + 3^2}{(-4+3)^2} = \dfrac{-16+9}{(-1)^2} = \dfrac{-7}{1} = -7$

51. $\dfrac{-4^2 - (-2)^5}{1 - 3^2} = \dfrac{-16 - (-32)}{1 - 9} = \dfrac{16}{-8} = -2$

53. $\dfrac{2(-6)^2 - 3(-5)^2}{(5-6)^3} = \dfrac{2(36) - 3(25)}{(-1)^3}$
$= \dfrac{72 - 75}{-1} = \dfrac{-3}{-1} = 3$

55. $\dfrac{-w^2}{(-w)^2} = \dfrac{-2^2}{(-2)^2} = \dfrac{-4}{4} = -1$

57. $\dfrac{-w^4 + x^2}{(-y+x)^2} = \dfrac{-2^4 + (-3)^2}{[-4 + (-3)]^2} = \dfrac{-16+9}{(-7)^2} = \dfrac{-7}{49} = -\dfrac{1}{7}$

59. $(w - x + y - z)^2 = [2 - (-3) + 4 - (-5)]^2$
$= (14)^2 = 196$

61. $\dfrac{x(-y)}{x - y} = \dfrac{-3(-4)}{-3 - 4} = \dfrac{12}{-7} = -\dfrac{12}{7}$

63. When we multiply two powers of the same base, we keep the base and add the exponents. When we raise a power to a power, we keep the base and multiply the exponents.

64. x^8 means 'x multiplied by itself 8 times'. $x^5 + x^3$ means 'multiply x by itself 5 times and then add x multiplied by itself 3 times'. These are not the same.

65. (a) $\dfrac{x^8}{x^8} = x^{8-8} = x^0$ (b) $\dfrac{x^4}{x^7} = x^{4-7} = x^{-3}$

66. (a) According to Exponent Rule 2, we must multiply the exponents, not add them.
    (b) According to Exponent Rule 1, we must add the exponents, not multiply them.
    (c) According to Exponent Rule 3, we must subtract the exponents, not divide them.
    (d) Since x^2 and x^3 are unlike terms, we cannot combine them when they are added.

67. $\dfrac{2}{x} \cdot \dfrac{8}{y} = \dfrac{2 \cdot 8}{x \cdot y} = \dfrac{16}{xy}$

69. $\dfrac{2}{x} + \dfrac{8}{y} = \dfrac{2 \cdot y}{x \cdot y} + \dfrac{8 \cdot x}{y \cdot x} = \dfrac{2y}{xy} + \dfrac{8x}{xy} = \dfrac{2y + 8x}{xy}$

71. $\dfrac{x}{3} - \dfrac{x}{4} = \dfrac{x \cdot 4}{3 \cdot 4} - \dfrac{x \cdot 3}{4 \cdot 3}$
$= \dfrac{4x}{12} - \dfrac{3x}{12} = \dfrac{x}{12}$

73. Let $m = $ number of miles driven.
$0.20m + 3(15) = 72$
$0.20m + 45 = 72$
$0.20m = 27$
$10(0.20m) = 10(27)$
$2m = 270$
$m = 135$
Thus, 135 miles were driven.

Exercises 7.2

1. (a) $2(-3) = -6$ (b) $x^2 x^{-3} = x^{2+(-3)} = x^{-1} = \dfrac{1}{x}$

    (c) $(x^2)^{-3} = x^{2(-3)} = x^{-6} = \dfrac{1}{x^6}$ (d) $2^{-3} = \dfrac{1}{2^3} = \dfrac{1}{8}$

3. (a) $3(-4) = -12$ (b) $x^3 x^{-4} = x^{3+(-4)} = x^{-1} = \dfrac{1}{x}$

    (c) $(x^3)^{-4} = x^{3(-4)} = x^{-12} = \dfrac{1}{x^{12}}$ (d) $3^{-4} = \dfrac{1}{3^4} = \dfrac{1}{81}$

5. (a) $3 - 4 = -1$ (b) $x^3 x^{-4} = x^{3+(-4)} = x^{-1} = \dfrac{1}{x}$

    (c) $3(-4) = -12$ (d) $(x^3)^{-4} = x^{3(-4)} = x^{-12} = \dfrac{1}{x^{12}}$

    (e) $3 - (-4) = 7$ (f) $\dfrac{x^3}{x^{-4}} = x^{3-(-4)} = x^7$ (g) $3^{-4} = \dfrac{1}{3^4} = \dfrac{1}{81}$

7. $8^0 = 1$ 9. $5 \cdot 4^0 = 5 \cdot 1 = 5$

11. $xy^0 = x \cdot 1 = x$ 13. $5^{-2} = \dfrac{1}{5^2} = \dfrac{1}{25}$

15. $\dfrac{1}{5^{-2}} = \dfrac{1}{\frac{1}{5^2}} = \dfrac{1}{\frac{1}{25}} = 1 \cdot \dfrac{25}{1} = 25$ 17. $x^{-4} x^4 = x^{-4+4} = x^0 = 1$

19. $x^{-4} x^{-6} = x^{-4+(-6)}$ 21. $a^2 a^{-4} a a^{-7} = a^{2+(-4)+1+(-7)}$

    $= x^{-10} = \dfrac{1}{x^{10}}$ $= a^{-8} = \dfrac{1}{a^8}$

23. $10^{-3} 10^8 = 10^{-3+8}$ 25. $10^6 10^{-5} 10^{-4} = 10^{6+(-5)+(-4)}$

    $= 10^5 = 100,000$ $= 10^{-3} = \dfrac{1}{10^3}$

                                                    $= \dfrac{1}{1000} = 0.001$

27. $10^7 10^{-7} = 10^{7+(-7)} = 10^0 = 1$ 29. $(xy)^4 = x^4 y^4$

31. $2a^{-3} = \dfrac{2}{1} \cdot \dfrac{1}{a^3} = \dfrac{2}{a^3}$ 33. $-3y^{-2} = \dfrac{-3}{1} \cdot \dfrac{1}{y^2} = -\dfrac{3}{y^2}$

35. $-(3y^{-2}) = -\left(\dfrac{3}{1} \cdot \dfrac{1}{y^2}\right)$ 37. $xy^{-1} = \dfrac{x}{1} \cdot \dfrac{1}{y} = \dfrac{x}{y}$

             $= -\left(\dfrac{3}{y^2}\right) = -\dfrac{3}{y^2}$

39. $(a^4 b^3)^{-2} = (a^4)^{-2} (b^3)^{-2}$ 41. $(a^{-4} b^3)^{-2} = (a^{-4})^{-2} (b^3)^{-2}$

                $= a^{4(-2)} b^{3(-2)} = a^{-8} b^{-6}$ $= a^{-4(-2)} b^{3(-2)} = a^8 b^{-6}$

                $= \dfrac{1}{a^8} \cdot \dfrac{1}{b^6} = \dfrac{1}{a^8 b^6}$ $= \dfrac{a^8}{1} \cdot \dfrac{1}{b^6} = \dfrac{a^8}{b^6}$

43. $(3x^{-2}y^3z^{-4})^2 = 3^2(x^{-2})^2(y^3)^2(z^{-4})^2$
$= 9x^{-2(2)}y^{3(2)}z^{-4(2)}$
$= 9x^{-4}y^6z^{-8}$
$= \dfrac{9}{1} \cdot \dfrac{1}{x^4} \cdot \dfrac{y^6}{1} \cdot \dfrac{1}{z^8} = \dfrac{9y^6}{x^4z^8}$

45. $4(x^{-1}y)^{-3} = 4(x^{-1})^{-3}y^{-3}$
$= 4x^{-1(-3)}y^{-3}$
$= 4x^3y^{-3}$
$= \dfrac{4}{1} \cdot \dfrac{x^3}{1} \cdot \dfrac{1}{y^3} = \dfrac{4x^3}{y^3}$

47. $\dfrac{x^5}{x^2} = x^{5-2} = x^3$

49. $\dfrac{-3a^{-3}}{9a^9} = \dfrac{\cancel{-3}^{1}}{\cancel{9}_{3}} \cdot \dfrac{a^{-3}}{a^9} = -\dfrac{1}{3}a^{-3-9}$
$= -\dfrac{1}{3}a^{-12}$
$= -\dfrac{1}{3} \cdot \dfrac{1}{a^{12}} = -\dfrac{1}{3a^{12}}$

51. $x^{-2} + y^{-1} = \dfrac{1}{x^2} + \dfrac{1}{y}$
$= \dfrac{y}{x^2y} + \dfrac{x^2}{x^2y} = \dfrac{y+x^2}{x^2y}$

53. $\dfrac{x^4x^{-10}}{x^{-2}x^{-5}} = \dfrac{x^{4+(-10)}}{x^{-2+(-5)}} = \dfrac{x^{-6}}{x^{-7}}$
$= x^{-6-(-7)} = x^1 = x$

55. $\dfrac{x^4y^{-10}}{x^{-2}y^{-5}} = \dfrac{x^4}{x^{-2}} \cdot \dfrac{y^{-10}}{y^{-5}}$
$= x^{4-(-2)}y^{-10-(-5)}$
$= x^6y^{-5} = \dfrac{x^6}{1} \cdot \dfrac{1}{y^5} = \dfrac{x^6}{y^5}$

57. $\dfrac{10^{-3}10^5}{10^610^{-10}} = \dfrac{10^{-3+5}}{10^{6+(-10)}}$
$= \dfrac{10^2}{10^{-4}} = 10^{2-(-4)}$
$= 10^6 = 1,000,000$

59. $\dfrac{12(10^{-3})}{4(10^{-7})} = \dfrac{\cancel{12}^{3}}{\cancel{4}_{1}} \cdot \dfrac{10^{-3}}{10^{-7}}$
$= 3 \cdot 10^{-3-(-7)} = 3 \cdot 10^4$
$= 3 \cdot 10,000 = 30,000$

61. $\left(\dfrac{a^{-2}}{a^3}\right)^{-3} = (a^{-2-3})^{-3} = (a^{-5})^{-3}$
$= a^{-5(-3)} = a^{15}$

63. $\dfrac{(x^2y^{-1})^{-1}}{(x^3y^{-2})^2} = \dfrac{(x^2)^{-1}(y^{-1})^{-1}}{(x^3)^2(y^{-2})^2}$
$= \dfrac{x^{2(-1)}y^{-1(-1)}}{x^{3(2)}y^{-2(2)}}$
$= \dfrac{x^{-2}y^1}{x^6y^{-4}}$
$= \dfrac{x^{-2}}{x^6} \cdot \dfrac{y^1}{y^{-4}}$
$= x^{-2-6}y^{1-(-4)}$
$= x^{-8}y^5 = \dfrac{1}{x^8} \cdot \dfrac{y^5}{1} = \dfrac{y^5}{x^8}$

65. $\left(\dfrac{2m^{-2}n^{-3}}{m^{-6}n^{-1}}\right)^{-2} = \left(\dfrac{2}{1} \cdot \dfrac{m^{-2}}{m^{-6}} \cdot \dfrac{n^{-3}}{n^{-1}}\right)^{-2}$
$= (2m^{-2-(-6)}n^{-3-(-1)})^{-2}$
$= (2m^4n^{-2})^{-2}$
$= 2^{-2}(m^4)^{-2}(n^{-2})^{-2}$
$= \dfrac{1}{2^2} \cdot m^{4(-2)}n^{-2(-2)}$
$= \dfrac{1}{4}m^{-8}n^4$
$= \dfrac{1}{4} \cdot \dfrac{1}{m^8} \cdot \dfrac{n^4}{1} = \dfrac{n^4}{4m^8}$

67. $\left(\dfrac{x^{-1}y^{-2}}{3x^{-2}y^{-3}}\right)^{-1} = \left(\dfrac{1}{3}\cdot\dfrac{x^{-1}}{x^{-2}}\cdot\dfrac{y^{-2}}{y^{-3}}\right)^{-1}$

$= \left(\dfrac{1}{3}x^{-1-(-2)}y^{-2-(-3)}\right)^{-1}$

$= \left(\dfrac{1}{3}xy\right)^{-1}$

$= \left(\dfrac{1}{3}\right)^{-1}x^{-1}y^{-1}$

$= \dfrac{1}{\frac{1}{3}}x^{-1}y^{-1}$

$= \dfrac{3}{1}\cdot\dfrac{1}{x}\cdot\dfrac{1}{y} = \dfrac{3}{xy}$

69. $\dfrac{(2m^{-2}n^{-3})^{-4}}{(m^{-6}n^{-1})^{-2}} = \dfrac{2^{-4}(m^{-2})^{-4}(n^{-3})^{-4}}{(m^{-6})^{-2}(n^{-1})^{-2}}$

$= \dfrac{2^{-4}m^{-2(-4)}n^{-3(-4)}}{m^{-6(-2)}n^{-1(-2)}}$

$= \dfrac{2^{-4}m^{8}n^{12}}{m^{12}n^{2}}$

$= \dfrac{1}{2^4}\cdot\dfrac{m^8}{m^{12}}\cdot\dfrac{n^{12}}{n^2}$

$= \dfrac{1}{16}m^{8-12}n^{12-2}$

$= \dfrac{1}{16}m^{-4}n^{10}$

$= \dfrac{1}{16}\cdot\dfrac{1}{m^4}\cdot\dfrac{n^{10}}{1}$

$= \dfrac{n^{10}}{16m^4}$

71. $\dfrac{(x^5y)^{-2}(x^{-2}y^3)^2}{(x^{-3}y^{-4})^{-2}} = \dfrac{(x^5)^{-2}y^{-2}\cdot(x^{-2})^2(y^3)^2}{(x^{-3})^{-2}(y^{-4})^{-2}} = \dfrac{x^{5(-2)}y^{-2}x^{-2(2)}y^{3(2)}}{x^{-3(-2)}y^{-4(-2)}} = \dfrac{x^{-10}y^{-2}x^{-4}y^6}{x^6y^8}$

$= \dfrac{x^{-10+(-4)}y^{-2+6}}{x^6y^8} = \dfrac{x^{-14}y^4}{x^6y^8} = \dfrac{x^{-14}}{x^6}\cdot\dfrac{y^4}{y^8} = x^{-14-6}y^{4-8} = x^{-20}y^{-4}$

$= \dfrac{1}{x^{20}}\cdot\dfrac{1}{y^4} = \dfrac{1}{x^{20}y^4}$

73. $x^{-3} = 2^{-3} = \dfrac{1}{2^3} = \dfrac{1}{8}$

75. $8x^{-1} = 8\cdot 2^{-1} = 8\cdot\dfrac{1}{2} = 4$

77. $x^{-1} + y^{-1} = 2^{-1} + (-3)^{-1} = \dfrac{1}{2} + \dfrac{1}{-3}$

$= \dfrac{1}{2} - \dfrac{1}{3} = \dfrac{1}{6}$

79. $\dfrac{x^6}{x^4}$ requires us to divide x^6 by x^4,

whereas $\dfrac{x^6}{x^{-4}} = \dfrac{x^6}{\left(\dfrac{1}{x^4}\right)} = x^6\cdot\dfrac{x^4}{1}$

asks us to multiply x^6 by x^4.

80. When -1 appears in the exponent, it tells us to take the reciprocal of the base. Thus, $3^{-1} = \dfrac{1}{3}$. When the minus sign appears in front of the 3, it tells us to take the opposite of 3. Put another way, 3^{-1} is the multiplicative inverse of 3, while -3 is the additive inverse of 3.

81. Let $t =$ number of hours Maria works.
Then $t - 2 =$ number of hours Francis works.
$5t + 7(t - 2) = 70$
$5t + 7t - 14 = 70$
$12t - 14 = 70$
$12t = 84$
$t = 7$
Seven hours after Maria started, it is 4 P.M.

Exercises 7.3

1. $4,530 = 4.53 \times 10^3$

3. $0.0453 = 4.53 \times 10^{-2}$

5. $0.00007 = 7 \times 10^{-5}$

7. $7,000,000 = 7 \times 10^6$

9. $85,370 = 8.537 \times 10^4$

11. $0.0085370 = 8.537 \times 10^{-3}$

13. $90 = 9 \times 10^1$ or 9×10

15. $9 = 9 \times 10^0$ or 9×1

17. $0.9 = 9 \times 10^{-1}$

19. $0.09 = 9 \times 10^{-2}$

21. $0.00000003 = 3 \times 10^{-8}$

23. $28 = 2.8 \times 10^1$ or 2.8×10

25. $47.5 = 4.75 \times 10^1$ or 4.75×10

27. $9727.3 = 9.7273 \times 10^3$

29. $56 \times 10^{-2} = (5.6 \times 10^1) \times 10^{-2}$
$= 5.6 \times (10^1 \cdot 10^{-2})$
$= 5.6 \times 10^{-1}$

31. $0.154 \times 10^4 = (1.54 \times 10^{-1}) \times 10^4$
$= 1.54 \times (10^{-1} \cdot 10^4)$
$= 1.54 \times 10^3$

33. $28.40 \times 10^6 = (2.84 \times 10^1) \times 10^6$
$= 2.84 \times (10^1 \cdot 10^6)$
$= 2.84 \times 10^7$

35. $2.8 \times 10^4 = 28,000$

37. $2.8 \times 10^{-4} = 0.00028$

39. $4.29 \times 10^7 = 42,900,000$

41. $4.29 \times 10^{-7} = 0.000000429$

43. $3.52 \times 10^{-3} = 0.00352$

45. $3.5286 \times 10^5 = 352,860$

47. $0.026 \times 10^{-3} = 0.000026$

49. $(0.004)(250) = (4 \times 10^{-3})(2.5 \times 10^2)$
$= (4)(2.5) \times (10^{-3} \cdot 10^2)$
$= 10 \times 10^{-3+2} = 10^1 \times 10^{-1}$
$= 10^{1+(-1)} = 10^0 = 1$

51. $\dfrac{0.003}{6,000} = \dfrac{3 \times 10^{-3}}{6 \times 10^3} = \dfrac{3}{6} \times \dfrac{10^{-3}}{10^3}$
$= 0.5 \times 10^{-3-3} = 0.5 \times 10^{-6}$
$= 5 \times 10^{-7}$

53. $\dfrac{(480)(0.008)}{(0.24)(4,000)} = \dfrac{(4.8 \times 10^2)(8 \times 10^{-3})}{(2.4 \times 10^{-1})(4 \times 10^3)}$
$= \dfrac{(\cancel{4.8})^2 (\cancel{8})^2}{(\cancel{2.4})_1 (\cancel{4})_1} \times \dfrac{10^2 \, 10^{-3}}{10^{-1} \, 10^3}$
$= 4 \times \dfrac{10^{2+(-3)}}{10^{-1+3}}$
$= 4 \times \dfrac{10^{-1}}{10^2}$
$= 4 \times 10^{-1-2}$
$= 4 \times 10^{-3} = 0.004$

55. $\dfrac{(0.0036)(0.005)}{(0.01)(0.06)} = \dfrac{(3.6 \times 10^{-3})(5 \times 10^{-3})}{(1 \times 10^{-2})(6 \times 10^{-2})}$
$= \dfrac{(\cancel{3.6})^{0.6} (5)}{(1)(\cancel{6})_1} \times \dfrac{10^{-3} \, 10^{-3}}{10^{-2} \, 10^{-2}}$
$= 3 \times \dfrac{10^{-3+(-3)}}{10^{-2+(-2)}}$
$= 3 \times \dfrac{10^{-6}}{10^{-4}}$
$= 3 \times 10^{-6-(-4)}$
$= 3 \times 10^{-2} = 0.03$

57. c 59. e 61. c
63. c 65. d 67. 5.98×10^{24} kg

69. $(80,000)(9.3 \times 10^{-23}) = (8 \times 10^4)(9.3 \times 10^{-23}) = (8)(9.3) \times (10^4 \cdot 10^{-23})$
$= 74.4 \times 10^{4+(-23)} = 74.4 \times 10^{-19} = 7.44 \times 10^{-18}$ grams

71. $0.00000001 = 1 \times 10^{-8}$ cm

73. $(153)(1 \times 10^{-8}) = 153 \times 10^{-8}$
$= 1.53 \times 10^{-6}$ cm

75. $4,250$ million $= (4,250)(1,000,000)$
$= (4.25 \times 10^3)(1 \times 10^6)$
$= (4.25)(1) \times 10^3 10^6$
$= 4.25 \times 10^{3+6}$
$= 4.25 \times 10^9$ miles

77. Let w = weight of the Earth in tons.
$$\frac{5.98 \times 10^{24}}{w} = \frac{888.9}{1}$$
$5.98 \times 10^{24} = (8.889 \times 10^2)w$
$$\frac{5.98 \times 10^{24}}{8.889 \times 10^2} = w$$
$$\left(\frac{5.98}{8.889}\right) \times \frac{10^{24}}{10^2} = w$$
$0.6727 \times 10^{24-2} = w$
$0.6727 \times 10^{22} = 6.727 \times 10^{21} = w$

Thus, the weight of the Earth is 6.727×10^{21} tons.

79. There are $(365)(24)(60)(60) = 31,536,000$ seconds in one year. Then one light year equals $(186,000)(31,536,000)$ miles.
$(186,000)(31,536,000) = (1.86 \times 10^5)(3.1536 \times 10^7)$
$= (1.86)(3.1536) \times 10^5 10^7$
$= 5.865696 \times 10^{5+7}$
$= 5.865696 \times 10^{12}$ miles.

81. $(5 \times 10^9)(5.865696 \times 10^{12})(1.6) = (5)(5.865696)(1.6) \times 10^9 10^{12}$
$= 46.925568 \times 10^{9+12}$
$= 46.925568 \times 10^{21}$
$= 4.6925568 \times 10^{22}$ kilometers

83. First multiply 3.74 by 6.38; then multiply 10^{-5} by 10^4. Take the product of these two results and express this product in scientific notation.
$(3.74)(6.38) = 23.8612$
$10^{-5} 10^4 = 10^{-5+4} = 10^{-1}$
$23.8612 \times 10^{-1} = 2.38612 \times 10^0$
$= 2.38612$

84. If the number is bigger than 1, the exponent cannot be negative; if the number is smaller than 1, the exponent must be negative.

Exercises 7.4

1. (a) one term: $3x^5$
   (b) degree of $3x^5$: 5
   (c) degree of polynomial: 5

3. (a) two terms: $3x$, 4
   (b) degree of $3x$: 1
   degree of 4: 0
   (c) degree of polynomial: 1

5. (a) two terms: x^2, y^3
   (b) degree of x^2: 2
   degree of y^3: 3
   (c) degree of polynomial: 3

7. (a) one term: x^2y^3
   (b) degree of x^2y^3: $5(=2+3)$
   (c) degree of polynomial: 5

9. (a) one term: 8
   (b) degree of 8: 0
   (c) degree of polynomial: 0

11. (a) three terms: $2x^3$, $-5x^2$, x
    (b) degree of $2x^3$: 3
    degree of $-5x^2$: 2
    degree of x: 1
    (c) degree of polynomial: 3

13. (a) two terms: $2x^3$, y^5
    (b) degree of $2x^3$: 3
    degree of y^5: 5
    (c) degree of polynomial: 5

15. (a) one term: $4x^3y^5$
    (b) degree of $4x^3y^5$: $8(=3+5)$
    (c) degree of polynomial: 8

17. (a) four terms: x^5, $-x^3y^4$, $-2x^2y^3$, y^6
    (b) degree of x^5: 5
    degree of $-x^3y^4$: $7(=3+4)$
    degree of $-2x^2y^3$: $5(=2+3)$
    degree of y^6: 6
    (c) degree of polynomial: 7

19. (a) degree of x^2: 2
    degree of $-5x$: 1
    degree of 6: 0
    (b) degree of polynomial: 2
    (c) The coefficient of x^2 is 1.
    The coefficient of $-5x$ is -5.
    6 is both a term and a coefficient.

21. (a) degree of x^2: 2
    degree of 4: 0
    (b) degree of polynomial: 2
    (c) Write the polynomial as $x^2 + 0x + 4$.
    Then the coefficient of x^2 is 1 and
    the coefficient of $0x$ is 0.
    4 is both a term and a coefficient.

23. (a) degree of x^3: 3
    degree of -1: 0
    (b) degree of polynomial: 3
    (c) Write the polynomial as $x^3 + 0x^2 + 0x - 1$.
    Then the coefficient of x^3 is 1 and
    the coefficients of $0x^2$ and $0x$ are 0.
    -1 is both a term and a coefficient.

25. (a) degree of 1: 0
    degree of $-x^5$: 5
    (b) degree of polynomial: 5
    (c) Write the polynomial as
    $-x^5 + 0x^4 + 0x^3 + 0x^2 + 0x + 1$.
    Then the coefficient of $-x^5$ is -1 and
    the coefficients of $0x^4$, $0x^3$, $0x^2$, and $0x$
    are all 0. 1 is both a term and a coefficient.

27. $(2x^2 - 5) + (3x^2 - 5) = 2x^2 + 3x^2 - 5 - 5$
    $= 5x^2 - 10$

29. $(3u^3 - 2u + 7) + (u^3 - u^2 + 7u)$
$= 3u^3 + u^3 - u^2 - 2u + 7u + 7$
$= 4u^3 - u^2 + 5u + 7$

31. $(3u^3 - 2u + 7) - (u^3 - u^2 + 7u)$
$= 3u^3 - 2u + 7 - u^3 + u^2 - 7u$
$= 3u^3 - u^3 + u^2 - 2u - 7u + 7$
$= 2u^3 + u^2 - 9u + 7$

33. $(4t^3 - t) + (t^2 + t) - (t^3 - t^2)$
$= 4t^3 - t + t^2 + t - t^3 + t^2$
$= 4t^3 - t^3 + t^2 + t^2 - t + t$
$= 3t^3 + 2t^2$

35. $(x^2y + 3xy - x^2y^2) + (x^2y - 5x^2y^2 - xy^2)$
$= x^2y + x^2y + 3xy - x^2y^2 - 5x^2y^2 - xy^2$
$= 2x^2y + 3xy - 6x^2y^2 - xy^2$

37. $(x^2y + 3xy - x^2y^2) - (x^2y - 5x^2y^2 - xy^2)$
$= x^2y + 3xy - x^2y^2 - x^2y + 5x^2y^2 + xy^2$
$= x^2y - x^2y + 3xy - x^2y^2 + 5x^2y^2 + xy^2$
$= 3xy + 4x^2y^2 + xy^2$

39. $2(y^2 - 4y + 1) + 3(2y^2 - y - 1)$
$= 2y^2 - 8y + 2 + 6y^2 - 3y - 3$
$= 2y^2 + 6y^2 - 8y - 3y + 2 - 3$
$= 8y^2 - 11y - 1$

41. $5(x^2 - 3x + 2) - 3(2x^2 - 5x - 2)$
$= 5x^2 - 15x + 10 - 6x^2 + 15x + 6$
$= 5x^2 - 6x^2 - 15x + 15x + 10 + 6$
$= -x^2 + 16$

43. $(x^2 + 3x - 7) + (5x - x^2) + (3x^2 - x - 2)$
$= x^2 - x^2 + 3x^2 + 3x + 5x - x - 7 - 2$
$= 3x^2 + 7x - 9$

45. $(2x^2 - 3x + 5) - (x^2 - 7x + 3)$
$= 2x^2 - 3x + 5 - x^2 + 7x - 3$
$= 2x^2 - x^2 - 3x + 7x + 5 - 3$
$= x^2 + 4x + 2$

47. $(a^3 - b^2) + (a^2b + 2b^2) - (a^3 - a^2 - b + b^2)$
$= a^3 - b^2 + a^2b + 2b^2 - a^3 + a^2 + b - b^2$
$= a^3 - a^3 - b^2 + 2b^2 - b^2 + a^2b + a^2 + b$
$= a^2b + a^2 + b$

49. $2x - 1 - [(3x + 6) + (5x - 8)]$
$= 2x - 1 - (8x - 2)$
$= 2x - 1 - 8x + 2$
$= 2x - 8x - 1 + 2 = -6x + 1$

51. $x^2 - x + 3$
$= (-5)^2 - (-5) + 3$
$= 25 + 5 + 3 = 33$

53. $y^4 + y^3 + y^2 + y + 1$
$= (-3)^4 + (-3)^3 + (-3)^2 + (-3) + 1$
$= 81 - 27 + 9 - 3 + 1 = 61$

55. $6x^2 - 7x + 3 = 6\left(-\dfrac{1}{2}\right)^2 - 7\left(-\dfrac{1}{2}\right) + 3$
$= 6\left(\dfrac{1}{4}\right) + \dfrac{7}{2} + 3$
$= \dfrac{3}{2} + \dfrac{7}{2} + 3 = 8$

57. $-\dfrac{1}{2}x^2 - \dfrac{1}{4}x - 2 = -\dfrac{1}{2}(8)^2 - \dfrac{1}{4}(8) - 2$
$= -\dfrac{1}{2}(64) - 2 - 2$
$= -32 - 4 = -36$

59. $\dfrac{1}{3}x^2 - \dfrac{2}{5}x - 3 = \dfrac{1}{3}\left(\dfrac{1}{2}\right)^2 - \dfrac{2}{5}\left(\dfrac{1}{2}\right) - 3$
$= \dfrac{1}{3}\left(\dfrac{1}{4}\right) - \dfrac{1}{5} - 3$
$= \dfrac{1}{12} - \dfrac{1}{5} - 3 = -\dfrac{187}{60}$

61. $-3x^2y + 5xy^2 = -3(2)^2(-1) + 5(2)(-1)^2$
    $= -3(4)(-1) + 5(2)(1)$
    $= 12 + 10 = 22$

63. $5x - 12 - (3x + 8) = 5x - 12 - 3x - 8$
    $= 5x - 3x - 12 - 8$
    $= 2x - 20$

65. $P = 100x - x^2$

    When $x = 30$,
    $P = 100(30) - (30)^2$
    $= 3000 - 900$
    $= \$2100.$

    When $x = 50$,
    $P = 100(50) - (50)^2$
    $= 5000 - 2500$
    $= \$2500.$

    When $x = 90$,
    $P = 100(90) - (90)^2$
    $= 9000 - 8100$
    $= \$900.$

67. $h = 120 - 28t - 16t^2$

    When $t = 1$,
    $h = 120 - 28(1) - 16(1)^2$
    $= 120 - 28 - 16$
    $= 76$ feet.

    When $t = 1.5$,
    $h = 120 - 28(1.5) - 16(1.5)^2$
    $= 120 - 42 - 36$
    $= 42$ feet.

    When $t = 2$,
    $h = 120 - 28(2) - 16(2)^2$
    $= 120 - 56 - 64$
    $= 0$ feet.

69. $h = 5 + 80t - 16t^2$

    When $t = 1$,
    $h = 5 + 80(1) - 16(1)^2$
    $= 5 + 80 - 16$
    $= 69$ feet.

    When $t = 2$,
    $h = 5 + 80(2) - 16(2)^2$
    $= 5 + 160 - 64$
    $= 101$ feet.

    When $t = 2.5$,
    $h = 5 + 80(2.5) - 16(2.5)^2$
    $= 5 + 200 - 100$
    $= 105$ feet.

71. $N = 2d - 0.01d^2$

    When $d = 30$,
    $N = 2(30) - 0.01(30)^2$
    $= 60 - 9$
    $= 51$ thousand
    $= 51,000$ people.

    When $d = 60$,
    $N = 2(60) - 0.01(60)^2$
    $= 120 - 36$
    $= 84$ thousand
    $= 84,000$ people.

    When $d = 90$,
    $N = 2(90) - 0.01(90)^2$
    $= 180 - 81$
    $= 99$ thousand
    $= 99,000$ people.

73. In a sum, the expressions to be added are called terms; in a product, the expressions to be multiplied are called factors.

74. 3 is not a factor of the expression $6x + 8$ because 3 does not exactly divide 8.

75. 2 is a factor of the expression $6x + 8$ because 2 does exactly divide both $6x$ and 8. Here, we can write $6x + 8 = 2(3x + 4)$.

77. $2(x + 3) - 5(x + 4) = x - 2$
    $2x + 6 - 5x - 20 = x - 2$
    $-3x - 14 = x - 2$
    $-4x = 12$
    $x = -3$

79. $4(x - 1) - 3x = x - 4$
    $4x - 4 - 3x = x - 4$
    $x - 4 = x - 4$
    Identity

81. Let r = rate of interest on second investment.
$$0.08(1800) + r(1400) = 284$$
$$144 + 1400r = 284$$
$$1400r = 140$$
$$r = 0.10$$

The rate of interest on the second investment must be 10%.

Exercises 7.5

1. $3x(5x^3)(4x^2) = (3)(5)(4)(x \cdot x^3 \cdot x^2)$
    $= 60x^6$

3. $3x(5x^3 + 4x^2) = 3x \cdot 5x^3 + 3x \cdot 4x^2$
    $= 15x^4 + 12x^3$

5. $4xy(3yz)(-5xz)$
    $= (4)(3)(-5)(xx)(yy)(zz)$
    $= -60x^2y^2z^2$

7. $4xy(3yz - 5xz)$
    $= 4xy \cdot 3yz - 4xy \cdot 5xz$
    $= 12xy^2z - 20x^2yz$

9. $3x^2(x + 3y) + 4xy(x - 3y)$
    $= 3x^2 \cdot x + 3x^2 \cdot 3y + 4xy \cdot x - 4xy \cdot 3y$
    $= 3x^3 + 9x^2y + 4x^2y - 12xy^2$
    $= 3x^3 + 13x^2y - 12xy^2$

11. $5xy^2(xy - y) - 2y(x^2y^2 - xy^2)$
    $= 5xy^2 \cdot xy - 5xy^2 \cdot y - 2y \cdot x^2y^2 + 2y \cdot xy^2$
    $= 5x^2y^3 - 5xy^3 - 2x^2y^3 + 2xy^3$
    $= 3x^2y^3 - 3xy^3$

13. $(x + 2)(x^2 - x + 3)$
    $= x(x^2 - x + 3) + 2(x^2 - x + 3)$
    $= x^3 - x^2 + 3x + 2x^2 - 2x + 6$
    $= x^3 + x^2 + x + 6$

15. $(y - 5)(y^2 + 2y - 6)$
    $= y(y^2 + 2y - 6) - 5(y^2 + 2y - 6)$
    $= y^3 + 2y^2 - 6y - 5y^2 - 10y + 30$
    $= y^3 - 3y^2 - 16y + 30$

17. $(3x - 2)(x^2 + 3x - 5)$
    $= 3x(x^2 + 3x - 5) - 2(x^2 + 3x - 5)$
    $= 3x^3 + 9x^2 - 15x - 2x^2 - 6x + 10$
    $= 3x^3 + 7x^2 - 21x + 10$

19. $(5z + 2)(3z^2 + 2z + 8)$
    $= 5z(3z^2 + 2z + 8) + 2(3z^2 + 2z + 8)$
    $= 15z^3 + 10z^2 + 40z + 6z^2 + 4z + 16$
    $= 15z^3 + 16z^2 + 44z + 16$

21. $(x + y)(x^2 - xy + y^2) = x(x^2 - xy + y^2) + y(x^2 - xy + y^2)$
    $= x^3 - x^2y + xy^2 + x^2y - xy^2 + y^3$
    $= x^3 + y^3$

23. $(x^2 + x + 1)(x^2 + x - 1) = x^2(x^2 + x - 1) + x(x^2 + x - 1) + 1(x^2 + x - 1)$
    $= x^4 + x^3 - x^2 + x^3 + x^2 - x + x^2 + x - 1$
    $= x^4 + 2x^3 + x^2 - 1$

25. $(x^3 + xy - y^2)(x^3 - 3xy + y^2) = x^3(x^3 - 3xy + y^2) + xy(x^3 - 3xy + y^2) - y^2(x^3 - 3xy + y^2)$
    $= x^6 - 3x^4y + x^3y^2 + x^4y - 3x^2y^2 + xy^3 - x^3y^2 + 3xy^3 - y^4$
    $= x^6 - 2x^4y - 3x^2y^2 + 4xy^3 - y^4$

27. $(x + 5)(x + 3) = x^2 + 3x + 5x + 15$
    $= x^2 + 8x + 15$

29. $(x - 5)(x - 3) = x^2 - 3x - 5x + 15$
    $= x^2 - 8x + 15$

31. $(x-5)(x+3) = x^2 + 3x - 5x - 15$
    $= x^2 - 2x - 15$

33. $(x+5)(x-3) = x^2 - 3x + 5x - 15$
    $= x^2 + 2x - 15$

35. $(x+2y)(x+3y) = x^2 + 3xy + 2xy + 6y^2$
    $= x^2 + 5xy + 6y^2$

37. $(a+8b)(a-5b) = a^2 - 5ab + 8ab - 40b^2$
    $= a^2 + 3ab - 40b^2$

39. $(3x-4)(4x-1) = 12x^2 - 3x - 16x + 4$
    $= 12x^2 - 19x + 4$

41. $(2r-s)(r+3s) = 2r^2 + 6rs - rs - 3s^2$
    $= 2r^2 + 5rs - 3s^2$

43. $(x^2+3)(x^2+2) = x^4 + 2x^2 + 3x^2 + 6$
    $= x^4 + 5x^2 + 6$

45. $(x+7)(x+7) = x^2 + 7x + 7x + 49$
    $= x^2 + 14x + 49$

47. $(x+7)(x-7) = x^2 - 7x + 7x - 49$
    $= x^2 - 49$

49. $(x-4)^2 = (x-4)(x-4)$
    $= x^2 - 4x - 4x + 16$
    $= x^2 - 8x + 16$

51. $2(x-3)(x+5) = 2(x^2 + 5x - 3x - 15)$
    $= 2(x^2 + 2x - 15)$
    $= 2x^2 + 4x - 30$

53. $6(a-8)(a+2) = 6(a^2 + 2a - 8a - 16)$
    $= 6(a^2 - 6a - 16)$
    $= 6a^2 - 36a - 96$

55. $(x+2)^3 = (x+2)(x+2)(x+2)$
    $= (x+2)(x^2 + 2x + 2x + 4)$
    $= (x+2)(x^2 + 4x + 4)$
    $= x(x^2 + 4x + 4) + 2(x^2 + 4x + 4)$
    $= x^3 + 4x^2 + 4x + 2x^2 + 8x + 8$
    $= x^3 + 6x^2 + 12x + 8$

57. $(3x-5)^2 = (3x-5)(3x-5)$
    $= 9x^2 - 15x - 15x + 25$
    $= 9x^2 - 30x + 25$

59. $(2a-9b)^2$
    $= (2a-9b)(2a-9b)$
    $= 4a^2 - 18ab - 18ab + 81b^2$
    $= 4a^2 - 36ab + 81b^2$

61. $2x^2(x+4)(x-8)$
    $= 2x^2(x^2 - 8x + 4x - 32)$
    $= 2x^2(x^2 - 4x - 32)$
    $= 2x^4 - 8x^3 - 64x^2$

63. $3x(5x-6)(3x-2)$
    $= 3x(15x^2 - 10x - 18x + 12)$
    $= 3x(15x^2 - 28x + 12)$
    $= 45x^3 - 84x^2 + 36x$

65. $(x+4)(x-3) + (x-6)(x-2)$
    $= x^2 - 3x + 4x - 12 + x^2 - 2x - 6x + 12$
    $= 2x^2 - 7x$

67. $(a-5)(a-4) - (a-3)(a-2)$
    $= a^2 - 4a - 5a + 20 - (a^2 - 2a - 3a + 6)$
    $= a^2 - 9a + 20 - (a^2 - 5a + 6)$
    $= a^2 - 9a + 20 - a^2 + 5a - 6$
    $= -4a + 14$

69. $(x-6)^2 - (x+6)^2$
    $= (x-6)(x-6) - (x+6)(x+6)$
    $= x^2 - 6x - 6x + 36 - (x^2 + 6x + 6x + 36)$
    $= x^2 - 12x + 36 - (x^2 + 12x + 36)$
    $= x^2 - 12x + 36 - x^2 - 12x - 36$
    $= -24x$

71. $(2x-3)^3 = (2x-3)(2x-3)(2x-3)$
    $= (2x-3)(4x^2 - 6x - 6x + 9)$
    $= (2x-3)(4x^2 - 12x + 9)$
    $= 2x(4x^2 - 12x + 9) - 3(4x^2 - 12x + 9)$
    $= 8x^3 - 24x^2 + 18x - 12x^2 + 36x - 27$
    $= 8x^3 - 36x^2 + 54x - 27$

73. $(3a + 4b)^3 = (3a + 4b)(3a + 4b)(3a + 4b)$
$= (3a + 4b)(9a^2 + 12ab + 12ab + 16b^2)$
$= (3a + 4b)(9a^2 + 24ab + 16b^2)$
$= 3a(9a^2 + 24ab + 16b^2) + 4b(9a^2 + 24ab + 16b^2)$
$= 27a^3 + 72a^2b + 48ab^2 + 36a^2b + 96ab^2 + 64b^3)$
$= 27a^3 + 108a^2b + 144ab^2 + 64b^3$

75. $(y - 2)^3 + (y - 2)^2 + y - 2 = (y - 2)(y - 2)(y - 2) + (y - 2)(y - 2) + y - 2$
$= (y - 2)(y^2 - 2y - 2y + 4) + y^2 - 2y - 2y + 4 + y - 2$
$= (y - 2)(y^2 - 4y + 4) + y^2 - 3y + 2$
$= y(y^2 - 4y + 4) - 2(y^2 - 4y + 4) + y^2 - 3y + 2$
$= y^3 - 4y^2 + 4y - 2y^2 + 8y - 8 + y^2 - 3y + 2$
$= y^3 - 5y^2 + 9y + 2$

77. $(x + 2)^3 - (x + 2)^2 - (x + 2) + 2 = (x + 2)(x + 2)(x + 2) - (x + 2)(x + 2) - (x + 2) + 2$
$= x^3 + 6x^2 + 12x + 8 - (x^2 + 4x + 4) - (x + 2) + 2$
(see Exercise 55 for details)
$= x^3 + 6x^2 + 12x + 8 - x^2 - 4x - 4 - x - 2 + 2$
$= x^3 + 5x^2 + 7x + 4$

79. Let W = width of the rectangle.
Then $2W + 3$ = length of the rectangle.
So $P = 2W + 2(2W + 3) = 2W + 4W + 6 = 6W + 6$ is the perimeter of the rectangle and
$A = W(2W + 3) = 2W^2 + 3W$ is the area of the rectangle.

81. Let s = length of the side of the original square.
Then $s + 4.5$ = length of the side of the new square.
So $s \cdot s = s^2$ is the area of the original square and $(s + 4.5)(s + 4.5) = s^2 + 4.5s + 4.5s + 20.25$
$= s^2 + 9s + 20.25$ is the area of the new square. Therefore, the change in the area is
$s^2 + 9s + 20.25 - s^2 = 9s + 20.25$.

83. Let a = width of original rectangle.
Then $5a - 8$ = length of original rectangle and $a + 3$ = width of new rectangle.
So $a(5a - 8) = 5a^2 - 8a$ is the area of the original rectangle and
$(a + 3)(5a - 8) = 5a^2 - 8a + 15a - 24 = 5a^2 + 7a - 24$ is the area of the new rectangle.
Therefore, the change in area is
$5a^2 + 7a - 24 - (5a^2 - 8a) = 5a^2 + 7a - 24 - 5a^2 + 8a = 15a - 24$.

85. Let x = length of side of the square to be cut out. The original rectangle has dimensions 8 by 10, so has an area of 80. The area of the square to be cut out is $x \cdot x = x^2$. Therefore, the remaining area is $80 - x^2$.

87. When x is increased by 500,
$R = (x + 500)[50 - 0.004(x + 500)]$
$= (x + 500)(50 - 0.004x - 2)$
$= (x + 500)(48 - 0.004x)$
$= 48x - 0.004x^2 + 24,000 - 2x$
$= 46x - 0.004x^2 + 24,000$

Then the increase in revenue is
$46x - 0.004x^2 + 24,000 - [x(50 - 0.004x)]$
$= 46x - 0.004x^2 + 24,000 - (50x - 0.004x^2)$
$= 46x - 0.004x^2 + 24,000 - 50x + 0.004x^2$
$= -4x + 24,000$

89. (a) Both examples require us to find the square of an expression involving x and y.
    (b) The first example asks us to square a product; the second asks us to square a sum.
    (c) $(xy)^2 = x^2y^2$, by Exponent Rule 4. $(x+y)^2 = x^2 + 2xy + y^2$ by the FOIL method.

90. $(x+y)^n = x^n + y^n$ only when $n = 1$.

91. (a) $-4x^2(2x - 7) = -8x^3 + 28x^2$
    $-4x^2(2x - 7)(3x + 1) = (-8x^3 + 28x^2)(3x + 1)$
    $= -24x^4 - 8x^3 + 84x^3 + 28x^2$
    $= -24x^4 + 76x^3 + 28x^2$

    (b) $-4x^2(3x + 1) = -12x^3 - 4x^2$
    $-4x^2(2x - 7)(3x + 1) = -4x^2(3x + 1)(2x - 7)$
    $= (-12x^3 - 4x^2)(2x - 7)$
    $= -24x^4 + 84x^3 - 8x^3 + 28x^2$
    $= -24x^4 + 76x^3 + 28x^2$

    (c) $(2x - 7)(3x + 1) = 6x^2 + 2x - 21x - 7$
    $= 6x^2 - 19x - 7$
    $-4x^2(2x - 7)(3x + 1) = -4x^2(6x^2 - 19x - 7)$
    $= -24x^4 + 76x^3 + 28x^2$

    The three answers are the same, and should be because of the commutative and associative laws of multiplication.

93. $8 - 3(x + 1) < 32$
    $8 - 3x - 3 < 32$
    $-3x + 5 < 32$
    $-3x < 27$
    $x > -9$

95. $-3 \leq 9 - x < 1$
    $-9 \quad -9 \quad\quad -9$
    $\overline{-12 \leq \quad -x \quad < -8}$
    $12 \geq \quad x \quad > 8$
    or
    $8 < \quad x \quad \leq 12$

97. Let $x =$ number of ounces of the 55% solution.
    Then $40 - x =$ number of ounces of the 35% solution.
    $0.55x + 0.35(40 - x) = 0.51(40)$
    $0.55x + 14 - 0.35x = 20.4$
    $0.20x + 14 = 20.4$
    $0.20x = 6.4$
    $2.0x = 64$
    $x = 32$

    Then $40 - x = 40 - 32 = 8$.
    So 32 ounces of the 55% solution should be mixed with 8 ounces of the 35% solution.

Exercises 7.6

1. $(x + 4)(x + 3) = x^2 + 3x + 4x + 12$
   $= x^2 + 7x + 12$

3. $(x - 4)(x - 3) = x^2 - 3x - 4x + 12$
   $= x^2 - 7x + 12$

5. $(x+4)(x-3) = x^2 - 3x + 4x - 12$
   $= x^2 + x - 12$

7. $x + 4(x-3) = x + 4x - 12$
   $= 5x - 12$

9. $(x-4)(x+3) = x^2 + 3x - 4x - 12$
   $= x^2 - x - 12$

11. $(x+6)(x+2) = x^2 + 2x + 6x + 12$
    $= x^2 + 8x + 12$

13. $(x-6)(x-2) = x^2 - 2x - 6x + 12$
    $= x^2 - 8x + 12$

15. $(x+6)(x-2) = x^2 - 2x + 6x - 12$
    $= x^2 + 4x - 12$

17. $(x-6)(x+2) = x^2 + 2x - 6x - 12$
    $= x^2 - 4x - 12$

19. $(x+12)(x+1) = x^2 + x + 12x + 12$
    $= x^2 + 13x + 12$

21. $(x-12)(x-1) = x^2 - x - 12x + 12$
    $= x^2 - 13x + 12$

23. $(x+12)(x-1) = x^2 - x + 12x - 12$
    $= x^2 + 11x - 12$

25. $(x-12)(x+1) = x^2 + x - 12x - 12$
    $= x^2 - 11x - 12$

27. $(x-20)(x+2) = x^2 + 2x - 20x - 40$
    $= x^2 - 18x - 40$

29. $(a+8)(a+8) = a^2 + 8a + 8a + 64$
    $= a^2 + 16a + 64$
    (Perfect square)

31. $(a-8)(a-8) = a^2 - 8a - 8a + 64$
    $= a^2 - 16a + 64$
    (Perfect square)

33. $(a+8)(a-8) = a^2 - 8a + 8a - 64$
    $= a^2 - 64$
    (Difference of two squares)

35. $(c-4)^2 = (c-4)(c-4)$
    $= c^2 - 4c - 4c + 16$
    $= c^2 - 8c + 16$
    (Perfect square)

37. $(c+4)^2 = (c+4)(c+4)$
    $= c^2 + 4c + 4c + 16$
    $= c^2 + 8c + 16$
    (Perfect square)

39. $(c+4)(c-4) = c^2 - 4c + 4c + 16$
    $= c^2 - 16$
    (Difference of two squares)

41. $(3x+4)(x+7) = 3x^2 + 21x + 4x + 28$
    $= 3x^2 + 25x + 28$

43. $(3x+7)(x+4) = 3x^2 + 12x + 7x + 28$
    $= 3x^2 + 19x + 28$

45. $(2y-3)(y-5) = 2y^2 - 10y - 3y + 15$
    $= 2y^2 - 13y + 15$

47. $(3x+4)(x-7) = 3x^2 - 21x + 4x - 28$
    $= 3x^2 - 17x - 28$

49. $(3x-4)(x+7) = 3x^2 + 21x - 4x - 28$
    $= 3x^2 + 17x - 28$

51. $(3x+4)(5x+7) = 15x^2 + 21x + 20x + 28$
    $= 15x^2 + 41x + 28$

53. $(3x+7)(5x+4) = 15x^2 + 12x + 35x + 28$
    $= 15x^2 + 47x + 28$

55. $(3x+4)(5x-7) = 15x^2 - 21x + 20x - 28$
    $= 15x^2 - x - 28$

57. $(3x - 4)(5x + 7) = 15x^2 + 21x - 20x - 28$
$= 15x^2 + x - 28$

59. $(2a + 5)^2 = (2a + 5)(2a + 5)$
$= 4a^2 + 10a + 10a + 25$
$= 4a^2 + 20a + 25$
(Perfect square)

61. $(2a + 5)(2a - 5) = 4a^2 - 10a + 10a - 25$
$= 4a^2 - 25$
(Difference of two squares)

63. $(2a - 5)^2 = (2a - 5)(2a - 5)$
$= 4a^2 - 10a - 10a + 25$
$= 4a^2 - 20a + 25$
(Perfect square)

65. $(x - 4)(x^2 + 6x - 7)$
$= x(x^2 + 6x - 7) - 4(x^2 + 6x - 7)$
$= x^3 + 6x^2 - 7x - 4x^2 - 24x + 28$
$= x^3 + 2x^2 - 31x + 28$

67. $(3xy)^2 = 3^2 x^2 y^2 = 9x^2 y^2$

69. $(x^3 + y^2)^2 = (x^3 + y^2)(x^3 + y^2)$
$= x^6 + x^3 y^2 + x^3 y^2 + y^4$
$= x^6 + 2x^3 y^2 + y^4$
(Perfect square)

71. $(2a + 5y)^2 = (2a + 5y)(2a + 5y)$
$= 4a^2 + 10ay + 10ay + 25y^2$
$= 4a^2 + 20ay + 25y^2$
(Perfect square)

73. $(2a + 5y)(2a - 5y) = 4a^2 - 10ay + 10ay - 25y^2$
$= 4a^2 - 25y^2$
(Difference of two squares)

75. $(x + 6)(x + 4) = x^2 + 4x + 6x + 24$
$= x^2 + 10x + 24$

$(x - 6)(x - 4) = x^2 - 4x - 6x + 24$
$= x^2 - 10x + 24$

The effect of switching both $+$ signs to $-$ signs is to change the sign of the middle term from $+$ to $-$.

76. $(x + 6)(x - 4) = x^2 - 4x + 6x - 24$
$= x^2 + 2x - 24$

$(x - 6)(x + 4) = x^2 + 4x - 6x - 24$
$= x^2 - 2x - 24$

The middle terms of the resulting trinomials have opposite signs.

77. (5) and (7) (6) and (8) (13) and (15) (14) and (16) (21) and (23)
(22) and (24) (41) and (43) (42) and (44) (49) and (51) (50) and (52)

The middle terms will always have opposite signs, since
$(x + a)(x - b) = x^2 + (a - b)x - ab$
$(x - a)(x + b) = x^2 + (-a + b)x - ab$
and $(a - b)$ and $(-a + b)$ are always opposites of one another.

79. $\dfrac{20xy}{9z} \div 18xyz = \dfrac{\overset{10}{\cancel{20}\cancel{x}\cancel{y}}}{9z} \cdot \dfrac{1}{\underset{9}{\cancel{18}\cancel{x}\cancel{y}z}} = \dfrac{10}{81z^2}$

81. $\dfrac{a}{3} - \dfrac{a}{2} = \dfrac{a}{5}$ LCD = 30

$30\left(\dfrac{a}{3} - \dfrac{a}{2}\right) = 30\left(\dfrac{a}{5}\right)$

$10a - 15a = 6a$

$-5a = 6a$

$0 = 11a$

$0 = a$

83. Let r = rate of interest on second investment.

$0.09(1400) + r(1100) = 203$

$126 + 1100r = 203$

$1100r = 77$

$r = 0.07$

So the rate of interest on the second investment is 7%.

Chapter 7 Review Exercises

1. $3^{-4} = \dfrac{1}{3^4} = \dfrac{1}{81}$

3. $\left(3^{-1} + 2^{-2}\right)^2 = \left(\dfrac{1}{3^1} + \dfrac{1}{2^2}\right)^2$

$= \left(\dfrac{1}{3} + \dfrac{1}{4}\right)^2 = \left(\dfrac{7}{12}\right)^2$

$= \dfrac{7^2}{12^2} = \dfrac{49}{144}$

5. $\dfrac{(xy^2)^3}{(x^2y)^4} = \dfrac{x^3(y^2)^3}{(x^2)^4 y^4} = \dfrac{x^3 y^6}{x^8 y^4}$

$= \dfrac{x^3}{x^8} \cdot \dfrac{y^6}{y^4} = x^{-5} y^2$

$= \dfrac{1}{x^5} \cdot \dfrac{y^2}{1} = \dfrac{y^2}{x^5}$

7. $\dfrac{(3x^3 y^2)^4}{9(x^2 y^4)^3} = \dfrac{3^4 (x^3)^4 (y^2)^4}{9(x^2)^3 (y^4)^3}$

$= \dfrac{81 x^{12} y^8}{9 x^6 y^{12}} = \dfrac{81}{9} \cdot \dfrac{x^{12}}{x^6} \cdot \dfrac{y^8}{y^{12}}$

$= 9 \cdot x^6 \cdot y^{-4} = \dfrac{9}{1} \cdot \dfrac{x^6}{1} \cdot \dfrac{1}{y^4} = \dfrac{9 x^6}{y^4}$

9. $\left(x^{-2}\right)^{-3} = x^6$

11. $\left(\dfrac{2 x^{-2} x^3}{x^{-3}}\right)^{-2} = \left(\dfrac{2x}{x^{-3}}\right)^{-2}$

$= \left(2 x^4\right)^{-2} = 2^{-2} \left(x^4\right)^{-2}$

$= \dfrac{1}{2^2} x^{-8} = \dfrac{1}{4} \cdot \dfrac{1}{x^8} = \dfrac{1}{4 x^8}$

13. $58{,}700{,}000 = 5.87 \times 10^7$

15. $0.000002 = 2 \times 10^{-6}$

17. $2.56 \times 10^{-3} = 0.00256$

19. $5.773 \times 10^8 = 577{,}300{,}000$

21. $(0.008)(250{,}000) = (8 \times 10^{-3})(2.5 \times 10^5)$

$= (8)(2.5) \times 10^{-3} 10^5$

$= 20 \times 10^2 = 2 \times 10^3$

$= 2000$

23. $\dfrac{0.001}{0.000025} = \dfrac{1 \times 10^{-3}}{2.5 \times 10^{-5}}$

$= \dfrac{1}{2.5} \times \dfrac{10^{-3}}{10^{-5}}$

$= 0.4 \times 10^2$

$= 4 \times 10^1 = 40$

25. (a) three terms: x^2, $3x$, -7
    (b) degree of x^2: 2
    degree of $3x$: 1
    degree of -7: 0
    (c) degree of polynomial: 2

27. (a) three terms: $3x^3y$, $-5y^2$, $6xy$
    (b) degree of $3x^3y$: $4 (= 3+1)$
    degree of $-5y^2$: 2
    degree of $6xy$: $2 (= 1+1)$
    (c) degree of polynomial: 4

29. (a) two terms: $8x$, -5
    (b) degree of $8x$: 1
    degree of -5: 0
    (c) degree of polynomial: 1

31. (a) one term: 9
    (b) degree of 9: 0
    (c) degree of polynomial: 0

33. (a) one term
    (b) degree of $(3x^5)(2x^3) = 6x^8$: 8
    (c) degree of polynomial: 8

35. $2x^3 - 7x^2 + 0x + 4$

37. $y^5 + 0y^4 + 0y^3 + y^2 - 2y - 1$

39. $(3x^2 - 5x + 7) + (5x - x^2 - 5)$
    $= 3x^2 - 5x + 7 + 5x - x^2 - 5$
    $= 2x^2 + 2$

41. $(3x^2 - 5x + 7) - (5x - x^2 - 5)$
    $= 3x^2 - 5x + 7 - 5x + x^2 + 5$
    $= 4x^2 - 10x + 12$

43. $2(x^2y - xy^2 - 5x^2y^2) + 3(xy^2 + x^2y + x^2y^2)$
    $= 2x^2y - 2xy^2 - 10x^2y^2 + 3xy^2 + 3x^2y + 3x^2y^2$
    $= 5x^2y + xy^2 - 7x^2y^2$

45. $2(x^2y - xy^2) - 5x^2y^2 - 3(xy^2 - x^2y + x^2y^2) = 2x^2y - 2xy^2 - 5x^2y^2 - 3xy^2 + 3x^2y - 3x^2y^2$
    $= 5x^2y - 5xy^2 - 8x^2y^2$

47. $2a^2(a - 3b) + 4a(a^2 + ab) - 2(a^3 - a^2b) = 2a^3 - 6a^2b + 4a^3 + 4a^2b - 2a^3 + 2a^2b = 4a^3$

49. $x^2 + 4x - (x^2 - 4x)$
    $= x^2 + 4x - x^2 + 4x$
    $= 8x$

51. $(x^2 + 4x - 3) + (2x^2 - x - 2) - (3x - 5)$
    $= x^2 + 4x - 3 + 2x^2 - x - 2 - 3x + 5$
    $= 3x^2$

53. $(x + 4)(x - 7) = x^2 - 7x + 4x - 28$
    $= x^2 - 3x - 28$

55. $(2x - 3)(4x - 5) = 8x^2 - 10x - 12x + 15$
    $= 8x^2 - 22x + 15$

57. $(3a - 4b)(2a + 5b)$
    $= 6a^2 + 15ab - 8ab - 20b^2$
    $= 6a^2 + 7ab - 20b^2$

59. $3a - 4b(2a + 5b)$
    $= 3a - 8ab - 20b^2$

61. $(x + 2)(x - 3)(x + 1) = (x + 2)(x^2 + x - 3x - 3) = (x + 2)(x^2 - 2x - 3)$
    $= x(x^2 - 2x - 3) + 2(x^2 - 2x - 3)$
    $= x^3 - 2x^2 - 3x + 2x^2 - 4x - 6 = x^3 - 7x - 6$

63. $(x + 5)(x + 7) = x^2 + 7x + 5x + 35$
    $= x^2 + 12x + 35$

65. $(x - 5)(x - 7) = x^2 - 7x - 5x + 35$
    $= x^2 - 12x + 35$

67. $(x + 5)(x - 7) = x^2 - 7x + 5x - 35$
    $= x^2 - 2x - 35$

69. $(x - 5)(x + 7) = x^2 + 7x - 5x - 35$
    $= x^2 + 2x - 35$

71. $(x-5)(x-5) = x^2 - 5x - 5x + 25$
    $= x^2 - 10x + 25$

73. $(x-5)(x+5) = x^2 + 5x - 5x - 25$
    $= x^2 - 25$

75. $(x+9y)(x-9y)$
    $= x^2 - 9xy + 9xy - 81y^2$
    $= x^2 - 81y^2$

77. $(2x+3)(x-7)$
    $= 2x^2 - 14x + 3x - 21$
    $= 2x^2 - 11x - 21$

79. $2x + 3(x-7) = 2x + 3x - 21$
    $= 5x - 21$

81. $(5x-2)(3x+4) = 15x^2 + 20x - 6x - 8$
    $= 15x^2 + 14x - 8$

83. $(x+6)^2$
    $= (x+6)(x+6)$
    $= x^2 + 6x + 6x + 36$
    $= x^2 + 12x + 36$

85. $(x-5)^3$
    $= (x-5)(x-5)(x-5)$
    $= (x-5)(x^2 - 5x - 5x + 25)$
    $= (x-5)(x^2 - 10x + 25)$
    $= x(x^2 - 10x + 25) - 5(x^2 - 10x + 25)$
    $= x^3 - 10x^2 + 25x - 5x^2 + 50x - 125$
    $= x^3 - 15x^2 + 75x - 125$

87. $3x^2(x-4)(x+2)$
    $= 3x^2(x^2 + 2x - 4x - 8)$
    $= 3x^2(x^2 - 2x - 8)$
    $= 3x^4 - 6x^3 - 24x^2$

89. $(x+8)(x-8)$
    $= x^2 - 8x + 8x - 64$
    $= x^2 - 64$

91. $(x+2)(x^2 - 3x + 4) = x(x^2 - 3x + 4) + 2(x^2 - 3x + 4)$
    $= x^3 - 3x^2 + 4x + 2x^2 - 6x + 8 = x^3 - x^2 - 2x + 8$

93. $(x^2 + 2x - 1)(x^2 + 2x + 1) = x^2(x^2 + 2x + 1) + 2x(x^2 + 2x + 1) - 1(x^2 + 2x + 1)$
    $= x^4 + 2x^3 + x^2 + 2x^3 + 4x^2 + 2x - x^2 - 2x - 1$
    $= x^4 + 4x^3 + 4x^2 - 1$

95. $(2x-3)(x+4) - (x-2)(x-1) = 2x^2 + 8x - 3x - 12 - (x^2 - x - 2x + 2)$
    $= 2x^2 + 5x - 12 - (x^2 - 3x + 2)$
    $= 2x^2 + 5x - 12 - x^2 + 3x - 2$
    $= x^2 + 8x - 14$

97. Let $w =$ width of the rectangle.
    Then $3w - 5 =$ length of the rectangle.
    Since the area of a rectangle, A, is the product of its length and width, we find
    $A = (3w - 5)w = 3w^2 - 5w$.

99. $(60)(5,000)(10^{-6})$ meters
    $= (6 \times 10^1)(5 \times 10^3)(10^{-6})$ meters
    $= 30 \times (10^1 \cdot 10^3 \cdot 10^{-6})$ meters
    $= 30 \times 10^{-2}$ meters
    $= 3.0 \times 10^{-1}$ or 0.3 meters

Chapter 7 Practice Test

1. (a) $5^0 + 2^{-2} + 4^{-1}$
   $= 1 + \dfrac{1}{2^2} + \dfrac{1}{4^1}$
   $= 1 + \dfrac{1}{4} + \dfrac{1}{4}$
   $= \dfrac{4}{4} + \dfrac{1}{4} + \dfrac{1}{4} = \dfrac{6}{4} = \dfrac{3}{2}$

   (b) $6x^0 - 8x^{-4} + x^{-1}$
   $= 6(2)^0 - 8(2)^{-4} + 2^{-1}$
   $= 6(1) - 8\left(\dfrac{1}{2^4}\right) + \dfrac{1}{2}$
   $= 6 - \dfrac{8}{16} + \dfrac{1}{2}$
   $= 6 - \dfrac{1}{2} + \dfrac{1}{2} = 6$

3. $x^{-4}x^{-5} = x^{-4+(-5)} = x^{-9} = \dfrac{1}{x^9}$

5. $\dfrac{(2x^{-3}y^4)^4}{4(x^{-2}y^{-1})^3} = \dfrac{2^4(x^{-3})^4(y^4)^4}{4(x^{-2})^3(y^{-1})^3} = \dfrac{16x^{-3(4)}y^{4(4)}}{4x^{-2(3)}y^{-1(3)}} = \dfrac{16x^{-12}y^{16}}{4x^{-6}y^{-3}} = \left(\dfrac{16}{4}\right)\left(\dfrac{x^{-12}}{x^{-6}}\right)\left(\dfrac{y^{16}}{y^{-3}}\right)$
   $= 4x^{-12-(-6)}y^{16-(-3)} = 4x^{-6}y^{19} = \dfrac{4}{1} \cdot \dfrac{1}{x^6} \cdot \dfrac{y^{19}}{1} = \dfrac{4y^{19}}{x^6}$

7. $-3x^2y(4x^2y)(-2x^3)$
   $= -3(4)(-2)x^2x^2x^3yy$
   $= 24x^{2+2+3}y^{1+1}$
   $= 24x^7y^2$

9. $2x(x^2 - y) - 3(x - xy) - (2x^3 - 3x)$
   $= 2x^3 - 2xy - 3x + 3xy - 2x^3 + 3x$
   $= 2x^3 - 2x^3 - 2xy + 3xy - 3x + 3x$
   $= xy$

11. $3x^2(2x - y) - xy(x + y)$
    $= 6x^3 - 3x^2y - x^2y - xy^2$
    $= 6x^3 - 4x^2y - xy^2$

13. $(a-1)^2 - (a+1)^2$
    $= (a-1)(a-1) - (a+1)(a+1)$
    $= a^2 - 2a + 1 - (a^2 + 2a + 1)$
    $= a^2 - 2a + 1 - a^2 - 2a - 1$
    $= -4a$

15. $\dfrac{(0.24)(5,000)}{0.006}$
    $= \dfrac{(2.4 \times 10^{-1})(5 \times 10^3)}{6 \times 10^{-3}}$
    $= \dfrac{(2.4)(5)}{6} \times \dfrac{10^{-1}10^3}{10^{-3}}$
    $= \dfrac{12}{6} \times \dfrac{10^2}{10^{-3}} = 2 \times 10^{2-(-3)}$
    $= 2 \times 10^5 = 200,000$

17. $\dfrac{8.32 \times 10^{14}}{5.86 \times 10^{12}}$
    $= \dfrac{8.32}{5.86} \times \dfrac{10^{14}}{10^{12}}$
    $= 1.4198 \times 10^{14-12}$
    $= 1.4198 \times 10^2$
    $= 141.98$ light years

Chapter 8
Factoring

Exercises 8.1

1. $4x$

3. $5x$

5. $6mnp$

7. $5x + 20 = 5 \cdot x + 5 \cdot 4 = 5(x + 4)$

9. $6x - 18 = 6 \cdot x - 6 \cdot 3 = 6(x - 3)$

11. $4y + 27$ is not factorable.

13. $28a - 42 = 14 \cdot 2a - 14 \cdot 3$
$ = 14(2a - 3)$

15. $12a + 9 = 3 \cdot 4a + 3 \cdot 3$
$ = 3(4a + 3)$

17. $3a + 6b - 8c$ is not factorable.

19. $x^2 + 3x = x \cdot x + x \cdot 3 = x(x + 3)$

21. $2t^2 + 8t = 2 \cdot t \cdot t + 2 \cdot t \cdot 4$
$ = 2t(t + 4)$

23. $26y^2 - 39y^3 = 13y^2 \cdot 2 - 13y^2 \cdot 3y$
$ = 13y^2(2 - 3y)$

25. $4x^5 + 2x^2 - 8x$
$= 2x \cdot 2x^4 + 2x \cdot x - 2x \cdot 4$
$= 2x(2x^4 + x - 4)$

27. $a^2 + a = a \cdot a + a \cdot 1$
$ = a(a + 1)$

29. $x^2 - 5x + xy = x \cdot x - x \cdot 5 + x \cdot y$
$ = x(x - 5 + y)$

31. $3c^6 - 6c^3 = 3c^3 \cdot c^3 - 3c^3 \cdot 2$
$ = 3c^3(c^3 - 2)$

33. $x^2y - xy^2 = xy \cdot x - xy \cdot y$
$ = xy(x - y)$

35. $24x^2 + 15x = 3x \cdot 8x + 3x \cdot 5$
$ = 3x(8x + 5)$

37. $38x^3y^2 - 75z^4$ is not factorable.

39. $12c^3d^5 + 4c^2d^3 = 4c^2d^3 \cdot 3cd^2 + 4c^2d^3 \cdot 1$
$ = 4c^2d^3(3cd^2 + 1)$

41. $x^2y^3 - y^2z^4 + x^3z^2$ is not factorable.

43. $2x^2yz^3 + 8xyz^2 - 10x^2y^2z^2$
$= 2xyz^2 \cdot xz + 2xyz^2 \cdot 4 - 2xyz^2 \cdot 5xy$
$= 2xyz^2(xz + 4 - 5xy)$

45. $6u^3v^2 + 18u^3v^3 - 12u^3v^5$
$= 6u^3v^2 \cdot 1 + 6u^3v^2 \cdot 3v - 6u^3v^2 \cdot 2v^3$
$= 6u^3v^2(1 + 3v - 2v^3)$

47. $x(x - 5) + 4(x - 5) = (x - 5)(x + 4)$

49. $y(y + 6) - 3(y + 6) = (y + 6)(y - 3)$

51. $x^2 + 8x + xy + 8y = (x^2 + 8x) + (xy + 8y)$
$ = x(x + 8) + y(x + 8)$
$ = (x + 8)(x + y)$

53. $m^2 + mn + 9m + 9n$
$= (m^2 + mn) + (9m + 9n)$
$= m(m + n) + 9(m + n)$
$= (m + n)(m + 9)$

55. $x^2 - xy - 4x + 4y$
$= (x^2 - xy) + (-4x + 4y)$
$= x(x - y) - 4(x - y)$
$= (x - y)(x - 4)$

57. $3x^2y + 6xy - 5x - 10$
$= (3x^2y + 6xy) + (-5x - 10)$
$= 3xy(x + 2) - 5(x + 2)$
$= (x + 2)(3xy - 5)$

59. (a) $(x + 2)(x + 3)$
(b) $x^2y^2(x + y)$

60. Step 1: $-x$ is replaced by $-5x + 4x$.
Step 2: The first two terms are grouped and the last two terms are grouped. The common factor is then removed from each group.
Step 3: The common factor of $(x - 5)$ is removed from the sum.

61. When we factor out -5 from the third and fourth terms, the remaining factor should be $x + 3$, not $x - 3$. This means that there is no common factor to remove.

63. $(x^{-3})^{-4} = x^{(-3)(-4)} = x^{12}$

65. $\dfrac{(x^{-2}y^3)^{-1}}{(x^3)^{-2}} = \dfrac{(x^{-2})^{-1}(y^3)^{-1}}{x^{-6}} = \dfrac{x^2 y^{-3}}{x^{-6}}$
$= x^{2-(-6)}y^{-3} = \dfrac{x^8}{y^3}$

67. Let $x = $ number of liters of pure water to be added.
$0x + 0.18(36) = 0.12(36 + x)$
$6.48 = 4.32 + 0.12x$
$2.16 = 0.12x$
$216 = 12x$
$18 = x$
So 18 liters of pure water must be added.

Exercises 8.2

1. $x^2 + 3x = x(x + 3)$

3. signs alike: $+$
factors of 2 whose sum is 3: 1 and 2
$x^2 + 3x + 2 = (x + 1)(x + 2)$

5. signs alike: $-$
factors of 2 whose sum is 3: 1 and 2
$x^2 - 3x + 2 = (x - 1)(x - 2)$

7. signs are different
factors of 2 whose difference is 3: none
$x^2 + 3x - 2$ is not factorable.

9. signs are different
factors of 2 whose difference is 1: 1 and 2
$x^2 + x - 2 = (x - 1)(x + 2)$

11. signs are different
factors of 2 whose difference is 1: 1 and 2
$x^2 - x - 2 = (x + 1)(x - 2)$

13. signs alike: $+$
factors of 12 whose sum is 8: 2 and 6
$a^2 + 8a + 12 = (a + 2)(a + 6)$

15. signs are different
factors of 12 whose difference is 1: 4 and 3
$a^2 - a - 12 = (a - 4)(a + 3)$

17. signs alike: $-$
factors of 12 whose sum is 1: none
$a^2 - a + 12$ is not factorable.

19. $a^2 - 12a = a(a - 12)$

21. signs are different
    factors of 12 whose difference is 1: 4 and 3
    $a - 12 + a^2 = a^2 + a - 12$
    $ = (a + 4)(a - 3)$

23. signs alike: +
    factors of 28 whose sum is 11: 4 and 7
    $y^2 + 11y + 28 = (y + 4)(y + 7)$

25. signs are different
    factors of 36 whose difference is 5: 9 and 4
    $x^2 - 5x - 36 = (x - 9)(x + 4)$

27. signs are different
    factors of 40 whose difference is 6: 4 and 10
    $a^2 + 6a - 40 = (a + 10)(a - 4)$

29. signs alike: −
    factors of 30 whose sum is 17: 2 and 15
    $z^2 - 17z + 30 = (z - 2)(z - 15)$

31. $x^2 - 9x = x(x - 9)$

33. signs are different
    factors of 9 whose difference is 0: −3 and 3
    $x^2 - 9 = (x - 3)(x + 3)$

35. signs are different
    factors of 10 whose difference is 9: 10 and 1
    $x^2 - 9x - 10 = (x - 10)(x + 1)$

37. signs alike: −
    factors of 2 whose sum is 3: 1 and 2
    $x^2 - 3xy + 2y^2 = (x - y)(x - 2y)$

39. signs alike: +
    factors of 24 whose sum is 10: 4 and 6
    $a^2 + 10a + 24 = (a + 4)(a + 6)$

41. signs alike: +
    factors of 24 whose sum is 10: 4 and 6
    $y^2 + 12y + 36 = (y + 6)(y + 6)$
    $ \text{or } (y + 6)^2$

43. signs are different
    factors of 36 whose difference is 0: −6 and 6
    $y^2 - 36 = (y - 6)(y + 6)$

45. signs are different
    factors of 18 whose difference is 7: 9 and 2
    $x^2 - 7x - 18 = (x - 9)(x + 2)$

47. signs are different
    factors of 10 whose difference is 3: 5 and 2
    $r^2 - 3rs - 10s^2 = (r - 5s)(r + 2s)$

49. signs alike: −
    factors of 5 whose sum is 6: 1 and 5
    $c^2 - 6c + 5 = (c - 1)(c - 5)$

51. $4x^2 + 8x + 4 = 4(x^2 + 2x + 1)$
    signs alike: +
    factors of 1 whose sum is 2: 1 and 1
    $ = 4(x + 1)(x + 1)$
    $ \text{or } 4(x + 1)^2$

53. signs are different
    factors of 30 whose difference is 1: 6 and 5
    $x^2 - 30 + x = x^2 + x - 30$
    $ = (x + 6)(x - 5)$

55. $2x^2 - 50 = 2(x^2 - 25)$
    signs are different
    factors of 25 whose difference is 0: −5 and 5
    $ = 2(x - 5)(x + 5)$

57. signs are different
    factors of 20 whose difference is 1: 5 and 4
    $x^2 - x - 20 = (x - 5)(x + 4)$

59. signs alike: −
    factors of 20 whose sum is 1: none
    $x^2 - x + 20$ is not factorable.

61. signs alike: −
    factors of 28 whose sum is 11: 4 and 7
    $y^2 - 11y + 28 = (y - 4)(y - 7)$

63. $2y^2 + 2y - 84 = 2(y^2 + y - 42)$
    signs are different
    factors of 42 whose difference is 1: 7 and 6
    $ = 2(y + 7)(y - 6)$

65. $49 - d^2 = (7-d)(7+d)$

67. $49 + d^2$ is not factorable.

69. $10x^2 - 40xy - 120y^2$
$= 10(x^2 - 4xy - 12y^2)$
$= 10(x - 6y)(x + 2y)$

71. $a^2 + 13 + 14a = a^2 + 14a + 13$
$= (a+13)(a+1)$

73. $6s^2 - 6s - 72 = 6(s^2 - s - 12)$
$= 6(s-4)(s+3)$

75. $4x^2 - 64 = 4(x^2 - 16)$
$= 4(x-4)(x+4)$

77. The factors of 10 are 1 and 10 and 2 and 5. Since the last term is positive, the signs in the parentheses must be the same.
Possibilities:
$(x+1)(x+10) = x^2 + 11x + 10$
$(x-1)(x-10) = x^2 - 11x + 10$
$(x+2)(x+5) = x^2 + 7x + 10$
$(x-2)(x-5) = x^2 - 7x + 10$
Therefore, the possible values of k are 11, -11, 7, and -7.

78. As in Exercise 77, the factors of 10 are 1 and 10 and 2 and 5. Since the last term is negative, the signs in the parentheses must be opposite.
Possibilities:
$(x+1)(x-10) = x^2 - 9x - 10$
$(x-1)(x+10) = x^2 + 9x - 10$
$(x+2)(x-5) = x^2 - 3x - 10$
$(x-2)(x+5) = x^2 + 3x - 10$
Therefore, the possible values of b are -9, 9, -3, and 3.

79. There are infinitely many such integers. We can always find such a "c" by choosing two integers that differ by 5 and forming the product.
$(x + \text{larger integer})(x - \text{smaller integer})$
For example, since 9 and 4 differ by 5, $(x+9)(x-4) = x^2 + 5x - 36$ will give the value $c = -36$. Clearly, we can find two integers that differ by 5 in infinitely many ways. (Interestingly, if we ask for all positive integers c with this property, there are only two answers: $c = 4$ and $c = 6$.)

81. $(5x^2 + y)^2 = (5x^2 + y)(5x^2 + y)$
$= 25x^4 + 5x^2y + 5x^2y + y^2$
$= 25x^4 + 10x^2y + y^2$

83. $(2x-7)(3x+4) = 6x^2 + 8x - 21x - 28$
$= 6x^2 - 13x - 28$

85. $28{,}700 = 2.87 \times 10^4$

87. $\dfrac{(2{,}400)(0.003)}{(0.02)(0.004)} = \dfrac{(2.4 \times 10^3)(3 \times 10^{-3})}{(2 \times 10^{-2})(4 \times 10^{-3})}$
$= \dfrac{(2.4)(3)}{(2)(4)} \times \dfrac{10^3 10^{-3}}{10^{-2} 10^{-3}}$
$= 0.9 \times 10^5 = 9 \times 10^4$
$= 90{,}000$

Exercises 8.3

1. $x^2 + 5x = x(x+5)$

3. $x^2 + 5x + 4 = (x+1)(x+4)$

5. $x^2 + 5x - 4$ is not factorable.

7. $AC = 3(4) = 12$
Product: 12;   Sum: 8;   6 and 2
$3x^2 + 8x + 4 = 3x^2 + 6x + 2x + 4$
$ = 3x(x+2) + 2(x+2)$
$ = (x+2)(3x+2)$

9. $AC = 2(12) = 24$
Product: 24;   Sum: 11;   8 and 3
$2x^2 + 11x + 12 = 2x^2 + 8x + 3x + 12$
$ = 2x(x+4) + 3(x+4)$
$ = (x+4)(2x+3)$

11. $2x^2 + 10x + 12 = 2(x^2 + 5x + 6)$
$ = 2(x+2)(x+3)$

13. $AC = 5(10) = 50$
Product: 50;   Sum: -27; -2 and -25
$5x^2 - 27x + 10 = 5x^2 - 2x - 25x + 10$
$ = x(5x-2) - 5(5x-2)$
$ = (5x-2)(x-5)$

15. $5x^2 - 15x + 10 = 5(x^2 - 3x + 2)$
$ = 5(x-1)(x-2)$

17. $AC = 2(-6) = -12$
Product: -12;   Sum: -1; -4 and 3
$2y^2 - y - 6 = 2y^2 - 4y + 3y - 6$
$ = 2y(y-2) + 3(y-2)$
$ = (y-2)(2y+3)$

19. $AC = 5(-18) = -90$
Product: -90;   Sum: 9; -6 and 15
$5a^2 + 9a - 18 = 5a^2 - 6a + 15a - 18$
$ = a(5a-6) + 3(5a-6)$
$ = (5a-6)(a+3)$

21. $AC = 2(6) = 12$
Product: 12;   Sum: 7;   3 and 4
$2t^2 + 7t + 6 = 2t^2 + 3t + 4t + 6$
$ = t(2t+3) + 2(2t+3)$
$ = (2t+3)(t+2)$

23. $2t^2 + 6t + 6 = 2(t^2 + 3t + 3)$

25. $3w^2 - 6w - 30 = 3(w^2 - 2w - 10)$

27. $AC = 3(2) = 6$
Product: 6;   Sum: -4;   none
$3x^2 - 4x + 2$ is not factorable.

29. $AC = 3(15) = 45$
Product: 45;   Sum: -14; -5 and -9
$3x^2 - 14xy + 15y^2 = 3x^2 - 5xy - 9xy + 15y^2$
$ = x(3x-5y) - 3y(3x-5y)$
$ = (3x-5y)(x-3y)$

31. $AC = 6(10) = 60$
Product: 60;   Sum: 17;   5 and 12
$6a^2 + 17a + 10 = 6a^2 + 5a + 12a + 10$
$ = a(6a+5) + 2(6a+5)$
$ = (6a+5)(a+2)$

33. $AC = 6(-10) = -60$
Product: -60;   Sum: 17; -3 and 20
$6a^2 + 17a - 10 = 6a^2 - 3a + 20a - 10$
$ = 3a(2a-1) + 10(2a-1)$
$ = (2a-1)(3a+10)$

35. $6a^2 - 18a - 24 = 6(a^2 - 3a - 4)$
$ = 6(a-4)(a+1)$

37. $x^2 - 36y^2 = (x-6y)(x+6y)$

39. $4x^2 - 36y^2 = 4(x^2 - 9y^2)$
$ = 4(x-3y)(x+3y)$

41. $\begin{aligned}x^3 + 5x^2 - 24x &= x(x^2 + 5x - 24) \\ &= x(x+8)(x-3)\end{aligned}$

43. $\begin{aligned}x^2 + 5x^2 - 24x &= 6x^2 - 24x \\ &= 6x(x-4)\end{aligned}$

45. $\begin{aligned}4x^4 - 24x^3 + 32x^2 &= 4x^2(x^2 - 6x + 8) \\ &= 4x^2(x-2)(x-4)\end{aligned}$

47. $6x^2y - 8xy^2 + 12xy = 2xy(3x - 4y + 6)$

49. $3x^2 - 7x - 48 = (3x - 16)(x + 3)$

51. $8x^2 - 32x = 8x(x-4)$

53. $\begin{aligned}2x - x^2 + 15 &= -x^2 + 2x + 15 \\ &= -(x^2 - 2x - 15) \\ &= -(x-5)(x+3)\end{aligned}$

55. $\begin{aligned}84xy - 16x^2y - 4x^3y &= 4xy(21 - 4x - x^2) \\ &= -4xy(x^2 + 4x - 21) \\ &= -4xy(x+7)(x-3)\end{aligned}$

57. $-x^2 + 25 = 25 - x^2 = (5-x)(5+x)$

59. The proposed factor $2x + 2$ has a common factor of 2, which would imply that the original trinomial $6x^2 - 5x - 4$ has a common factor of 2. This is not the case. We can eliminate eight possible factorizations of $6x^2 - 5x - 4$ in this way. In addition to $(3x-2)(2x+2)$, we can eliminate

$(3x-1)(2x+4)$ $(3x+1)(2x-4)$ $(3x+2)(2x-2)$ $(6x-2)(x+2)$
$(6x+2)(x-2)$ $(6x-4)(x+1)$ $(6x+4)(x-1)$

this leaves four possibilities:
$(6x-1)(x+4)$, $(6x+1)(x-4)$, $(3x+4)(2x-1)$, and $(3x-4)(2x+1)$,
the last of which is the correct one.

61. Let $f =$ number of franks sold.
Then $f + 20 =$ number of knishes sold.
$$\begin{aligned}1.25f + 0.80(f + 20) &= 169.75 \\ 1.25f + 0.80f + 16 &= 169.75 \\ 2.05f + 16 &= 169.75 \\ 2.05f &= 153.75 \\ f &= 75\end{aligned}$$

Then $f + 20 = 75 + 20 = 95$.
So the vendor sold 75 franks and 95 knishes on the day in question.

Exercises 8.4

1. $\begin{aligned}(x-2)(x+3) &= 0 \\ x - 2 = 0 \text{ or } x + 3 &= 0 \\ x = 2 \qquad x &= -3\end{aligned}$

3. $\begin{aligned}(x-2)(x+3) &= 6 \\ x^2 + x - 6 &= 6 \\ x^2 + x - 12 &= 0 \\ (x+4)(x-3) &= 0 \\ x + 4 = 0 \text{ or } x - 3 &= 0 \\ x = -4 \qquad x &= 3\end{aligned}$

5. $\begin{aligned}x - 2(x+3) &= 6 \\ x - 2x - 6 &= 6 \\ -x - 6 &= 6 \\ -x &= 12 \\ x &= -12\end{aligned}$

7. $\begin{aligned}y(y-4) &= 0 \\ y = 0 \text{ or } y - 4 &= 0 \\ y &= 4\end{aligned}$

9. $$\begin{aligned} y(y-4) &= 12 \\ y^2 - 4y &= 12 \\ y^2 - 4y - 12 &= 0 \\ (y-6)(y+2) &= 0 \\ y - 6 = 0 \text{ or } y + 2 &= 0 \\ y = 6 \qquad y &= -2 \end{aligned}$$

11. $$\begin{aligned} x^2 - x - 6 &= 0 \\ (x-3)(x+2) &= 0 \\ x - 3 = 0 \text{ or } x + 2 &= 0 \\ x = 3 \qquad x &= -2 \end{aligned}$$

13. $$\begin{aligned} x^2 - 3x &= 10 \\ x^2 - 3x - 10 &= 0 \\ (x-5)(x+2) &= 0 \\ x - 5 = 0 \text{ or } x + 2 &= 0 \\ x = 5 \qquad x &= -2 \end{aligned}$$

15. $$\begin{aligned} -m^2 + 2m + 8 &= 0 \\ m^2 - 2m - 8 &= 0 \\ (m-4)(m+2) &= 0 \\ m - 4 = 0 \text{ or } m + 2 &= 0 \\ m = 4 \qquad m &= -2 \end{aligned}$$

17. $$\begin{aligned} -m^2 &= 8 - 9m \\ -m^2 + 9x - 8 &= 0 \\ m^2 - 9x + 8 &= 0 \\ (m-1)(m-8) &= 0 \\ m - 1 = 0 \text{ or } m - 8 &= 0 \\ m = 1 \qquad m &= 8 \end{aligned}$$

19. $$\begin{aligned} p^2 + 3p &= p(p+4) \\ p^2 + 3p &= p^2 + 4p \\ 3p &= 4p \\ 0 &= p \end{aligned}$$

21. $$\begin{aligned} y^2 &= 4y \\ y^2 - 4y &= 0 \\ y(y-4) &= 0 \\ y = 0 \text{ or } y - 4 &= 0 \\ y &= 4 \end{aligned}$$

23. $$\begin{aligned} 5w^2 &= 8w \\ 5w^2 - 8w &= 0 \\ w(5w - 8) &= 0 \\ w = 0 \text{ or } 5w - 8 &= 0 \\ 5w &= 8 \\ w &= \frac{8}{5} \end{aligned}$$

25. $$\begin{aligned} 2a^2 &= 11a - 12 \\ 2a^2 - 11a + 12 &= 0 \\ (2a-3)(a-4) &= 0 \\ 2a - 3 = 0 \text{ or } a - 4 &= 0 \\ 2a = 3 \qquad a &= 4 \\ a = \frac{3}{2} \end{aligned}$$

27. $$\begin{aligned} 2a(a+3) &= 0 \\ 2a = 0 \text{ or } a + 3 &= 0 \\ a = 0 \qquad a &= -3 \end{aligned}$$

29. $$\begin{aligned} 2a(a+3) &= 20 \\ 2a^2 + 6a &= 20 \\ 2a^2 + 6a - 20 &= 0 \\ 2(a^2 + 3a - 10) &= 0 \\ a^2 + 3a - 10 &= 0 \\ (a+5)(a-2) &= 0 \\ a + 5 = 0 \text{ or } a - 2 &= 0 \\ a = -5 \qquad a &= 2 \end{aligned}$$

31. $$\begin{aligned} 2x^2 + 5x - 4 &= x^2 + 3x - 7 \\ x^2 + 2x + 3 &= 0 \end{aligned}$$
The quadratic expression $x^2 + 2x + 3$ cannot be factored.

33. $$\begin{aligned} a^2 - 4a - 2 &= 2a^2 - 9a - 16 \\ -4a - 2 &= a^2 - 9a - 16 \\ -2 &= a^2 - 5a - 16 \\ 0 &= a^2 - 5a - 14 \\ 0 &= (a-7)(a+2) \\ 0 = a - 7 \text{ or } 0 &= a + 2 \\ 7 = a \qquad -2 &= a \end{aligned}$$

35.
$$5x^2 = 45$$
$$5x^2 - 45 = 0$$
$$5(x^2 - 9) = 0$$
$$5(x - 3)(x + 3) = 0$$
$$x - 3 = 0 \text{ or } x + 3 = 0$$
$$x = 3 \qquad x = -3$$

37.
$$(x + 3)^2 = 3x^2 - 10$$
$$x^2 + 6x + 9 = 3x^2 - 10$$
$$6x + 9 = 2x^2 - 10$$
$$9 = 2x^2 - 6x - 10$$
$$0 = 2x^2 - 6x - 19$$
The quadratic expression $2x^2 - 6x - 19$ cannot be factored.

39.
$$4y = 4y^2 + 1$$
$$0 = 4y^2 - 4y + 1$$
$$0 = (2y - 1)(2y - 1)$$
$$0 = 2y - 1$$
$$1 = 2y$$
$$\frac{1}{2} = y$$

41.
$$(x + 2)^2 = 25$$
$$(x + 2)(x + 2) = 25$$
$$x^2 + 4x + 4 = 25$$
$$x^2 + 4x - 21 = 0$$
$$(x + 7)(x - 3) = 0$$
$$x + 7 = 0 \text{ or } x - 3 = 0$$
$$x = -7 \qquad x = 3$$

43.
$$(x - 4)(x + 1) = (x - 3)(x - 2)$$
$$x^2 - 3x - 4 = x^2 - 5x + 6$$
$$-3x - 4 = -5x + 6$$
$$2x - 4 = 6$$
$$2x = 10$$
$$x = 5$$

45.
$$(2x - 4)(x + 1) = (x - 3)(x - 2)$$
$$2x^2 - 2x - 4 = x^2 - 5x + 6$$
$$x^2 - 2x - 4 = -5x + 6$$
$$x^2 + 3x - 4 = 6$$
$$x^2 + 3x - 10 = 0$$
$$(x + 5)(x - 2) = 0$$
$$x + 5 = 0 \text{ or } x - 2 = 0$$
$$x = -5 \qquad x = 2$$

47. Let L = length of the rectangle.
Then $3L - 2$ = width of the rectangle.
$$L(3L - 2) = 33$$
$$3L^2 - 2L = 33$$
$$3L^2 - 2L - 33 = 0$$
$$(3L - 11)(L + 3) = 0$$
$$3L - 11 = 0 \text{ or } L + 3 = 0$$
$$3L = 11 \qquad L = -3$$
$$L = \frac{11}{3} \qquad \text{(reject)}$$

When $L = \frac{11}{3}$,
$$3L - 2 = 3\left(\frac{11}{3}\right) - 2 = 11 - 2 = 9.$$
Thus, the dimensions of the rectangle are $\frac{11}{3}$ or $3\frac{2}{3}$ ft. by 9 ft.

49. Let x = length of the side of the square painting.
Then $x + 2$ = length of the side of the framed painting.
$$(x + 2)^2 - x^2 = 28$$
$$x^2 + 4x + 4 - x^2 = 28$$
$$4x + 4 = 28$$
$$4x = 24$$
$$x = 6$$
The square painting is 6 in. by 6 in.

51. Let x = width of the path.
$$(2x + 5)(2x + 7) = 63$$
$$4x^2 + 24x + 35 = 63$$
$$4x^2 + 24x - 28 = 0$$
$$4(x^2 + 6x - 7) = 0$$
$$x^2 + 6x - 7 = 0$$
$$(x - 1)(x + 7) = 0$$
$$x - 1 = 0 \text{ or } x + 7 = 0$$
$$x = 1 \qquad x = -7$$
$$\qquad\qquad \text{(reject)}$$
Thus, the width of the path is 1 ft.

53.
$$y^2 - 65y = 1200$$
$$y^2 - 65y - 1200 = 0$$
$$(y - 80)(y + 15) = 0$$
$$y - 80 = 0 \quad \text{or} \quad y + 15 = 0$$
$$y = 80 \qquad y = -15$$
$$\text{(reject)}$$
Thus, 80 sq. yd of carpeting must be sold.

55.
$$240t - 16t^2 = 800$$
$$240t = 16t^2 + 800$$
$$0 = 16t^2 - 240t + 800$$
$$0 = 16(t^2 - 15t + 50)$$
$$0 = t^2 - 15t + 50$$
$$0 = (t - 5)(t - 10)$$
$$0 = t - 5 \quad \text{or} \quad 0 = t - 10$$
$$5 = t \qquad 10 = t$$
The object will be 800 ft. high after 5 seconds (when it is rising) and again after 10 seconds (when it is falling).

57.
$$n^2 - n = 90$$
$$n^2 - n - 90 = 0$$
$$(n - 10)(n + 9) = 0$$
$$n - 10 = 0 \quad \text{or} \quad n + 9 = 0$$
$$n = 10 \qquad n = -9 \text{ (reject)}$$
Thus, the league has 10 teams.

59. (a) We cannot set each of the factors equal to 7. The zero-product rule requires that the product of the factors be equal to 0.

(b) $x = 3$ is not a possible solution. The first factor on the left side of the equation can never be equal to zero. We can either ignore its presence or divide both sides of the equation by it. (This logic is valid for constant factors of a zero product, but not for variable factors.)

60. Since the square of any real number must be non-negative, x^2 must be at least zero. Therefore, $x^2 + 4$ must be at least 4, which means that $x^2 + 4$ cannot ever equal 0. Thus, the equation $x^2 + 4 = 0$ cannot have any solution in the real number system.

Exercises 8.5

1. $\dfrac{3x + 12}{6} = \dfrac{\cancel{3}x}{\cancel{6}} + \dfrac{\cancel{12}^{\,2}}{\cancel{6}} = \dfrac{x}{2} + 2$

   or $\dfrac{x + 4}{2}$

3. $\dfrac{t^2 - 6t}{6t} = \dfrac{\cancel{t^2}^{\,t}}{\cancel{6t}} - \dfrac{\cancel{6t}}{\cancel{6t}} = \dfrac{t}{6} - 1$

   or $\dfrac{t - 6}{6}$

5. $\dfrac{3x^2y - 9xy^2}{3xy} = \dfrac{\cancel{3x^2y}^{\,x}}{\cancel{3xy}} - \dfrac{\cancel{9xy^2}^{\,3y}}{\cancel{3xy}} = x - 3y$

7. $\dfrac{3x^2y - 9xy^2}{6x^2y^2} = \dfrac{\cancel{3x^2y}^{\,}}{\cancel{6x^2y^2}^{\,2y}} - \dfrac{\cancel{9xy^2}^{\,3}}{\cancel{6x^2y^2}^{\,2x}}$

   $= \dfrac{1}{2y} - \dfrac{3}{2x}$ or $\dfrac{x - 3y}{2xy}$

9. $\dfrac{10a^2b^3c - 15ab^2c^2 - 20a^3b^2c^3}{5ab^2c} = \dfrac{\cancel{10a^2b^3c}^{\,2ab}}{\cancel{5ab^2c}} - \dfrac{\cancel{15ab^2c^2}^{\,3c}}{\cancel{5ab^2c}} - \dfrac{\cancel{20a^3b^2c^3}^{\,4a^2c^2}}{\cancel{5ab^2c}}$

   $= 2ab - 3c - 4a^2c^2$

11.
$$\begin{array}{r}x-5\\x+2\overline{\smash{)}x^2-3x+2}\\-(x^2+2x)\\\hline-5x+2\\-(-5x-10)\\\hline 12\end{array}$$

13.
$$\begin{array}{r}t+2\\t-5\overline{\smash{)}t^2-3t-10}\\-(t^2-5t)\\\hline 2t-10\\-(2t-10)\\\hline 0\end{array}$$

15.
$$\begin{array}{r}w+1\\w+3\overline{\smash{)}w^2+4w-21}\\-(w^2+3w)\\\hline w-21\\-(\,w+3)\\\hline -24\end{array}$$

17.
$$\begin{array}{r}2x-1\\x-1\overline{\smash{)}2x^2-3x+7}\\-(2x^2-2x)\\\hline -x+7\\-(-x+1)\\\hline 6\end{array}$$

19.
$$\begin{array}{r}y^2+3y+7\\y-2\overline{\smash{)}y^3+y^2+y-14}\\-(y^3-2y^2)\\\hline 3y^2+y\\-(3y^2-6y)\\\hline 7y-14\\-(7y-14)\\\hline 0\end{array}$$

21.
$$\begin{array}{r}2a^2-a-2\\a+1\overline{\smash{)}2a^3+a^2-3a+2}\\-(2a^3+2a^2)\\\hline -a^2-3a\\-(-a^2-a)\\\hline -2a+2\\-(-2a-2)\\\hline 4\end{array}$$

23.
$$\begin{array}{r}x^2-4x+12\\x+3\overline{\smash{)}x^3-x^2+0x+36}\\-(x^3+3x^2)\\\hline -4x^2+0x\\-(-4x^2-12x)\\\hline 12x+36\\-(12x+36)\\\hline 0\end{array}$$

25.
$$\begin{array}{r}x^3+2x^2+4x+8\\x-2\overline{\smash{)}x^4+0x^3+0x^2+0x-16}\\-(x^4-2x^3)\\\hline 2x^3+0x^2\\-(2x^3-4x^2)\\\hline 4x^2+0x\\-(4x^2-8x)\\\hline 8x-16\\-(8x-16)\\\hline 0\end{array}$$

27.
$$\begin{array}{r}x^2+4x-2\\3x+2\overline{\smash{)}3x^3+14x^2+2x-4}\\-(3x^3+2x^2)\\\hline 12x^2+2x\\-(12x^2+8x)\\\hline -6x-4\\-(-6x-4)\\\hline 0\end{array}$$

29.
$$\begin{array}{r}2t^2+5t-4\\2t-5\overline{\smash{)}4t^3+0t^2-33t+24}\\-(4t^3-10t^2)\\\hline 10t^2-33t\\-(10t^2-25t)\\\hline -8t+24\\-(-8t+20)\\\hline 4\end{array}$$

31. To divide $5x^4 - 8x^3 + 7x^2 - 2x + 1$ by $x^2 + x - 1$, we would begin by asking what x^2 must be multiplied by in order to give a product of $5x^4$. When we determine this to be $5x^2$, we multiply $5x^2$ and $x^2 + x - 1$ and subtract the resulting product from $5x^4 - 8x^3 + 7x^2 - 2x + 1$. We continue this process, as shown below.

$$\begin{array}{r} 5x^2 - 13x + 25 \\ x^2+x-1 \overline{)\, 5x^4 - 8x^3 + 7x^2 - 2x + 1} \\ -(5x^4 + 5x^3 - 5x^2) \\ \hline -13x^3 + 12x^2 - 2x \\ -(-13x^3 - 13x^2 + 13x) \\ \hline 25x^2 - 15x + 1 \\ -(25x^2 + 25x - 25) \\ \hline -40x + 26 \end{array}$$

The long division process ends when the degree of the remainder (in this case, 1) is less than the degree of the divisor (in this case, 2).

33. $-x^2 = -(-3)^2 = -9$

35. $(xy)^2 = (-3 \cdot 5)^2 = (-15)^2 = 225$

37. $x^{-2} = (-3)^{-2} = \dfrac{1}{(-3)^2} = \dfrac{1}{9}$

39. $(xy)^{-1} = (-3 \cdot 5)^{-1} = (-15)^{-1}$
$= \dfrac{1}{-15} = -\dfrac{1}{15}$

Chapter 8 Review Exercises

1. $x^2 + 7x + 12 = (x+3)(x+4)$

3. $x^2 + 7x = x(x+7)$

5. $x^2 - 13x + 12 = (x-1)(x-12)$

7. $x^2 - 6xy - 27y^2 = (x+3y)(x-9y)$

9. $x^2 - 64 = (x-8)(x+8)$

11. $2x^2 + 9x + 10 = (2x+5)(x+2)$

13. $3x^2 - 6x - 24 = 3(x^2 - 2x - 8)$
$= 3(x-4)(x+2)$

15. $6a^2 + 36a + 48 = 6(a^2 + 6a + 8)$
$= 6(a+2)(a+4)$

17. $5x^3 y - 80xy^3 = 5xy(x^2 - 16y^2)$
$= 5xy(x-4y)(x+4y)$

19. $x^2 + 9x = x(x+9)$

21. $25t^2 - 1 = (5t-1)(5t+1)$

23. $30 - x^2 + x = -x^2 + x + 30$
$= -(x^2 - x - 30)$
$= -(x-6)(x+5)$

25. $12x - 3x^2 - 9 = -3x^2 + 12x - 9$
$= -3(x^2 - 4x + 3)$
$= -3(x-1)(x-3)$

27. $x(x-7) + 3(x-7) = (x-7)(x+3)$

29. $m^2 + 2mn + 9m + 18n$
$= m(m+2n) + 9(m+2n)$
$= (m+2n)(m+9)$

31. $(x-5)(x+4) = 0$
$x - 5 = 0$ or $x + 4 = 0$
$x = 5 \qquad x = -4$

33. $(x-5)(x+4) = 36$
$x^2 - x - 20 = 36$
$x^2 - x - 56 = 0$
$(x-8)(x+7) = 0$
$x - 8 = 0$ or $x + 7 = 0$
$x = 8 \qquad x = -7$

35. $3x^2 + 8x = 35$
$3x^2 + 8x - 35 = 0$
$(3x - 7)(x + 5) = 0$
$3x - 7 = 0$ or $x + 5 = 0$
$3x = 7 \qquad x = -5$
$x = \dfrac{7}{3}$

37. $(2x-3)(x+2) = (x+6)(x-1)$
$2x^2 + x - 6 = x^2 + 5x - 6$
$\underline{-x^2 - 5x + 6 = -x^2 - 5x + 6}$
$x^2 - 4x = 0$
$x(x - 4) = 0$
$x = 0$ or $x - 4 = 0$
$x = 0 \qquad x = 4$

39. $5x^2 = 80$
$x^2 = 16$
$x^2 - 16 = 0$
$(x-4)(x+4) = 0$
$x - 4 = 0$ or $x + 4 = 0$
$x = 4 \qquad x = -4$

41. Let $w =$ width of the rectangle.
Then $2w - 3 =$ length of the rectangle.
$w(2w - 3) = 14$
$2w^2 - 3w = 14$
$2w^2 - 3w - 14 = 0$
$(2w - 7)(w + 2) = 0$
$2w - 7 = 0$ or $w + 2 = 0$
$2w = 7 \qquad w = -2$
$w = \dfrac{7}{2}$ \qquad (reject)
When $w = \dfrac{7}{2}$,
$2w - 3 = \cancel{2}\left(\dfrac{7}{\cancel{2}}\right) - 3 = 7 - 3 = 4$.
The dimensions of the rectangle are 4 cm. by $\dfrac{7}{2}$ or $3\dfrac{1}{2}$ cm.

43. Let $x =$ length of the side
of the square garden.
$(x + 4)^2 = 144$
$x^2 + 8x + 16 = 144$
$x^2 + 8x - 128 = 0$
$(x - 8)(x + 16) = 0$
$x - 8 = 0$ or $x + 16 = 0$
$x = 8 \qquad x = -2$
\qquad (reject)
The dimensions of the garden
are 8 ft. by 8 ft.

45. $n^2 - n = 56$
$n^2 - n - 56 = 0$
$(n - 8)(n + 7) = 0$
$n - 8 = 0$ or $n + 7 = 0$
$n = 8 \qquad n = -7$
\qquad (reject)
There are 8 teams in the league.

47. $\dfrac{6r^2t - 4rt^3 + 10r^2t^4}{2rt^2} = \dfrac{\cancel{2rt}(3r - 2t^2 + 5rt^3)}{\cancel{2rt^2}_{t}}$
$= \dfrac{3r - 2t^2 + 5rt^2}{t}$

49.
$$\begin{array}{r} y^2 + y + 3 \\ y-2 \overline{\smash{\big)}\, y^3 - y^2 + y - 1} \\ \underline{-(y^3 - 2y^2)} \\ y^2 + y \\ \underline{-(y^2 - 2y)} \\ 3y - 1 \\ \underline{-(3y - 6)} \\ 5 \end{array}$$

51.
$$\begin{array}{r} 2x^2 - x + 2 \\ 3x-5 \overline{\smash{\big)}\, 6x^3 - 13x^2 + 11x - 10} \\ \underline{-(6x^3 - 10x^2)} \\ -3x^2 + 11x \\ \underline{-(-3x^2 + 5x)} \\ 6x - 10 \\ \underline{-(6x - 10)} \\ 0 \end{array}$$

53.
$$\begin{array}{r} 16x^3 + 32x^2 + 64x + 128 \\ x-2 \overline{\smash{\big)}\, 16x^4 + 0x^3 + 0x^2 + 0x - 64} \\ \underline{-(16x^4 - 32x^3)} \\ 32x^3 + 0x^2 \\ \underline{-(32x^3 - 64x^2)} \\ 64x^2 - 0x \\ \underline{-(64x^2 - 128x)} \\ 128x - 64 \\ \underline{-(128x - 256)} \\ 192 \end{array}$$

Chapter 8 Practice Test

1. $6x^3 + 12x^2 - 15x$
$= 3x(2x^2 + 4x - 5)$

3. $x^2 + 9x + 8$
$= (x+1)(x+8)$

5. $4x^2 - 20x$
$= 4x(x-5)$

7. $6x^2 + 24x + 18$
$= 6(x^2 + 4x + 3)$
$= 6(x+1)(x+3)$

9. $x^2 + 4x + 4$
$= (x+2)(x+2)$

11. $x^2y^2 - 9$
$= (xy - 3)(xy + 3)$

13. $\dfrac{12r^3t^2 - 18r^2t^2 + 20r^3t^4}{4r^3t^3} = \dfrac{\cancel{12r^3t^2}^{3}}{\cancel{4r^3t^3}_{t}} - \dfrac{\cancel{18r^2t^2}^{9}}{\cancel{4r^3t^3}_{2rt}} + \dfrac{\cancel{20r^3t^4}^{5t}}{\cancel{4r^3t^3}} = \dfrac{3}{t} - \dfrac{9}{2rt} + 5t$ or $\dfrac{6r - 9 + 10rt^2}{2rt}$

15. $x^2 - 8x = 20$
$x^2 - 8x - 20 = 0$
$(x-10)(x+2) = 0$
$x - 10 = 0$ or $x + 2 = 0$
$x = 10 x = -2$

17. $(x-8)(x-6) = 3$
$x^2 - 14x + 48 = 3$
$x^2 - 14x + 45 = 0$
$(x-5)(x-9) = 0$
$x - 5 = 0$ or $x - 9 = 0$
$x = 5 x = 9$

19. Let $x =$ width of the path.
$(2x + 9)^2 - 81 = 40$
$4x^2 + 36x + 81 - 81 = 40$
$4x^2 + 36x = 40$
$4x^2 + 36x - 40 = 0$
$x^2 + 9x - 10 = 0$
$(x-1)(x+10) = 0$
$x - 1 = 0$ or $x + 10 = 0$
$x = 1 x = -10$ (reject)

The path is 1 ft. wide.

Chapter 9
More Rational Expressions

Exercises 9.1

1. $\dfrac{\cancel{8x^3}^{4}\cancel{y^{10}}^{y^5}}{\cancel{10x^6}_{5x^3}\cancel{y^5}} = \dfrac{4y^5}{5x^3}$

3. $\dfrac{5x - 7x}{x^2 - 7x^2} = \dfrac{\cancel{-2x}^{1}}{\cancel{-6x^2}_{3x}} = \dfrac{1}{3x}$

5. $\dfrac{\cancel{6x^2}^{2}\cancel{(x+4)^5}^{(x+4)^4}}{\cancel{9x^3}_{3x}\cancel{(x+4)}} = \dfrac{2(x+4)^4}{3x}$

7. $\dfrac{12a^2b + 6c^3}{8a^2b + 4c^3} = \dfrac{\cancel{6}^{3}\cancel{(2a^2b + c^3)}}{\cancel{4}_{2}\cancel{(2a^2b + c^3)}} = \dfrac{3}{2}$

9. $\dfrac{3x - 6}{5x - 10} = \dfrac{3\cancel{(x-2)}}{5\cancel{(x-2)}} = \dfrac{3}{5}$

11. $\dfrac{3x - 6}{6x - 12} = \dfrac{\cancel{3}^{1}\cancel{(x-2)}}{\cancel{6}_{2}\cancel{(x-2)}} = \dfrac{1}{2}$

13. $\dfrac{3x - 6}{6x + 12} = \dfrac{\cancel{3}(x - 2)}{\cancel{6}_{2}(x + 2)} = \dfrac{x - 2}{2(x + 2)}$

15. $\dfrac{5y}{10y + 20} = \dfrac{\cancel{5}y}{\cancel{10}_{2}(y + 2)} = \dfrac{y}{2(y + 2)}$

17. $\dfrac{6x + 18}{x^2 - 9} = \dfrac{6\cancel{(x+3)}}{(x - 3)\cancel{(x+3)}}$
    $= \dfrac{6}{x - 3}$

19. $\dfrac{6x^2 + 18}{x^2 - 9} = \dfrac{6(x^2 + 3)}{x^2 - 9}$
    (cannot be reduced)

21. $\dfrac{t^2 + 3t}{t^2 + 3t - 10} = \dfrac{t(t + 3)}{(t + 5)(t - 2)}$
    (cannot be reduced)

23. $\dfrac{2x^2 + x + x}{4x^3 + 4x^2} = \dfrac{2x^2 + 2x}{4x^3 + 4x^2}$
    $= \dfrac{\cancel{2x}^{1}\cancel{(x+1)}}{\cancel{4x^2}_{2x}\cancel{(x+1)}} = \dfrac{1}{2x}$

25. $\dfrac{y^2 - 5y - 6}{y^2 - 12y + 36} = \dfrac{\cancel{(y-6)}(y + 1)}{\cancel{(y-6)}(y - 6)}$
    $= \dfrac{y + 1}{y - 6}$

27. $\dfrac{s^2 - 2s - 15}{s^2 - 6s + 5} = \dfrac{\cancel{(s-5)}(s + 3)}{\cancel{(s-5)}(s - 1)}$
    $= \dfrac{s + 3}{s - 1}$

29. $\dfrac{x^2(x + 3)(x - 4)}{x^4 - 4x^3} = \dfrac{\cancel{x^2}(x + 3)\cancel{(x-4)}}{\cancel{x^4}_{x}\cancel{(x-4)}}$
    $= \dfrac{x + 3}{x}$

31. $\dfrac{3a^2 + a - 2}{a^2 - a - 2} = \dfrac{(3a - 2)\cancel{(a+1)}}{(a - 2)\cancel{(a+1)}}$
    $= \dfrac{3a - 2}{a - 2}$

33. $\dfrac{4x^2 + 7x - 2}{x^2 + 4x + 4} = \dfrac{(4x-1)\cancel{(x+2)}}{(x+2)\cancel{(x+2)}}$
$= \dfrac{4x-1}{x+2}$

35. $\dfrac{x^2 - x + 3x - 8}{x^2 + 4x} = \dfrac{x^2 + 2x - 8}{x^2 + 4x}$
$= \dfrac{(x-2)\cancel{(x+4)}}{x\cancel{(x+4)}}$
$= \dfrac{x-2}{x}$

37. $\dfrac{6x^2 - 12x - 18}{3x^2 - 9x - 30} = \dfrac{6(x^2 - 2x - 3)}{3(x^2 - 3x - 10)}$
$= \dfrac{\overset{2}{\cancel{6}}(x-3)(x+1)}{\cancel{3}(x-5)(x+2)}$
$= \dfrac{2(x-3)(x+1)}{(x-5)(x+2)}$

39. $\dfrac{6x^2 - 5x^2 - 4}{x^2 - 6x + 8} = \dfrac{x^2 - 4}{x^2 - 6x + 8}$
$= \dfrac{\cancel{(x-2)}(x+2)}{\cancel{(x-2)}(x-4)}$
$= \dfrac{x+2}{x-4}$

41. $\dfrac{x^2 - 7x + 10}{x^2 - 7x + 12} = \dfrac{(x-2)(x-5)}{(x-3)(x-4)}$
(cannot be reduced)

43. $\dfrac{y^3 - y^2 - 2y}{6y^2 - 24} = \dfrac{y(y^2 - y - 2)}{6(y^2 - 4)}$
$= \dfrac{y\cancel{(y-2)}(y+1)}{6\cancel{(y-2)}(y+2)} = \dfrac{y(y+1)}{6(y+2)}$

45. $\dfrac{c^2 - 9c}{c^3 - 9c^2} = \dfrac{\overset{1}{\cancel{c}}\cancel{(c-9)}}{\underset{c}{\cancel{c^2}}\cancel{(c-9)}} = \dfrac{1}{c}$

47. The student was incorrect in each case, because he or she tried to cancel terms instead of factors. Neither of the fractions can be reduced because neither contains a common factor in its numerator and denominator. (in part (a), we could write
$\dfrac{x^2 + 5}{x} = \dfrac{\overset{x}{\cancel{x^2}}}{\underset{1}{\cancel{x}}} + \dfrac{5}{x} = x + \dfrac{5}{x}$ if we wished.)

49. $\dfrac{9x^2}{25y^6} \div \dfrac{xy}{15} = \dfrac{9\overset{x}{\cancel{x^2}}}{\underset{5}{\cancel{25}}y^6} \cdot \dfrac{\overset{3}{\cancel{15}}}{\cancel{x}y}$
$= \dfrac{27x}{5y^7}$

51. $\dfrac{5}{x} \div 10 = \dfrac{\overset{1}{\cancel{5}}}{x} \cdot \dfrac{1}{\underset{2}{\cancel{10}}}$
$= \dfrac{1}{2x}$

53. $\dfrac{2a}{5} = \dfrac{a-2}{3}$ LCD = 15
$\overset{3}{\cancel{15}}\left(\dfrac{2a}{\cancel{5}}\right) = \overset{5}{\cancel{15}}\left(\dfrac{a-2}{\cancel{3}}\right)$
$6a = 5a - 10$
$a = -10$

Exercises 9.2

1. $\dfrac{\overset{4}{\cancel{8}}\overset{y^2}{\cancel{x^2y^3}}}{\underset{3}{\cancel{9y^5}}\underset{z^2}{}} \cdot \dfrac{\overset{4\,a}{\cancel{12a^2z}}}{\underset{1}{\cancel{2ax^2}}} = \dfrac{16ay^2}{3z^2}$

3. $\dfrac{3st^2}{5p} \div \dfrac{15t^2}{s^2} = \dfrac{3st^{\cancel{2}}}{5p} \cdot \dfrac{s^2}{\underset{5}{\cancel{15}}t^{\cancel{2}}} = \dfrac{s^3}{25p}$

5. $\dfrac{12x^2y^3}{5z} \cdot 30xyz = \dfrac{12x^2y^3}{\cancel{5z}_1} \cdot \dfrac{\overset{6}{\cancel{30xyz}}}{1}$
$= 72x^3y^4$

7. $28a^2b^3z \div \dfrac{4a}{7b} = \dfrac{\overset{7\,a}{\cancel{28a^2}}b^3z^4}{1} \cdot \dfrac{7b}{\underset{1}{\cancel{4a}}}$
$= 49ab^4z^4$

9. $\dfrac{x^2+4x}{x^2} \cdot \dfrac{x}{x^2+6x+5}$
$= \dfrac{\cancel{x}(x+4)}{\underset{\cancel{x}\,1}{\cancel{x^2}}} \cdot \dfrac{\overset{1}{\cancel{x}}}{(x+1)(x+5)}$
$= \dfrac{x+4}{(x+1)(x+5)}$

11. $\dfrac{x^2+4x}{x^2+4x+4} \cdot \dfrac{x^2-4}{x^2-4x+4}$
$= \dfrac{x(x+4)}{(x+2)\cancel{(x+2)}} \cdot \dfrac{\cancel{(x-2)}\overset{1}{\cancel{(x+2)}}}{\cancel{(x-2)}(x-2)}$
$= \dfrac{x(x+4)}{(x+2)(x-2)}$

13. $\dfrac{r^2-4r-5}{2r-10} \div \dfrac{r^2-3r+2}{4r^2}$
$= \dfrac{r^2-4r-5}{2r-10} \cdot \dfrac{4r^2}{r^2-3r+2}$
$= \dfrac{\cancel{(r-5)}(r+1)}{\underset{1}{\cancel{2}}\cancel{(r-5)}} \cdot \dfrac{\overset{2}{\cancel{4}}r^2}{(r-1)(r-2)}$
$= \dfrac{2r^2(r+1)}{(r-1)(r-2)}$

15. $\dfrac{m^2}{m^2+3m} \div \dfrac{3m^2}{m^2+6m}$
$= \dfrac{m^2}{m^2+3m} \cdot \dfrac{m^2+6m}{3m^2}$
$= \dfrac{\overset{1}{\cancel{m^2}}}{\cancel{m}(m+3)} \cdot \dfrac{\cancel{m}(m+6)}{3\cancel{m^2}}$
$= \dfrac{m+6}{3(m+3)}$

17. $\dfrac{x^2+3x+2}{x^2+2x} \cdot \dfrac{x}{x^2+2}$
$= \dfrac{(x+1)\cancel{(x+2)}}{\underset{1}{\cancel{x}\cancel{(x+2)}}} \cdot \dfrac{\overset{1}{\cancel{x}}}{x^2+2}$
$= \dfrac{x+1}{x^2+2}$

19. $\dfrac{y^2-3y-4}{y^2-2y-8} \cdot \dfrac{y^2+4y+4}{y^2-8y+16}$
$= \dfrac{\cancel{(y-4)}(y+1)}{\cancel{(y-4)}\cancel{(y+2)}} \cdot \dfrac{\cancel{(y+2)}(y+2)}{(y-4)(y-4)}$
$= \dfrac{(y+1)(y+2)}{(y-4)(y-4)}$ or $\dfrac{(y+1)(y+2)}{(y-4)^2}$

21. $\dfrac{x^2+2x}{x^2-x-2} \cdot \dfrac{x-2}{x}$
$= \dfrac{\cancel{x}(x+2)}{\cancel{(x-2)}(x+1)} \cdot \dfrac{\overset{1}{\cancel{x-2}}}{\underset{1}{\cancel{x}}}$
$= \dfrac{x+2}{x+1}$

23. $\dfrac{2x^2+x-15}{x^2-9} \cdot \dfrac{6x^2+7x+1}{2x^2-3x-5}$
$= \dfrac{\cancel{(2x-5)}\cancel{(x+3)}}{(x-3)\cancel{(x+3)}} \cdot \dfrac{(6x+1)\cancel{(x+1)}}{\cancel{(2x-5)}\cancel{(x+1)}}$
$= \dfrac{6x+1}{x-3}$

25. $\dfrac{4a}{a+4} \cdot \dfrac{a+5}{5a} \div \dfrac{a^2+6a+5}{a^2+5a+4} = \left(\dfrac{4\cancel{a}}{a+4} \cdot \dfrac{a+5}{5\cancel{a}}\right) \div \dfrac{a^2+6a+5}{a^2+5a+4} = \dfrac{4(a+5)}{5(a+4)} \cdot \dfrac{a^2+5a+4}{a^2+6a+5}$
$= \dfrac{4\cancel{(a+5)}}{5\cancel{(a+4)}} \cdot \dfrac{\cancel{(a+1)}\overset{1}{\cancel{(a+4)}}}{\cancel{(a+1)}\cancel{(a+5)}} = \dfrac{4}{5}$

27. $\dfrac{x}{2} \div \dfrac{2}{x} \cdot \dfrac{x^2 - 16}{x^2 - 4x}$

$= \left(\dfrac{x}{2} \div \dfrac{2}{x}\right) \cdot \dfrac{x^2 - 16}{x^2 - 4x}$

$= \left(\dfrac{x}{2} \cdot \dfrac{x}{2}\right) \cdot \dfrac{x^2 - 16}{x^2 - 4x}$

$= \dfrac{\cancel{x^2}^{\,x}}{4} \cdot \dfrac{\cancel{(x-4)}(x+4)}{\cancel{x}\cancel{(x-4)}_{\,1}} = \dfrac{x(x+4)}{4}$

29. $\dfrac{t^2 + 2t}{t^2 + 2t + 1} \div \dfrac{2t^2 + 7t + 6}{t^2 + t}$

$= \dfrac{t^2 + 2t}{t^2 + 2t + 1} \cdot \dfrac{t^2 + t}{2t^2 + 7t + 6}$

$= \dfrac{t\cancel{(t+2)}}{(t+1)\cancel{(t+1)}} \cdot \dfrac{t\cancel{(t+1)}}{(2t+3)\cancel{(t+2)}}$

$= \dfrac{t^2}{(t+1)(2t+3)}$

31. $\dfrac{2x^2 + 6x + 4x}{x^2 - 25} \cdot \dfrac{(x+5)^2}{4x - 2x}$

$= \dfrac{2x^2 + 10x}{x^2 - 25} \cdot \dfrac{(x+5)^2}{2x}$

$= \dfrac{\cancel{2x}\cancel{(x+5)}}{(x-5)\cancel{(x+5)}} \cdot \dfrac{(x+5)(x+5)}{\cancel{2x}_{\,1}}$

$= \dfrac{(x+5)(x+5)}{x-5}$ or $\dfrac{(x+5)^2}{x-5}$

33. $\dfrac{x^3 y - xy^3}{8x^2 y + 4xy^2} \div \dfrac{(x-y)^2}{2x^2 + 3xy + y^2}$

$= \dfrac{x^3 y - xy^3}{8x^2 y + 4xy^2} \cdot \dfrac{2x^2 + 3xy + y^2}{(x-y)^2}$

$= \dfrac{xy(x^2 - y^2)}{4xy(2x+y)} \cdot \dfrac{(2x+y)(x+y)}{(x-y)^2}$

$= \dfrac{\cancel{xy}\cancel{(x-y)}(x+y)}{4\cancel{xy}\cancel{(2x+y)}} \cdot \dfrac{\cancel{(2x+y)}(x+y)}{\cancel{(x-y)}(x-y)}$

$= \dfrac{(x+y)(x+y)}{4(x-y)}$ or $\dfrac{(x+y)^2}{4(x-y)}$

35. $\left(\dfrac{x}{3} \cdot \dfrac{x}{4}\right) \cdot \dfrac{12}{x^2 - 3x}$

$= \dfrac{x^2}{12} \cdot \dfrac{12}{x^2 - 3x}$

$= \dfrac{\cancel{x^2}^{\,x}}{\cancel{12}_{\,1}} \cdot \dfrac{\cancel{12}^{\,1}}{\cancel{x}(x-3)}$

$= \dfrac{x}{x-3}$

37. $\left(\dfrac{c}{2} + \dfrac{c}{5}\right) \div \dfrac{c^2 + 7c}{10}$

$= \left(\dfrac{c(5)}{2(5)} + \dfrac{c(2)}{5(2)}\right) \div \dfrac{c^2 + 7c}{10}$

$= \left(\dfrac{5c}{10} + \dfrac{2c}{10}\right) \div \dfrac{c^2 + 7c}{10}$

$= \dfrac{7c}{10} \div \dfrac{c^2 + 7c}{10} = \dfrac{7c}{10} \cdot \dfrac{10}{c^2 + 7c}$

$= \dfrac{7\cancel{c}}{\cancel{10}} \cdot \dfrac{\cancel{10}^{\,1}}{\cancel{c}(c+7)} = \dfrac{7}{c+7}$

39. $\left(\dfrac{w}{4} - \dfrac{9}{w}\right) \cdot \left(\dfrac{w+10}{2w} - \dfrac{5}{w}\right) = \left(\dfrac{w(w)}{4(w)} - \dfrac{9(4)}{w(4)}\right) \cdot \left(\dfrac{w+10}{2w} - \dfrac{5(2)}{w(2)}\right)$

$= \left(\dfrac{w^2}{4w} - \dfrac{36}{4w}\right) \cdot \left(\dfrac{w+10}{2w} - \dfrac{10}{2w}\right)$

$= \left(\dfrac{w^2 - 36}{4w}\right) \cdot \left(\dfrac{w+10-10}{2w}\right)$

$= \dfrac{(w-6)(w+6)}{4w} \cdot \dfrac{\cancel{w}^{\,1}}{2\cancel{w}} = \dfrac{(w-6)(w+6)}{8w}$

41. Two cars leave from the same point and travel in opposite directions. The slower car, traveling at the rate of 20 kph, leaves two hours later than the faster one, which travels at 30 kph. How long has the slower car been traveling when the two cars are 310 km apart? (Let t = time traveled by the slower car.)

42. Wilma invested a sum of money at an 8% interest rate and a second sum, $500 more than the first, at a 12% interest rate. How much money did she invest at each rate if her total interest on the two investments is $260? (Let x = amount of money Wilma invested at 8%.)

43. $\dfrac{5}{6x} + \dfrac{3}{8x^2}$ LCD $= 24x^2$

$= \dfrac{5(4x)}{6x(4x)} + \dfrac{3(3)}{8x^2(3)}$

$= \dfrac{20x}{24x^2} + \dfrac{9}{24x^2}$

$= \dfrac{20x + 9}{24x^2}$

45. $9 - \dfrac{6}{x}$

$= \dfrac{9x}{x} - \dfrac{6}{x}$

$= \dfrac{9x - 6}{x}$

47. Let x = cost of an orchestra seat.
Then $x - 4$ = cost of a balcony seat.
$250x + 150(x - 4) = 3400$
$250x + 150x - 600 = 3400$
$400x - 600 = 3400$
$400x = 4000$
$x = 10$
They should charge $10 for an orchestra seat.

Exercises 9.3

1. $\dfrac{5x - 2}{4x + 8} + \dfrac{3x + 2}{4x + 8}$

$= \dfrac{5x - 2 + 3x + 2}{4x + 8}$

$= \dfrac{8x}{4x + 8} = \dfrac{\overset{2}{\cancel{8}}x}{\cancel{4}(x + 2)}$

$= \dfrac{2x}{x + 2}$

3. $\dfrac{5x - 2}{4x + 8} - \dfrac{3x + 2}{4x + 8} = \dfrac{5x - 2 - (3x + 2)}{4x + 8}$

$= \dfrac{5x - 2 - 3x - 2}{4x + 8}$

$= \dfrac{2x - 4}{4x + 8} = \dfrac{\cancel{2}(x - 2)}{\underset{2}{\cancel{4}}(x + 2)}$

$= \dfrac{x - 2}{2(x + 2)}$

5. $\dfrac{x + 7}{x + 2} + \dfrac{x + 3}{x + 2} + \dfrac{x - 2}{x + 2}$

$= \dfrac{(x + 7) + (x + 3) + (x - 2)}{x + 2}$

$= \dfrac{x + 7 + x + 3 + x - 2}{x + 2}$

$= \dfrac{3x + 8}{x + 2}$

7. $\dfrac{x + 7}{x + 2} - \dfrac{x + 3}{x + 2} + \dfrac{x - 2}{x + 2}$

$= \dfrac{(x + 7) - (x + 3) + (x - 2)}{x + 2}$

$= \dfrac{x + 7 - x - 3 + x - 2}{x + 2} = \dfrac{\cancel{x + 2}}{\underset{1}{\cancel{x + 2}}} = \dfrac{1}{1} = 1$

9. $\dfrac{y^2}{4y} + \dfrac{2y}{4y} = \dfrac{y^2 + 2y}{4y}$

$= \dfrac{y(y + 2)}{4y} = \dfrac{y + 2}{4}$

11. $\dfrac{5}{2x} + \dfrac{4}{x + 2} = \dfrac{5(x + 2)}{2x(x + 2)} + \dfrac{4(2x)}{(x + 2)(2x)}$

$= \dfrac{5x + 10}{2x(x + 2)} + \dfrac{8x}{2x(x + 2)}$

$= \dfrac{5x + 10 + 8x}{2x(x + 2)} = \dfrac{13x + 10}{2x(x + 2)}$

13. $\dfrac{5}{\cancel{2}x} \cdot \dfrac{\cancel{4}^{2}}{x+2} = \dfrac{10}{x(x+2)}$

15. $\dfrac{2}{x+2} + \dfrac{3}{x+3}$

$= \dfrac{2(x+3)}{(x+2)(x+3)} + \dfrac{3(x+2)}{(x+2)(x+3)}$

$= \dfrac{2x+6}{(x+2)(x+3)} + \dfrac{3x+6}{(x+2)(x+3)}$

$= \dfrac{2x+6+3x+6}{(x+2)(x+3)}$

$= \dfrac{5x+12}{(x+2)(x+3)}$

17. $\dfrac{7}{a+7} - \dfrac{5}{a+5}$

$= \dfrac{7(a+5)}{(a+7)(a+5)} - \dfrac{5(a+7)}{(a+5)(a+7)}$

$= \dfrac{7a+35}{(a+7)(a+5)} - \dfrac{5a+35}{(a+7)(a+5)}$

$= \dfrac{7a+35-(5a+35)}{(a+7)(a+5)}$

$= \dfrac{2a}{(a+7)(a+5)}$

19. $\dfrac{4}{3x^2} - \dfrac{2}{x^2+3x}$

$= \dfrac{4}{3x^2} - \dfrac{2}{x(x+3)}$

$= \dfrac{4(x+3)}{3x^2(x+3)} - \dfrac{2(3x)}{x(x+3)(3x)}$

$= \dfrac{4x+12}{3x^2(x+3)} - \dfrac{6x}{3x^2(x+3)}$

$= \dfrac{4x+12-6x}{3x^2(x+3)}$

$= \dfrac{-2x+12}{3x^2(x+3)} = \dfrac{2(-x+6)}{3x^2(x+3)}$

21. $\dfrac{3}{4a^2} + \dfrac{1}{6a-18}$

$= \dfrac{3}{4a^2} + \dfrac{1}{6(a-3)}$

$= \dfrac{3 \cdot 3(a-3)}{12a^2(a-3)} + \dfrac{1 \cdot 2a^2}{12a^2(a-3)}$

$= \dfrac{9a-27}{12a^2(a-3)} + \dfrac{2a^2}{12a^2(a-3)}$

$= \dfrac{9a-27+2a^2}{12a^2(a-3)}$

$= \dfrac{2a^2+9a-27}{12a^2(a-3)}$

23. $\dfrac{4}{x^2+4x} - \dfrac{2}{x^2-4x}$

$= \dfrac{4}{x(x+4)} - \dfrac{2}{x(x-4)}$

$= \dfrac{4(x-4)}{x(x+4)(x-4)} - \dfrac{2(x+4)}{x(x-4)(x+4)}$

$= \dfrac{4x-16}{x(x+4)(x-4)} - \dfrac{2x+8}{x(x+4)(x-4)}$

$= \dfrac{4x-16-(2x+8)}{x(x+4)(x-4)}$

$= \dfrac{4x-16-2x-8}{x(x+4)(x-4)}$

$= \dfrac{2x-24}{x(x+4)(x-4)} = \dfrac{2(x-12)}{x(x+4)(x-4)}$

25. $2 - \dfrac{x}{x-1}$

$= \dfrac{2}{1} - \dfrac{x}{x-1}$

$= \dfrac{2(x-1)}{1(x-1)} - \dfrac{x}{x-1}$

$= \dfrac{2x-2}{x-1} - \dfrac{x}{x-1}$

$= \dfrac{2x-2-x}{x-1} = \dfrac{x-2}{x-1}$

27. $2 \cdot \dfrac{x}{x-1}$

$= \dfrac{2}{1} \cdot \dfrac{x}{x-1}$

$= \dfrac{2x}{x-1}$

29. $\dfrac{3}{x^2-16} - \dfrac{3}{2x^2+8x}$

$= \dfrac{3}{(x-4)(x+4)} - \dfrac{3}{2x(x+4)}$

$= \dfrac{3(2x)}{(x-4)(x+4)(2x)} - \dfrac{3(x-4)}{2x(x+4)(x-4)}$

$= \dfrac{6x}{2x(x+4)(x-4)} - \dfrac{3x-12}{2x(x+4)(x-4)}$

$= \dfrac{6x-(3x-12)}{2x(x+4)(x-4)}$

$= \dfrac{6x-3x+12}{2x(x+4)(x-4)}$

$= \dfrac{3x+12}{2x(x+4)(x-4)}$

$= \dfrac{3(x+4)}{2x(x+4)(x-4)} = \dfrac{3}{2x(x-4)}$

31. $\dfrac{12}{x^2+x-2} - \dfrac{4}{x^2-x}$

$= \dfrac{12}{(x-1)(x+2)} - \dfrac{4}{x(x-1)}$

$= \dfrac{12(x)}{(x-1)(x+2)(x)} - \dfrac{4(x+2)}{x(x-1)(x+2)}$

$= \dfrac{12x}{x(x-1)(x+2)} - \dfrac{4x+8}{x(x-1)(x+2)}$

$= \dfrac{12x-(4x+8)}{x(x-1)(x+2)}$

$= \dfrac{12x-4x-8}{x(x-1)(x+2)}$

$= \dfrac{8x-8}{x(x-1)(x+2)}$

$= \dfrac{8(x-1)}{x(x-1)(x+2)} = \dfrac{8}{x(x+2)}$

33. $\dfrac{x}{x^2+6x+9} + \dfrac{1}{x^2+4x+3}$

$= \dfrac{x}{(x+3)(x+3)} + \dfrac{1}{(x+1)(x+3)}$

$= \dfrac{x(x+1)}{(x+3)(x+3)(x+1)} + \dfrac{1(x+3)}{(x+1)(x+3)(x+3)}$

$= \dfrac{x^2+x}{(x+1)(x+3)(x+3)} + \dfrac{x+3}{(x+1)(x+3)(x+3)}$

$= \dfrac{x^2+x+x+3}{(x+1)(x+3)(x+3)}$

$= \dfrac{x^2+2x+3}{(x+1)(x+3)(x+3)}$ or $\dfrac{x^2+2x+3}{(x+1)(x+3)^2}$

35. $\dfrac{3a+6}{3a+2} \cdot \dfrac{a+1}{a+2}$

$= \dfrac{3(a+2)}{3a+2} \cdot \dfrac{a+1}{(a+2)}$

$= \dfrac{3(a+1)}{3a+2}$ or $\dfrac{3a+3}{3a+2}$

37. $5 - \dfrac{1}{3x-6} + \dfrac{3}{x^2-2x} = \dfrac{5}{1} - \dfrac{1}{3(x-2)} + \dfrac{3}{x(x-2)}$

$= \dfrac{5(3x)(x-2)}{1(3x)(x-2)} - \dfrac{1(x)}{3(x-2)(x)} + \dfrac{3(3)}{x(x-2)(3)}$

$= \dfrac{15x^2-30x}{3x(x-2)} - \dfrac{x}{3x(x-2)} + \dfrac{9}{3x(x-2)}$

$= \dfrac{15x^2-30x-x+9}{3x(x-2)} = \dfrac{15x^2-31x+9}{3x(x-2)}$

39. $\dfrac{5}{x^2+x+6} - \dfrac{3}{x^2+3x} + \dfrac{2}{x^2-2x} = \dfrac{5}{(x+3)(x-2)} - \dfrac{3}{x(x+3)} + \dfrac{2}{x(x-2)}$

$= \dfrac{5(x)}{(x+3)(x-2)(x)} - \dfrac{3(x-2)}{x(x+3)(x-2)} + \dfrac{2(x+3)}{x(x-2)(x+3)}$

$= \dfrac{5x}{x(x+3)(x-2)} - \dfrac{3x-6}{x(x+3)(x-2)} + \dfrac{2x+6}{x(x+3)(x-2)}$

$= \dfrac{5x-(3x-6)+2x+6}{x(x+3)(x-2)}$

$= \dfrac{5x-3x+6+2x+6}{x(x+3)(x-2)}$

$= \dfrac{4x+12}{x(x+3)(x-2)} = \dfrac{4\cancel{(x+3)}}{x\cancel{(x+3)}(x-2)} = \dfrac{4}{x(x-2)}$

41. $\left(\dfrac{x}{2}-\dfrac{2}{x}\right) \div \dfrac{x^2-2x}{x^2}$

$= \left(\dfrac{x\cdot x}{2x}-\dfrac{2\cdot 2}{2x}\right)\cdot \dfrac{x^2}{x^2-2x}$

$= \left(\dfrac{x^2}{2x}-\dfrac{4}{2x}\right)\cdot \dfrac{x^2}{x(x-2)}$

$= \left(\dfrac{x^2-4}{2x}\right)\cdot \dfrac{x^2}{x(x-2)}$

$= \dfrac{\cancel{(x-2)}(x+2)}{2\cancel{x}} \cdot \dfrac{\cancel{x^2}}{\cancel{x}\cancel{(x-2)}}$

$= \dfrac{x+2}{2}$

43. $\dfrac{\dfrac{1}{x}-1}{\dfrac{2}{x}} = \dfrac{x\left(\dfrac{1}{x}-1\right)}{x\left(\dfrac{2}{x}\right)}$

$= \dfrac{\dfrac{\cancel{x}}{1}\cdot\dfrac{1}{\cancel{x}} - x\cdot 1}{\dfrac{\cancel{x}}{1}\cdot\dfrac{2}{\cancel{x}}}$

$= \dfrac{1-x}{2}$

45. $\dfrac{1-\dfrac{1}{a}}{\dfrac{1}{a}-\dfrac{1}{a^2}} = \dfrac{a^2\left(1-\dfrac{1}{a}\right)}{a^2\left(\dfrac{1}{a}-\dfrac{1}{a^2}\right)}$

$= \dfrac{a^2\cdot 1 - \dfrac{\cancel{a^2}^{a}}{1}\cdot\dfrac{1}{\cancel{a}}}{\dfrac{\cancel{a^2}}{1}\cdot\dfrac{1}{\cancel{a}} - \dfrac{\cancel{a^2}}{1}\cdot\dfrac{1}{\cancel{a^2}}}$

$= \dfrac{a^2-a}{a-1} = \dfrac{a\cancel{(a-1)}}{\cancel{a-1}} = \dfrac{a}{1} = a$

47. $\dfrac{\dfrac{a}{2}-\dfrac{b}{4}}{\dfrac{4}{b^2}-\dfrac{1}{a^2}} = \dfrac{4a^2b^2\left(\dfrac{a}{2}-\dfrac{b}{4}\right)}{4a^2b^2\left(\dfrac{4}{b^2}-\dfrac{1}{a^2}\right)}$

$= \dfrac{\dfrac{\cancel{4a^2b^2}^{2}}{1}\cdot\dfrac{a}{\cancel{2}} - \dfrac{\cancel{4}a^2b^2}{1}\cdot\dfrac{b}{\cancel{4}}}{\dfrac{4a^2\cancel{b^2}}{1}\cdot\dfrac{4}{\cancel{b^2}} - \dfrac{4\cancel{a^2}b^2}{1}\cdot\dfrac{1}{\cancel{a^2}}}$

$= \dfrac{2a^3b^2-a^2b^3}{16a^2-4b^2} = \dfrac{a^2b^2(2a-b)}{4(4a^2-b^2)}$

$= \dfrac{a^2b^2\cancel{(2a-b)}}{4\cancel{(2a-b)}(2a+b)} = \dfrac{a^2b^2}{4(2a+b)}$

49. $\dfrac{\dfrac{1}{x}+\dfrac{1}{x+2}}{\dfrac{x+1}{x+2}} = \dfrac{x(x+2)\left(\dfrac{1}{x}+\dfrac{1}{x+2}\right)}{x(x+2)\left(\dfrac{x+1}{x+2}\right)} = \dfrac{\dfrac{\cancel{x}(x+2)}{1}\cdot\dfrac{1}{\cancel{x}} + \dfrac{x\cancel{(x+2)}}{1}\cdot\dfrac{1}{\cancel{(x+2)}}}{\dfrac{x\cancel{(x+2)}}{1}\cdot\dfrac{x+1}{\cancel{(x+2)}}}$

$= \dfrac{x+2+x}{x(x+1)} = \dfrac{2x+2}{x(x+1)} = \dfrac{2\cancel{(x+1)}}{x\cancel{(x+1)}} = \dfrac{2}{x}$

51. $\dfrac{\dfrac{1}{t+1}-\dfrac{1}{t+2}}{1-\dfrac{1}{t+2}} = \dfrac{(t+1)(t+2)\left(\dfrac{1}{t+1}-\dfrac{1}{t+2}\right)}{(t+1)(t+2)\left(1-\dfrac{1}{t+2}\right)}$

$= \dfrac{\dfrac{\cancel{(t+1)}(t+2)}{1}\cdot\dfrac{1}{\cancel{(t+1)}} - \dfrac{(t+1)\cancel{(t+2)}}{1}\cdot\dfrac{1}{\cancel{(t+2)}}}{(t+1)(t+2)\cdot 1 - \dfrac{(t+1)\cancel{(t+2)}}{1}\cdot\dfrac{1}{\cancel{(t+2)}}}$

$= \dfrac{t+2-(t+1)}{(t+1)(t+2)-(t+1)} = \dfrac{t+2-t-1}{t^2+2t+t+2-t-1)}$

$= \dfrac{1}{t^2+2t+1} \text{ or } \dfrac{1}{(t+1)^2}$

53. $\dfrac{X-a}{\dfrac{s}{\sqrt{n}}} = \dfrac{78-70}{\dfrac{6.2}{\sqrt{20}}} = \dfrac{8}{\dfrac{6.2}{\sqrt{20}}} = 8\cdot\dfrac{\sqrt{20}}{6.2} = \dfrac{160}{6.2} = 25.8$

55. In the solution to Example 12, replace the 150-mile distance between cities A and B by x. Then the time going from A to B is $\dfrac{x}{60}$ and the time returning from B to A is $\dfrac{x}{40}$. Therefore:

Average rate for entire trip $= \dfrac{\text{Total distance}}{\text{Total time}}$

$= \dfrac{x+x}{\dfrac{x}{60}+\dfrac{x}{40}} = \dfrac{2x}{\dfrac{2x}{120}+\dfrac{3x}{120}} = \dfrac{2x}{\dfrac{5x}{120}} = 2\cancel{x}\cdot\dfrac{\overset{24}{\cancel{120}}}{\cancel{5x}} = 48$

Thus, the average rate of speed for the entire trip is 48 mph. independent of the distance between the cities.

56. (a) $\left(x+\dfrac{2}{x}\right)\left(x-\dfrac{3}{x}\right) = x^2 - \dfrac{\cancel{x}}{1}\left(\dfrac{3}{\cancel{x}}\right) + \dfrac{\cancel{x}}{1}\left(\dfrac{2}{\cancel{x}}\right) - \left(\dfrac{2}{x}\right)\left(\dfrac{3}{x}\right)$

$= x^2 - 3 + 2 - \dfrac{6}{x^2} = x^2 - 1 + \dfrac{6}{x^2}$

(b) $x+\dfrac{2}{x} = \dfrac{x}{1}+\dfrac{2}{x} = \dfrac{x(x)}{1(x)}+\dfrac{2}{x} = \dfrac{x^2}{x}+\dfrac{2}{x} = \dfrac{x^2+2}{x}$

$x-\dfrac{3}{x} = \dfrac{x}{1}-\dfrac{3}{x} = \dfrac{x(x)}{1(x)}-\dfrac{3}{x} = \dfrac{x^2}{x}-\dfrac{3}{x} = \dfrac{x^2-3}{x}$

Then $\left(x + \dfrac{2}{x}\right)\left(x - \dfrac{3}{x}\right) = \dfrac{x^2 + 2}{x} \cdot \dfrac{x^2 - 3}{x} = \dfrac{(x^2 + 2)(x^2 - 3)}{x^2}$

The second method is easier. If the instructions ask for the answer in the form of a single fraction, we get such an answer directly in part (b). In part (a), there would be further work to do.

57. $\dfrac{x-1}{2} = \dfrac{x}{3}$ LCD = 6

$\cancel{6}^{3}\left(\dfrac{x-1}{\cancel{2}}\right) = \cancel{6}^{2}\left(\dfrac{x}{\cancel{3}}\right)$

$3(x-1) = 2x$

$3x - 3 = 2x$

$-3 = -x$

$3 = x$

59. $\dfrac{a}{5} - \dfrac{a}{6} = \dfrac{1}{2}$ LCD = 30

$30\left(\dfrac{a}{5} - \dfrac{a}{6}\right) = 30\left(\dfrac{1}{2}\right)$

$\cancel{30}^{6}\left(\dfrac{a}{\cancel{5}}\right) - \cancel{30}^{5}\left(\dfrac{a}{\cancel{6}}\right) = \cancel{30}^{15}\left(\dfrac{1}{\cancel{2}}\right)$

$6a - 5a = 15$

$a = 15$

Exercises 9.4

1. $\dfrac{x}{2} + \dfrac{x}{3} = 10$

$6\left(\dfrac{x}{2} + \dfrac{x}{3}\right) = 6 \cdot 10$

$\dfrac{\cancel{6}^{3}}{1} \cdot \dfrac{x}{\cancel{2}_{1}} + \dfrac{\cancel{6}^{2}}{1} \cdot \dfrac{x}{\cancel{3}_{1}} = 60$

$3x + 2x = 60$

$5x = 60$

$x = 12$

3. $\dfrac{y}{6} - \dfrac{y}{4} = 2$

$12\left(\dfrac{y}{6} - \dfrac{y}{4}\right) = 12 \cdot 2$

$\dfrac{\cancel{12}^{2}}{1} \cdot \dfrac{y}{\cancel{6}_{1}} - \dfrac{\cancel{12}^{3}}{1} \cdot \dfrac{y}{\cancel{4}_{1}} = 24$

$2y - 3y = 24$

$-y = 24$

$y = -24$

5. $\dfrac{a+1}{3} + \dfrac{a+3}{4} = 4$

$12\left(\dfrac{a+1}{3} + \dfrac{a+3}{4}\right) = 12 \cdot 4$

$\dfrac{\cancel{12}^{4}}{1} \cdot \dfrac{a+1}{\cancel{3}_{1}} + \dfrac{\cancel{12}^{3}}{1} \cdot \dfrac{a+3}{\cancel{4}_{1}} = 48$

$4(a+1) + 3(a+3) = 48$

$4a + 4 + 3a + 9 = 48$

$7a + 13 = 48$

$7a = 35$

$a = 5$

7. $\dfrac{u-2}{2} - \dfrac{u+4}{8} = 1$

$8\left(\dfrac{u-2}{2} - \dfrac{u+4}{8}\right) = 8 \cdot 1$

$\dfrac{\cancel{8}^{4}}{1} \cdot \dfrac{u-2}{\cancel{2}_{1}} - \dfrac{\cancel{8}^{1}}{1} \cdot \dfrac{u+4}{\cancel{8}_{1}} = 8$

$4(u-2) - (u+4) = 8$

$4u - 8 - u - 4 = 8$

$3u - 12 = 8$

$3u = 20$

$u = \dfrac{20}{3}$

9. $\dfrac{x}{2} + \dfrac{x}{3} + \dfrac{x}{4}$ LCD = 12

$= \dfrac{x(6)}{2(6)} + \dfrac{x(4)}{3(4)} + \dfrac{x(3)}{4(3)}$

$= \dfrac{6x}{12} + \dfrac{4x}{12} + \dfrac{3x}{12}$

$= \dfrac{6x + 4x + 3x}{12} = \dfrac{13x}{12}$

11. $\dfrac{x}{2} \cdot \dfrac{x}{3} \cdot \dfrac{x}{4} = \dfrac{x \cdot x \cdot x}{2 \cdot 3 \cdot 4}$

$= \dfrac{x^3}{24}$

13. $\dfrac{x+1}{2} - \dfrac{3}{x} = \dfrac{x}{2}$

$2x\left(\dfrac{x+1}{2} - \dfrac{3}{x}\right) = 2x\left(\dfrac{x}{2}\right)$

$\dfrac{\cancel{2}x}{1} \cdot \dfrac{x+1}{\cancel{2}} - \dfrac{2\cancel{x}}{1} \cdot \dfrac{3}{\cancel{x}} = \dfrac{\cancel{2}x}{1} \cdot \dfrac{x}{\cancel{2}}$

$x(x+1) - 6 = x^2$

$x^2 + x - 6 = x^2$

$x - 6 = 0$

$x = 6$

15. $\dfrac{x+1}{2} \cdot \dfrac{3}{x} = \dfrac{2}{x}$

$\dfrac{(x+1) \cdot 3}{2 \cdot x} = \dfrac{2}{x}$

$\dfrac{2\cancel{x}}{1}\left[\dfrac{(x+1) \cdot 3}{2\cancel{x}}\right] = \dfrac{2\cancel{x}}{1} \cdot \dfrac{2}{\cancel{x}}$

$(x+1) \cdot 3 = 2 \cdot 2$

$3x + 3 = 4$

$3x = 1$

$x = \dfrac{1}{3}$

17. $\dfrac{x}{5} + \dfrac{x-1}{4} + \dfrac{x-3}{2} = 3$

$20\left(\dfrac{x}{5} + \dfrac{x-1}{4} + \dfrac{x-3}{2}\right) = 20 \cdot 3$

$\dfrac{\cancel{20}^4}{1} \cdot \dfrac{x}{\cancel{5}} + \dfrac{\cancel{20}^5}{1} \cdot \dfrac{x-1}{\cancel{4}} + \dfrac{\cancel{20}^{10}}{1} \cdot \dfrac{x-3}{\cancel{2}} = 60$

$4x + 5(x-1) + 10(x-3) = 60$

$4x + 5x - 5 + 10x - 30 = 60$

$19x - 35 = 60$

$19x = 95$

$x = 5$

19. $\dfrac{a+5}{2} = \dfrac{a+2}{5}$

$\dfrac{\cancel{10}^5}{1}\left(\dfrac{a+5}{\cancel{2}}\right) = \dfrac{\cancel{10}^2}{1}\left(\dfrac{a+2}{\cancel{5}}\right)$

$5(a+5) = 2(a+2)$

$5a + 25 = 2a + 4$

$3a + 25 = 4$

$3a = -21$

$a = -7$

21. $\dfrac{x+2}{3} + 1 = \dfrac{x+3}{2} - 1$

$6\left(\dfrac{x+2}{3} + 1\right) = 6\left(\dfrac{x+3}{2} - 1\right)$

$\dfrac{\cancel{6}^2}{1}\left(\dfrac{x+2}{\cancel{3}}\right) + 6 \cdot 1 = \dfrac{\cancel{6}^3}{1}\left(\dfrac{x+3}{\cancel{2}}\right) - 6 \cdot 1$

$2(x+2) + 6 = 3(x+3) - 6$

$2x + 4 + 6 = 3x + 9 - 6$

$2x + 10 = 3x + 3$

$10 = x + 3$

$7 = x$

23.
$$\frac{2r+1}{3} - \frac{r+1}{5} = \frac{r+8}{6}$$
$$30\left(\frac{2r+1}{3} - \frac{r+1}{5}\right) = 30\left(\frac{r+8}{6}\right)$$
$$\frac{\cancel{30}^{10}}{1} \cdot \frac{2r+1}{\cancel{3}_1} - \frac{\cancel{30}^{6}}{1} \cdot \frac{r+1}{\cancel{5}_1} = \frac{\cancel{30}^{5}}{1} \cdot \frac{r+8}{\cancel{6}_1}$$
$$10(2r+1) - 6(r+1) = 5(r+8)$$
$$20r + 10 - 6r - 6 = 5r + 40$$
$$14r + 4 = 5r + 40$$
$$9r + 4 = 40$$
$$9r = 36$$
$$r = 4$$

25.
$$\frac{3}{x} - \frac{2}{3} = \frac{2}{x}$$
$$3x\left(\frac{3}{x} - \frac{2}{3}\right) = 3x\left(\frac{2}{x}\right)$$
$$\frac{3\cancel{x}}{1} \cdot \frac{3}{\cancel{x}_1} - \frac{\cancel{3}x}{1} \cdot \frac{2}{\cancel{3}_1} = \frac{3\cancel{x}}{1} \cdot \frac{2}{\cancel{x}_1}$$
$$9 - 2x = 6$$
$$-2x = -3$$
$$x = \frac{3}{2}$$

27.
$$\frac{4}{t-2} + \frac{3}{t} = \frac{1}{2t}$$
$$2t(t-2)\left(\frac{4}{t-2} + \frac{3}{t}\right) = 2t(t-2)\left(\frac{1}{2t}\right)$$
$$\frac{2t\cancel{(t-2)}}{1} \cdot \frac{4}{\cancel{(t-2)}_1} + \frac{2\cancel{t}(t-2)}{1} \cdot \frac{3}{\cancel{t}_1} = \frac{\cancel{2t}(t-2)}{1} \cdot \frac{1}{\cancel{2t}_1}$$
$$8t + 6(t-2) = t - 2$$
$$8t + 6t - 12 = t - 2$$
$$14t - 12 = t - 2$$
$$13t - 12 = -2$$
$$13t = 10$$
$$t = \frac{10}{13}$$

29.
$$\frac{5}{2x} - \frac{8}{x+2} \quad \text{LCD} = 2x(x+2)$$
$$= \frac{5(x+2)}{2x(x+2)} - \frac{8(2x)}{2x(x+2)}$$
$$= \frac{5x+10}{2x(x+2)} - \frac{16x}{2x(x+2)}$$
$$= \frac{5x+10-16x}{2x(x+2)}$$
$$= \frac{-11x+10}{2x(x+2)}$$

31.
$$\frac{4}{x-1} - \frac{5}{8} = \frac{3}{2x-2}$$
$$\frac{4}{x-1} - \frac{5}{8} = \frac{3}{2(x-1)}$$
$$8(x-1)\left(\frac{4}{x-1} - \frac{5}{8}\right) = 8(x-1)\left(\frac{3}{2(x-1)}\right)$$
$$\frac{8\cancel{(x-1)}}{1} \cdot \frac{4}{\cancel{x-1}_1} - \frac{\cancel{8}(x-1)}{1} \cdot \frac{5}{\cancel{8}_1} = \frac{\cancel{8}^4\cancel{(x-1)}}{1} \cdot \frac{3}{\cancel{2(x-1)}_1}$$

$$32 - 5(x-1) = 12$$
$$32 - 5x + 5 = 12$$
$$-5x + 37 = 12$$
$$-5x = -25$$
$$x = 5$$

33.
$$\frac{8}{x-2} + 3 = \frac{x+6}{x-2}$$
$$(x-2)\left(\frac{8}{x-2} + 3\right) = (x-2)\left(\frac{x+6}{x-2}\right)$$
$$\frac{\cancel{x-2}}{1} \cdot \frac{8}{\cancel{x-2}} - \frac{x-2}{1} \cdot \frac{3}{1} = \frac{\cancel{x-2}}{1} \cdot \frac{x+6}{\cancel{x-2}}$$
$$8 + 3(x-2) = x+6$$
$$8 + 3x - 6 = x+6$$
$$3x + 2 = x + 6$$
$$2x + 2 = 6$$
$$2x = 4$$
$$x = 2$$

$x = 2$ is not a solution since it causes a denominator in the original equation to equal 0. Hence, no solution.

35.
$$\frac{4}{x+2} - \frac{3}{x+6} \quad \text{LCD} = (x+2)(x+6)$$
$$= \frac{4(x+6)}{(x+2)(x+6)} - \frac{3(x+2)}{(x+2)(x+6)}$$
$$= \frac{4x+24}{(x+2)(x+6)} - \frac{3x+6}{(x+2)(x+6)}$$
$$= \frac{4x+24-(3x+6)}{(x+2)(x+6)}$$
$$= \frac{4x+24-3x-6}{(x+2)(x+6)}$$
$$= \frac{x+18}{(x+2)(x+6)}$$

37.
$$\frac{3}{x+4} + \frac{2}{5} = \frac{x-1}{x+4}$$
$$5(x+4)\left(\frac{3}{x+4} + \frac{2}{5}\right) = 5(x+4)\left(\frac{x-1}{x+4}\right)$$
$$\frac{5(x+4)}{1} \cdot \frac{3}{x+4} + \frac{5(x+4)}{1} \cdot \frac{2}{5} = \frac{5(x+4)}{1} \cdot \frac{x-1}{x+4}$$

$$15 + 2(x+4) = 5(x-1)$$
$$15 + 2x + 8 = 5x - 5$$
$$2x + 23 = 5x - 5$$
$$-3x + 23 = -5$$
$$-3x = -28$$
$$x = \frac{28}{3}$$

39.
$$\frac{7}{x+3} - \frac{1}{4} = \frac{x+10}{x+3}$$
$$4(x+3)\left(\frac{7}{x+3} - \frac{1}{4}\right) = 4(x+3)\left(\frac{x+10}{x+3}\right)$$
$$\frac{4(x+3)}{1} \cdot \frac{7}{x+3} - \frac{4(x+3)}{1} \cdot \frac{1}{4} = \frac{4(x+3)}{1} \cdot \frac{x+10}{x+3}$$

$$28 - (x+3) = 4(x+10)$$
$$28 - x - 3 = 4x + 40$$
$$-x + 25 = 4x + 40$$
$$-5x + 25 = 40$$
$$-5x = 15$$
$$x = -3$$

$x = -3$ is not a solution since it causes a denominator in the original equation to equal 0. Hence, no solution.

41.
$$\frac{m+3}{6} - \frac{m+4}{8} = m$$
$$24\left(\frac{m+3}{6} - \frac{m+4}{8}\right) = 24m$$
$$\frac{\cancel{24}^4}{1} \cdot \frac{m+3}{\cancel{6}} - \frac{\cancel{24}^3}{1} \cdot \frac{m+4}{\cancel{8}} = 24m$$

$$4(m+3) - 3(m+4) = 24m$$
$$4m + 12 - 3m - 12 = 24m$$
$$m = 24m$$
$$-23m = 0$$
$$m = 0$$

43.
$$\frac{5}{x^2 - x} - \frac{1}{2x - 2} = \frac{1}{x}$$
$$\frac{5}{x(x-1)} - \frac{1}{2(x-1)} = \frac{1}{x}$$
$$2x(x-1)\left(\frac{5}{x(x-1)} - \frac{1}{2(x-1)}\right) = 2x(x-1)\left(\frac{1}{x}\right)$$
$$\frac{2\cancel{x}(\cancel{x-1})}{1} \cdot \frac{5}{\cancel{x}(\cancel{x-1})} - \frac{\cancel{2}x(\cancel{x-1})}{1} \cdot \frac{1}{\cancel{2}(\cancel{x-1})} = \frac{2\cancel{x}(x-1)}{1} \cdot \frac{1}{\cancel{x}}$$

$$10 - x = 2(x-1)$$
$$10 - x = 2x - 2$$
$$10 = 3x - 2$$
$$12 = 3x$$
$$4 = x$$

45.
$$\frac{x}{x-5} + \frac{3}{2} = \frac{5}{x-5}$$
$$2(x-5)\left(\frac{x}{x-5} + \frac{3}{2}\right) = 2(x-5)\left(\frac{5}{x-5}\right)$$
$$\frac{2(\cancel{x-5})}{1} \cdot \frac{x}{\cancel{x-5}} + \frac{\cancel{2}(x-5)}{1} \cdot \frac{3}{\cancel{2}} = \frac{2(\cancel{x-5})}{1} \cdot \frac{5}{\cancel{x-5}}$$

$$2x + (x-5) \cdot 3 = 10$$
$$2x + 3x - 15 = 10$$
$$5x - 15 = 10$$
$$5x = 25$$
$$x = 5$$

$x = 5$ is not a solution since it causes a denominator in the original equation to equal 0. Hence, no solution.

47.
$$\frac{a}{2} + \frac{2}{a-2} = \frac{a-4}{2}$$
$$2(a-2)\left(\frac{a}{2} + \frac{2}{a-2}\right) = 2(a-2)\left(\frac{a-4}{2}\right)$$
$$\frac{\cancel{2}(a-2)}{1} \cdot \frac{a}{\cancel{2}} + \frac{2(\cancel{a-2})}{1} \cdot \frac{2}{(\cancel{a-2})} = \frac{\cancel{2}(a-2)}{1} \cdot \frac{a-4}{\cancel{2}}$$
$$(a-2) \cdot a + 2 \cdot 2 = (a-2)(a-4)$$

$$a^2 - 2a + 4 = a^2 - 4a - 2a + 8$$
$$a^2 - 2a + 4 = a^2 - 6a + 8$$
$$-2a + 4 = -6a + 8$$
$$4a + 4 = 8$$
$$4a = 4$$
$$a = 1$$

49. $\dfrac{2}{x^2 - x} + \dfrac{8}{x^2 - 1}$ LCD $= x(x-1)(x+1)$

$$= \frac{2}{x(x-1)} + \frac{8}{(x-1)(x+1)}$$
$$= \frac{2(x+1)}{x(x-1)(x+1)} + \frac{8(x)}{x(x-1)(x+1)}$$
$$= \frac{2x + 2}{x(x-1)(x+1)} + \frac{8x}{x(x-1)(x+1)}$$
$$= \frac{2x + 2 + 8x}{x(x-1)(x+1)} = \frac{10x + 2}{x(x-1)(x+1)}$$

51. $\dfrac{1}{x+3} \div \dfrac{x+1}{x^2 - 9}$

$$= \frac{1}{x+3} \cdot \frac{x^2 - 9}{x+1}$$
$$= \frac{1}{\cancel{(x+3)}} \cdot \frac{\cancel{(x+3)}(x-3)}{x+1}$$
$$= \frac{x-3}{x+1}$$

53.
$$\frac{5}{x^2 - 2x - 3} = \frac{4}{x^2 - 3x - 4}$$
$$\frac{5}{(x-3)(x+1)} = \frac{4}{(x-4)(x+1)}$$
$$\frac{(x-3)(x-4)(x+1)}{1}\left(\frac{5}{(x-3)(x+1)}\right) = \frac{(x-3)(x-4)(x+1)}{1}\left(\frac{4}{(x-4)(x+1)}\right)$$
$$5(x-4) = 4(x-3)$$
$$5x - 20 = 4x - 12$$
$$x - 20 = -12$$
$$x = 8$$

55.
$$\frac{9}{x^2 - 3x + 2} - \frac{2}{x-1} = \frac{1}{x-2}$$
$$\frac{9}{(x-1)(x-2)} - \frac{2}{x-1} = \frac{1}{x-2}$$
$$(x-1)(x-2)\left(\frac{9}{(x-1)(x-2)} - \frac{2}{x-1}\right) = (x-1)(x-2)\left(\frac{1}{x-2}\right)$$
$$\frac{(x-1)(x-2)}{1} \cdot \frac{9}{(x-1)(x-2)} - \frac{(x-1)(x-2)}{1} \cdot \frac{2}{x-1} = \frac{(x-1)(x-2)}{1} \cdot \frac{1}{x-2}$$
$$9 - 2(x-2) = x - 1$$
$$9 - 2x + 4 = x - 1$$
$$13 - 2x = x - 1$$
$$14 = 3x$$
$$\frac{14}{3} = x$$

57. Let w = width of the rectangle.
Then $2w + 3$ = length of the rectangle.
$$\frac{w}{2w+3} = \frac{2}{5}$$
$$5(2w+3)\left(\frac{w}{2w+3}\right) = 5(2w+3)\left(\frac{2}{5}\right)$$
$$5w = 2(2w+3)$$
$$5w = 4w + 6$$
$$w = 6$$
Then $2w + 3 = 2(6) + 3 = 15$.
The dimensions of the rectangle are 6 by 15.

59. The line passing through the points $(1, c)$ and $(-5, 3)$ has
slope $= \dfrac{c-3}{1-(-5)} = \dfrac{c-3}{6}$.
The line passing through the points $(4, 3)$ and $(7, -2)$ has
slope $= \dfrac{-2-3}{7-4} = \dfrac{-5}{3}$.
Since parallel lines have equal slopes,
$$\frac{c-3}{6} = \frac{-5}{3}$$
$$6\left(\frac{c-3}{6}\right) = 6\left(\frac{-5}{3}\right)$$
$$c - 3 = -10$$
$$c = -7.$$

61. $C = \dfrac{85n + 10000}{n}$

$90 = \dfrac{85n + 10000}{n}$

$n \cdot 90 = \not{n} \cdot \dfrac{85n + 10000}{\not{n}}$

$90n = 85n + 10000$

$5n = 10000$

$n = 2000$

2000 CD players should be produced.

63. $n = \dfrac{100x}{0.6x + 5}$

$160 = \dfrac{100x}{0.6x + 5}$

$160(0.6x + 5) = \dfrac{100x}{\cancel{0.6x+5}}(\cancel{0.6x+5})$

$160(0.6x) + 160(5) = 100x$

$96x + 800 = 100x$

$800 = 4x$

$200 = x$

It would take 200 acres to support 160 deer.

65. Using the given formula, we construct the following table:

n	C
5000	87
8000	86.25
12000	85.83
15000	85.67

As the number of CD players passes 12,000, the average cost levels off between $85 and $86.

66. All of the given values satisfy the equation.

$\dfrac{x+3}{2} - \dfrac{2x+5}{4} = \dfrac{1}{4}$

$\dfrac{2(x+3)}{2(2)} - \dfrac{2x+5}{4} = \dfrac{1}{4}$

$\dfrac{2x+6}{4} - \dfrac{2x+5}{4} = \dfrac{1}{4}$

$\dfrac{2x+6-(2x+5)}{4} = \dfrac{1}{4}$

$\dfrac{2x+6-2x-5}{4} = \dfrac{1}{4}$

$\dfrac{1}{4} = \dfrac{1}{4}$

This shows that the given equation is always true, no matter what value of x is chosen. Such an equation is called an identity.

67. The length of a rectangle is five less than three times its width. Find the dimensions of the rectangle if its perimeter is 62 cm. (Let $x = $ width of the rectangle.)

69. Let $x = $ the number of students who take longer than four years to graduate. Then $2600 + x = $ the number of graduates altogether.

$\dfrac{2600}{x} = \dfrac{13}{3}$

$\dfrac{3\not{x}}{1}\left(\dfrac{2600}{\not{x}}\right) = \dfrac{\not{3}x}{1}\left(\dfrac{13}{\not{3}}\right)$

$7800 = 13x$

$600 = x$

Then $2600 + x = 2600 + 600 = 3200$.

There were 3200 graduates during the last five years.

Exercises 9.5

1. $4x - 3y = 12$
$ + 3y +3y$
$\overline{4x = 12 + 3y}$
$\dfrac{\cancel{4}x}{\cancel{4}} = \dfrac{12+3y}{4}$
$x = \dfrac{12+3y}{4}$

3. $3x + 6y = 18$
$-3x -3x$
$\overline{6y = 18 - 3x}$
$\dfrac{\cancel{6}y}{\cancel{6}} = \dfrac{18-3x}{6}$
$y = \dfrac{18-3x}{6} = \dfrac{\cancel{3}(6-x)}{\underset{2}{\cancel{6}}}$
$y = \dfrac{6-x}{2}$

5. $a - 3b = 2a - b + 5$
$+3b +3b$
$\overline{a = 2a + 2b + 5}$
$-2a -2a$
$\overline{-a = 2b + 5}$
$\dfrac{\cancel{-}a}{\cancel{-}1} = \dfrac{2b+5}{-1}$
$a = -(2b+5)$
$a = -2b - 5$

7. $3(m + 2p) = 4(p - m)$
$3m + 6p = 4p - 4m$
$+4m +4m$
$\overline{7m + 6p = 4p}$
$-6p -6p$
$\overline{7m = -2p}$
$\dfrac{\cancel{7}m}{\cancel{7}} = \dfrac{-2p}{7}$
$m = \dfrac{-2p}{7} = -\dfrac{2p}{7}$

9. $5x - 7z + 12 = 3y - 7x + 2z$
$-5x -5x$
$\overline{-7z + 12 = 3y - 12x + 2z}$
$-2z -2z$
$\overline{-9z + 12 = 3y - 12x}$
$-12 -12$
$\overline{-9z = 3y - 12x - 12}$
$\dfrac{\cancel{-9}z}{\cancel{-9}} = \dfrac{3y - 12x - 12}{-9}$
$z = \dfrac{\cancel{3}(y - 4x - 4)}{\underset{-3}{\cancel{-9}}}$
$z = -\dfrac{y - 4x - 4}{3}$

11. $2r + 3(r - 5) = r - 5s + 1$
$2r + 3r - 15 = r - 5s + 1$
$5r - 15 = r - 5s + 1$
$-r -r$
$\overline{4r = -5s + 1}$
$+15 +15$
$\overline{4r = -5s + 16}$
$\dfrac{\cancel{4}r}{\cancel{4}} = \dfrac{-5s + 16}{4}$
$r = \dfrac{-5s + 16}{4}$

13. $a(b + c) = d$
$\dfrac{a\cancel{(b+c)}}{\cancel{(b+c)}} = \dfrac{d}{b+c}$
$a = \dfrac{d}{b+c}$

15. $\dfrac{x}{2} + \dfrac{y}{3} = \dfrac{x}{3} + \dfrac{y}{4} - \dfrac{1}{6}$
$12\left(\dfrac{x}{2} + \dfrac{y}{3}\right) = 12\left(\dfrac{x}{3} + \dfrac{y}{4} - \dfrac{1}{6}\right)$
$\dfrac{\overset{6}{\cancel{12}}}{1} \cdot \dfrac{x}{\underset{1}{\cancel{2}}} + \dfrac{\overset{4}{\cancel{12}}}{1} \cdot \dfrac{y}{\underset{1}{\cancel{3}}} = \dfrac{\overset{4}{\cancel{12}}}{1} \cdot \dfrac{x}{\underset{1}{\cancel{3}}} + \dfrac{\overset{3}{\cancel{12}}}{1} \cdot \dfrac{y}{\underset{1}{\cancel{4}}} - \dfrac{\overset{2}{\cancel{12}}}{1} \cdot \dfrac{1}{\underset{1}{\cancel{6}}}$

15. con't.
$$\begin{aligned} 6x + 4y &= 4x + 3y - 2 \\ -4x & \quad -4x \\ \hline 2x + 4y &= 3y - 2 \\ -4y &\quad -4y \\ \hline 2x &= -y - 2 \\ \frac{\cancel{2}x}{\cancel{2}} &= \frac{-y-2}{2} \\ x &= \frac{-y-2}{2} \end{aligned}$$

17.
$$\begin{aligned} 4(x-y) - \frac{x-y}{2} &= 3 - (y-x) \\ 2\left[4(x-y) - \frac{x-y}{2}\right] &= 2[3 - (y-x)] \\ 8(x-y) - \frac{\cancel{2}^1}{1} \cdot \frac{x-y}{\cancel{2}_1} &= 6 - 2(y-x) \\ 8x - 8y - (x-y) &= 6 - 2y + 2x \\ 8x - 8y - x + y &= 6 - 2y + 2x \\ 7x - 7y &= 6 - 2y + 2x \\ +7y &\quad +7y \\ \hline 7x &= 6 + 5y + 2x \\ -2x &\quad 6 \quad -2x \\ \hline 5x &= 6 + 5y \\ \frac{\cancel{5}x}{\cancel{5}} &= \frac{6+5y}{5} \\ x &= \frac{6+5y}{5} \end{aligned}$$

19.
$$\begin{aligned} ax + b &= 2x + d \\ -b &\quad -b \\ \hline ax &= 2x + d - b \\ -2x &\quad -2x \\ \hline ax - 2x &= d - b \\ x(a-2) &= d - b \\ \frac{x\cancel{(a-2)}}{\cancel{(a-2)}} &= \frac{d-b}{a-2} \\ x &= \frac{d-b}{a-2} \end{aligned}$$

21.
$$\begin{aligned} 2pn - \frac{y}{3} &= 3n + ay + p \\ 3\left(2pn - \frac{y}{3}\right) &= 3(3n + ay + p) \\ 6pn - \frac{\cancel{3}^1}{1} \cdot \frac{y}{\cancel{3}_1} &= 9n + 3ay + 3p \\ 6pn - y &= 9n + 3ay + 3p \\ +y &\quad +y \\ \hline 6pn &= 9n + 3ay + 3p + y \\ -9n &\quad = -9n \\ \hline 6pn - 9n &= 3ay + 3p + y \\ -3p &\quad -3p \\ \hline 6pn - 9n - 3p &= 3ay \quad + y \\ 6pn - 9n - 3p &= (3a+1)y \\ \frac{6pn - 9n - 3p}{3a+1} &= \frac{\cancel{(3a+1)}y}{\cancel{3a+1}} \\ \frac{6pn - 9n - 3p}{3a+1} &= y \end{aligned}$$

23.
$$\begin{aligned} y &= \frac{u+1}{u-1} \\ (u-1)y &= \frac{\cancel{(u-1)}}{1} \cdot \frac{(u+1)}{\cancel{(u-1)}} \\ (u-1)y &= u+1 \end{aligned}$$

$$\begin{aligned} uy - y &= u + 1 \\ -u &\quad -u \\ \hline uy - y - u &= 1 \\ +y &\quad +y \\ \hline uy \quad - u &= 1 + y \end{aligned}$$

$$\begin{aligned} u(y-1) &= 1 + y \\ \frac{u\cancel{(y-1)}}{\cancel{y-1}} &= \frac{1+y}{y-1} \\ u &= \frac{1+y}{y-1} \end{aligned}$$

189

25. $(x+a)(y+b) = c$

$\dfrac{(x+a)\cancel{(y+b)}}{\cancel{(y+b)}} = \dfrac{c}{y+b}$

$x + a = \dfrac{c}{y+b}$

$\dfrac{-a \qquad\qquad -a}{x = \dfrac{c}{y+b} - a}$

27. $y = mx + b$

$\dfrac{-b \qquad -b}{y - b = mx}$

$\dfrac{y-b}{m} = \dfrac{m\cancel{x}}{\cancel{m}}$

$\dfrac{y-b}{m} = m$

29. $A = \dfrac{1}{2}h(b_1 + b_2)$

$2A = 2\left[\dfrac{1}{2}h(b_1 + b_2)\right]$

$2A = h(b_1 + b_2)$

$\dfrac{2A}{b_1 + b_2} = h$

31. $A = P(1 + rt)$

$\dfrac{A}{1 + rt} = \dfrac{P\cancel{(1+rt)}}{\cancel{1+rt}}$

$\dfrac{A}{1 + rt} = P$

33. $C = \dfrac{5}{9}(F - 32)$

$9C = 9\left[\dfrac{5}{9}(F - 32)\right]$

$9C = 5(F - 32)$

$\dfrac{9C}{5} = F - 32$

$\dfrac{9C}{5} + 32 = F$

35. $S = S_0 + v_0 t + \dfrac{1}{2}gt^2$

$\dfrac{-S_0 \qquad\qquad\qquad -S_0}{S - S_0 = v_0 t + \dfrac{1}{2}gt^2}$

$\dfrac{\qquad -\dfrac{1}{2}gt^2 \qquad -\dfrac{1}{2}gt^2}{S - S_0 - \dfrac{1}{2}gt^2 = v_0 t}$

$\dfrac{S - S_0 - \dfrac{1}{2}gt^2}{t} = \dfrac{v_0 \cancel{t}}{\cancel{t}}$

$\dfrac{S - S_0 - \dfrac{1}{2}gt^2}{t} = v_0$

37. $\dfrac{x - \mu}{s} < 2$

$\cancel{s}\left(\dfrac{x-\mu}{\cancel{s}}\right) < s(2)$ (since s is positive)

$x - \mu < 2s$

$x < 2s + \mu$

39. $\dfrac{1}{f} = \dfrac{1}{f_1} + \dfrac{1}{f_2}$

$ff_1f_2\left(\dfrac{1}{f}\right) = ff_1f_2\left(\dfrac{1}{f_1} + \dfrac{1}{f_2}\right)$

$\dfrac{f f_1 f_2}{1} \cdot \dfrac{1}{\cancel{f}} = \dfrac{f \cancel{f_1} f_2}{1} \cdot \dfrac{1}{\cancel{f_1}} + \dfrac{f f_1 \cancel{f_2}}{1} \cdot \dfrac{1}{\cancel{f_2}}$

$f_1 f_2 = f f_2 + f f_1$

$f_1 f_2 = f(f_1 + f_2)$

$\dfrac{f_1 f_2}{f_1 + f_2} = f$

41. The ratio of men to women on the faculty of a local college is 9 to 10. If there are 600 women on the faculty, how many men are on the faculty? (Let x = the number of men.)

43. $x^3 x^4 (x^3)^4 = x^3 x^4 x^{12}$
$= x^{3+4+12}$
$= x^{19}$

45. $2^{-3} + 3^{-2} = \dfrac{1}{2^3} + \dfrac{1}{3^2} = \dfrac{1}{8} + \dfrac{1}{9}$
$= \dfrac{9 \cdot 1}{9 \cdot 8} + \dfrac{1 \cdot 8}{9 \cdot 8}$
$= \dfrac{9}{72} + \dfrac{8}{72} = \dfrac{17}{72}$

Exercises 9.6

1. Let $w =$ the width of the rectangle.
Since $\dfrac{\text{width}}{\text{length}} = \dfrac{3}{7}$, let $\dfrac{7}{3}w =$ length.
$$2w + 2\left(\dfrac{7}{3}w\right) = 100$$
$$3\left[2w + 2\left(\dfrac{7}{3}w\right)\right] = 3(100)$$
$$6w + \dfrac{\cancel{3}}{1} \cdot \dfrac{2}{1}\left(\dfrac{7}{\cancel{3}}w\right) = 300$$
$$6w + 14w = 300$$
$$20w = 300$$
$$w = 15$$
Then $\dfrac{7}{3}w = \dfrac{7}{3}(15) = 35$. Thus, the width is 15 m and the length is 35 m.
CHECK: $\dfrac{15}{35} \stackrel{\vee}{=} \dfrac{3}{7}$ and
$2(15) + 2(35) \stackrel{\vee}{=} 100$.

3. Let $x =$ number of people preferring Brand X.
Then $200 - x =$ number of people who did not.
$$\dfrac{x}{200 - x} = \dfrac{13}{12}$$
$$\dfrac{12(200 - x)}{1} \cdot \dfrac{x}{200 - x} = \dfrac{\cancel{12}(200 - x)}{1} \cdot \dfrac{13}{\cancel{12}}$$
$$12x = 13(200 - x)$$
$$12x = 2600 - 13x$$
$$25x = 2600$$
$$x = 104$$
Then $200 - x = 200 - 104 = 96$.
Thus, 104 people preferred Brand X.
CHECK: $\dfrac{104}{96} = \dfrac{\cancel{8}(13)}{\cancel{8}(12)} \stackrel{\vee}{=} \dfrac{13}{12}$
and $104 + 96 \stackrel{\vee}{=} 200$.

5. Let $x =$ the number of hits Joe must get in his next 50 at-bats.
$$\dfrac{120 + x}{400 + 50} = 0.400 = \dfrac{400}{1000} = \dfrac{2}{5}$$
$$\dfrac{120 + x}{450} = \dfrac{2}{5}$$
$$\dfrac{\cancel{450}}{1} \cdot \dfrac{120 + x}{\cancel{450}} = \dfrac{\overset{90}{\cancel{450}}}{1} \cdot \dfrac{2}{\cancel{5}}$$
$$120 + x = 180$$
$$x = 60$$
This is impossible, since the largest possible value of x is 50. Therefore, Joe cannot raise his average to 0.400 in his next 50 at-bats.

7. Let $x =$ the number of hours for whole job.
$$x\left(\dfrac{1}{3}\right) + x\left(\dfrac{1}{2}\right) = 1$$
$$6\left[x\left(\dfrac{1}{3}\right) + x\left(\dfrac{1}{2}\right)\right] = 6(1)$$
$$\dfrac{\overset{2}{\cancel{6}}}{1} \cdot x \cdot \dfrac{1}{\cancel{3}} + \dfrac{\overset{3}{\cancel{6}}}{1} \cdot x \cdot \dfrac{1}{\cancel{2}} = 6$$
$$2x + 3x = 6$$
$$5x = 6$$
$$x = \dfrac{6}{5}$$
It would take them $\dfrac{6}{5}$ hours to complete the job working together.

7. con't. CHECK: In one hour, Bill completes $\frac{1}{3}$ of the job, so in $\frac{6}{5}$ hours, he completes $\frac{6}{5} \cdot \frac{1}{3} = \frac{2}{5}$ of the job. In one hour, Sandy completes $\frac{1}{2}$ of the job, so in $\frac{6}{5}$ hours, she completes $\frac{6}{5} \cdot \frac{1}{2} = \frac{3}{5}$ of the job. $\frac{2}{5} + \frac{3}{5} = 1$ (the entire job).

9. Let x = number of hours for the electrician to complete the job alone. Then $2x$ = the number of hours for the apprentice to complete the job alone.

$$6\left(\frac{1}{x}\right) + 6\left(\frac{1}{2x}\right) = 1$$

$$\frac{6}{1} \cdot \frac{1}{x} + \frac{\cancel{6}^{3}}{1} \cdot \frac{1}{\cancel{2}x} = 1$$

$$\frac{6}{x} + \frac{3}{x} = 1$$

$$\frac{9}{x} = 1$$

$$\frac{\cancel{x}}{1} \cdot \frac{9}{\cancel{x}} = x \cdot 1$$

$$9 = x$$

Thus, the electrician can complete the job in 9 hours, working alone. The apprentice can complete the job in $2x = 2(9) = 18$ hours, working alone.

CHECK: In one hour, the electrician completes $\frac{1}{9}$ of the job, so in 6 hours, he completes $\frac{6}{9} = \frac{2}{3}$ of the job. In one hour, the apprentice completes $\frac{1}{18}$ of the job, so in 6 hours, he completes $\frac{6}{18} = \frac{1}{3}$ of the job. $\frac{2}{3} + \frac{1}{3} = 1$ (the entire job).

11. Let r = rate of the train (in kph). Then $r + 100$ = rate of the plane (in kph).

$$\frac{300}{r} = \frac{500}{r + 100}$$

$$\frac{\cancel{r}(r + 100)}{1} \cdot \frac{300}{\cancel{r}} = \frac{r(\cancel{r + 100})}{1} \cdot \frac{500}{\cancel{r + 100}}$$

$$300(r + 100) = 500r$$

$$300r + 30000 = 500r$$

$$30000 = 200r$$

$$150 = r$$

Thus, the train travels at the rate of 150 kph and the plane travels at the rate of 250 kph.

CHECK: At the rate of 150 kph, the train covers 300 kilometers in $\frac{300}{150} = 2$ hours. At the rate of 250 kph, the plane covers 500 kilometers in $\frac{500}{250} = 2$ hours, the same amount of time.

13. Let x = speed of the current

$$\frac{8}{4 - x} = \frac{24}{4 + x}$$

$$\frac{(\cancel{4 - x})(4 + x)}{1} \cdot \frac{8}{(\cancel{4 - x})} = \frac{(4 - x)(\cancel{4 + x})}{1} \cdot \frac{24}{(\cancel{4 + x})}$$

$$8(4 + x) = 24(4 - x)$$

$$32 + 8x = 96 - 24x$$

$$32 + 32x = 96$$

$$32x = 64$$

$$x = 2$$

Thus, the speed of the current is 2 mph.

CHECK: time upstream = $\frac{8}{2}$ = 4 hours

time downstream = $\frac{24}{6}$ = 4 hours

The two times are equal.

15. Let $t =$ number of hours that Ronnie walks.
Then $3 - t =$ number of hours that Ronnie jogs.
$$6t = 14(3 - t)$$
$$6t = 42 - 14t$$
$$20t = 42$$
$$t = \frac{42}{20} = \frac{21}{10} = 2.1$$

Then the distance to Ronnie's friend's house is $6(2.1) = 12.6$ kilometers.
CHECK: Ronnie walks $6(2.1) = 12.6$ kilometers to his friend's house and jogs $14(3 - 2.1) = 14(0.9) = 12.6$ kilometers back, the same distance.

17. Let $x =$ smaller number.
Then $3x =$ larger number.
$$\frac{1}{x} + \frac{1}{3x} = \frac{5}{3}$$
$$3x\left(\frac{1}{x} + \frac{1}{3x}\right) = 3x\left(\frac{5}{3}\right)$$
$$\frac{3\not{x}}{1} \cdot \frac{1}{\not{x}} + \frac{\not{3x}}{1} \cdot \frac{1}{\not{3x}} = \frac{\not{3}x}{1} \cdot \frac{5}{\not{3}}$$
$$3 + 1 = 5x$$
$$4 = 5x$$
$$\frac{4}{5} = x$$

Then $3x = 3\left(\frac{4}{5}\right) = \frac{12}{5}$.
Thus, the numbers are $\frac{4}{5}$ and $\frac{12}{5}$.
CHECK: $\frac{12}{5}$ is three times $\frac{4}{5}$, and
$$\frac{1}{\left(\frac{12}{5}\right)} + \frac{1}{\left(\frac{4}{5}\right)} = \frac{5}{12} + \frac{5}{4}$$
$$= \frac{5}{12} + \frac{15}{12} = \frac{20}{12} = \frac{5}{3}.$$

19. Let $x =$ number to be added.
$$\frac{3 + x}{5 + x} = \frac{5}{6}$$
$$\frac{6(5 + x)}{1} \cdot \frac{3 + x}{5 + x} = \frac{\not{6}(5 + x)}{1} \cdot \frac{5}{\not{6}}$$
$$6(3 + x) = 5(5 + x)$$
$$18 + 6x = 25 + 5x$$
$$18 + x = 25$$
$$x = 7$$

Thus, the number to be added to both the numerator and denominator is 7.
CHECK: $\frac{3 + (7)}{5 + (7)} = \frac{10}{12} = \frac{5}{6}$.

21. Let $x =$ numerator of the fraction.
Then $x + 2 =$ denominator of the fraction.
$$\frac{x + 2}{x} = \frac{x}{x + 2}$$
$$\frac{\not{x}(x + 2)}{1} \cdot \frac{x + 2}{\not{x}} = \frac{x(\not{x + 2})}{1} \cdot \frac{x}{\not{x + 2}}$$
$$(x + 2)(x + 2) = x^2$$
$$x^2 + 4x + 4 = x^2$$
$$4x + 4 = 0$$
$$4x = -4$$
$$x = -1$$

Then $x + 2 = (-1) + 2 = 1$. Thus, the fraction is $\frac{-1}{1}$.
CHECK: 1 is two more than -1, and the reciprocal of $\frac{-1}{1}$ is $\frac{1}{-1}$, which is equal to the original fraction.

23. $S = 2\pi rh + 2\pi r^2 = 2\pi(6)(15) + 2\pi(6)^2 = 180\pi + 72\pi = 252\pi$ square centimeters

25.
$$\frac{1}{R_T} = \frac{1}{R_1} + \frac{1}{R_2}$$
$$\frac{1}{R_T} = \frac{1}{20} + \frac{1}{30}$$
$$60R_T\left(\frac{1}{R_T}\right) = 60R_T\left(\frac{1}{20} + \frac{1}{30}\right)$$
$$\frac{60\not{R_T}}{1} \cdot \frac{1}{\not{R_T}} = \frac{\overset{3}{\not{60}}R_T}{1} \cdot \frac{1}{\not{20}} + \frac{\overset{2}{\not{60}}R_T}{1} \cdot \frac{1}{\not{30}}$$
$$60 = 3R_T + 2R_T$$
$$60 = 5R_T$$
$$12 = R_T$$

The total resistance is 12 ohms.

27.
$$\frac{1}{f} = \frac{1}{d_s} + \frac{1}{d_i}$$
$$\frac{1}{f} = \frac{1}{6} + \frac{1}{3}$$
$$6f\left(\frac{1}{f}\right) = 6f\left(\frac{1}{6} + \frac{1}{3}\right)$$
$$\frac{6f}{1} \cdot \frac{1}{f} = \frac{\cancel{6}f}{1} \cdot \frac{1}{\cancel{6}} + \frac{\cancel{6}^2 f}{1} \cdot \frac{1}{\cancel{3}}$$
$$6 = f + 2f$$
$$6 = 3f$$
$$2 = f$$

The focal length of the lens is 2 cm.

29. Let h = number to hours Taisha needs to finish painting the room.
$$2\left(\frac{1}{5}\right) + h\left(\frac{1}{8}\right) = 1$$
$$\frac{2}{5} + \frac{h}{8} = 1$$
$$40\left(\frac{2}{5} + \frac{h}{8}\right) = 40 \cdot 1$$
$$\frac{\cancel{40}^8}{1} \cdot \frac{2}{\cancel{5}} + \frac{\cancel{40}^5}{1} \cdot \frac{h}{\cancel{8}} = 40$$
$$16 + 5h = 40$$
$$5h = 24$$
$$h = \frac{24}{5} \text{ or } 4\frac{4}{5}$$

Thus, it takes $2 + 4\frac{4}{5} = 6\frac{4}{5}$ hours to paint the room.

31. Let x = number of hours Tina works.
Then $x - 1$ = number of hours Latifa works.
$$x\left(\frac{1}{5}\right) + (x-1)\left(\frac{1}{7}\right) = 1$$
$$35\left[x\left(\frac{1}{5}\right) + (x-1)\left(\frac{1}{7}\right)\right] = 35 \cdot 1$$
$$\frac{\cancel{35}^7}{1} \cdot \frac{1}{\cancel{5}}x + \frac{\cancel{35}^5}{1} \cdot \frac{1}{\cancel{7}}(x-1) = 35$$
$$7x + 5(x-1) = 35$$
$$7x + 5x - 5 = 35$$
$$12x - 5 = 35$$
$$12x = 40$$
$$x = \frac{10}{3}$$

Thus, they will finish $\frac{10}{3}$ hours = 3 hr 20 min after Tina starts at 9:00 a.m. The time will be 12:20 p.m.

33. Let x = number of hours for Sean to complete the job alone.
Then $2x$ = number of hours Laura would need working alone.
$$8\left(\frac{1}{x}\right) + 5\left(\frac{1}{2x}\right) = 1$$
$$2x\left[8\left(\frac{1}{x}\right) + 5\left(\frac{1}{2x}\right)\right] = 2x \cdot 1$$
$$\frac{2\cancel{x}}{1} \cdot \frac{8}{\cancel{x}} + \frac{\cancel{2x}}{1} \cdot \frac{5}{\cancel{2x}} = 2x$$
$$16 + 5 = 2x$$
$$21 = 2x$$
$$\frac{21}{2} = x$$

It would take Sean $\frac{21}{2}$ hours = 10.5 hours working alone.

35. If someone can do a job in h hours, then he or she can complete $\frac{1}{h}$ of the job per hour. This is the rate at which the person works. When this rate is multiplied by the number of hours that the person works, we get the fraction part of the entire job that he or she accomplishes.

Chapter 9 Review Exercises

1. $\dfrac{x^2}{x^2+2}$ cannot be reduced.

3. $\dfrac{x^2+3x-4}{x^2-16} = \dfrac{\cancel{(x+4)}(x-1)}{(x-4)\cancel{(x+4)}} = \dfrac{x-1}{x-4}$

5. $\dfrac{a^2+8a+16}{a^2+6a+8} = \dfrac{(a+4)\cancel{(a+4)}}{(a+2)\cancel{(a+4)}}$
$= \dfrac{a+4}{a+2}$

7. $\dfrac{3z^2-12}{3z^2+9z+6} = \dfrac{3(z^2-4)}{3(z^2+3z+2)}$
$= \dfrac{\cancel{3}(z-2)\cancel{(z+2)}}{\cancel{3}(z+1)\cancel{(z+2)}} = \dfrac{z-2}{z+1}$

9. $\dfrac{x}{x+2} + \dfrac{x}{2} = \dfrac{x(2)}{(x+2)(2)} + \dfrac{x(x+2)}{2(x+2)}$
$= \dfrac{2x}{2(x+2)} + \dfrac{x^2+2x}{2(x+2)}$
$= \dfrac{2x+x^2+2x}{2(x+2)}$
$= \dfrac{x^2+4x}{2(x+2)} = \dfrac{x(x+4)}{2(x+2)}$

11. $\dfrac{3}{2x+4} + \dfrac{6}{x^2+2x} = \dfrac{3}{2(x+2)} + \dfrac{6}{x(x+2)}$
$= \dfrac{3(x)}{2(x+2)(x)} + \dfrac{6(2)}{x(x+2)(2)}$
$= \dfrac{3x}{2x(x+2)} + \dfrac{12}{2x(x+2)}$
$= \dfrac{3x+12}{2x(x+2)} = \dfrac{3(x+4)}{2x(x+2)}$

13. $\dfrac{x^2-5x-6}{2x-12} \div \dfrac{x^2+2x+1}{8x^2} = \dfrac{x^2-5x-6}{2x-12} \cdot \dfrac{8x^2}{x^2+2x+1}$
$= \dfrac{\cancel{(x-6)}\cancel{(x+1)}}{\cancel{2}\cancel{(x-6)}} \cdot \dfrac{\overset{4}{\cancel{8}}x^2}{\cancel{(x+1)}(x+1)} = \dfrac{4x^2}{x+1}$

15. $\dfrac{5}{z^2+z-6} - \dfrac{3}{z^2+3z}$
$= \dfrac{5}{(z+3)(z-2)} - \dfrac{3}{z(z+3)}$
$= \dfrac{5(z)}{(z+3)(z-2)(z)} - \dfrac{3(z-2)}{z(z+3)(z-2)}$
$= \dfrac{5z}{z(z+3)(z-2)} - \dfrac{3z-6}{z(z+3)(z-2)}$
$= \dfrac{5z-(3z-6)}{z(z+3)(z-2)} = \dfrac{5z-3z+6}{z(z+3)(z-2)}$
$= \dfrac{2z+6}{z(z+3)(z-2)} = \dfrac{2\cancel{(z+3)}}{z\cancel{(z+3)}(z-2)}$
$= \dfrac{2}{z(z-2)}$

17. $2 + \dfrac{3}{x+2} - \dfrac{1}{x}$
$= \dfrac{2}{1} + \dfrac{3}{x+2} - \dfrac{1}{x}$
$= \dfrac{2[x(x+2)]}{1[x(x+2)]} + \dfrac{3(x)}{(x+2)(x)} - \dfrac{1(x+2)}{x(x+2)}$
$= \dfrac{2x^2+4x}{x(x+2)} + \dfrac{3x}{x(x+2)} - \dfrac{x+2}{x(x+2)}$
$= \dfrac{2x^2+4x+3x-(x+2)}{x(x+2)}$
$= \dfrac{2x^2+4x+3x-x-2}{x(x+2)}$
$= \dfrac{2x^2+6x-2}{x(x+2)} = \dfrac{2(x^2+3x-1)}{x(x+2)}$

19.
$$\frac{5}{x} - \frac{2}{3x} = \frac{13}{6}$$

$$6x\left(\frac{5}{x} - \frac{2}{3x}\right) = 6x\left(\frac{13}{6}\right)$$

$$\frac{6\not{x}}{1} \cdot \frac{5}{\not{x}} - \frac{\overset{2}{\not{6}\not{x}}}{1} \cdot \frac{2}{\not{3}\not{x}} = \frac{\not{6}x}{1} \cdot \frac{13}{\not{6}}$$

$$30 - 4 = 13x$$
$$26 = 13x$$
$$2 = x$$

21.
$$\frac{y+2}{y} + \frac{4}{y+2} = \frac{y}{y+2}$$

$$y(y+2)\left(\frac{y+2}{y} + \frac{4}{y+2}\right) = y(y+2)\left(\frac{y}{y+2}\right)$$

$$\frac{\not{y}(y+2)}{1} \cdot \frac{y+2}{\not{y}} + \frac{y(\not{y+2})}{1} \cdot \frac{4}{\not{y+2}} = \frac{y(\not{y+2})}{1} \cdot \frac{y}{\not{y+2}}$$

$$(y+2)(y+2) + 4y = y^2$$
$$y^2 + 4y + 4 + 4y = y^2$$
$$y^2 + 8y + 4 = y^2$$
$$8y + 4 = 0$$
$$8y = -4$$
$$y = \frac{-4}{8} = -\frac{1}{2}$$

23.
$$\frac{x+2}{x-3} + \frac{4}{3} = \frac{5}{x-3}$$

$$3(x-3)\left(\frac{x+2}{x-3} + \frac{4}{3}\right) = 3(x-3)\left(\frac{5}{x-3}\right)$$

$$\frac{3(\not{x-3})}{1} \cdot \frac{x+2}{\not{x-3}} - \frac{\not{3}(x-3)}{1} \cdot \frac{4}{\not{3}} = \frac{3(\not{x-3})}{1} \cdot \frac{5}{\not{x-3}}$$

$$3(x+2) + 4(x-3) = 15$$
$$3x + 6 + 4x - 12 = 15$$
$$7x - 6 = 15$$
$$7x = 21$$
$$x = 3$$

This value does not check in the original equation, since it produces fractions with denominators of 0. Thus, the equation has no solution.

25.
$$\begin{aligned} 3x - 4y + 7 &= 8x - 7y + 3 \\ -8x \qquad\qquad &\quad -8x \\ \hline -5x - 4y + 7 &= \qquad -7y + 3 \\ +4y \qquad\qquad &\quad +4y \\ \hline -5x \qquad + 7 &= \qquad -3y + 3 \\ -7 \qquad &\quad -7 \\ \hline -5x \qquad\qquad &= \qquad -3y - 4 \\ \frac{\not{-5}x}{\not{-5}} &= \frac{-3y-4}{-5} \\ x &= \frac{-3y-4}{-5} \\ &\text{or } \frac{3y+4}{5} \end{aligned}$$

27. Let x = number of black marbles in the bag.

Then $x + 80$ = number of red marbles in the bag.

$$\frac{x+80}{x} = \frac{7}{5}$$

$$\frac{5\not{x}}{1} \cdot \frac{x+80}{\not{x}} = \frac{\not{5}x}{1} \cdot \frac{7}{\not{5}}$$

$$5(x+80) = 7x$$
$$5x + 400 = 7x$$
$$400 = 2x$$
$$200 = x$$

Then $x + 80 = 200 + 80 = 280$.
Thus, there are 200 black marbles and 280 red marbles in the bag, or 480 marbles altogether.
CHECK: 280 is 80 more than 200, and $\frac{280}{200} = \frac{7}{5}$.

29. Let $x =$ number of hours John needs to do the job alone.

$$4\left(\frac{1}{x}\right) + 4\left(\frac{1}{6}\right) = 1$$

$$\frac{4}{x} + \frac{4}{6} = 1$$

$$6x\left(\frac{4}{x} + \frac{4}{6}\right) = 6x \cdot 1$$

$$\frac{6\cancel{x}}{1} \cdot \frac{4}{\cancel{x}} + \frac{\cancel{6}x}{1} \cdot \frac{4}{\cancel{6}} = 6x$$

$$24 + 4x = 6x$$

$$24 = 2x$$

$$12 = x$$

Thus, John needs 12 hours to do the job alone.

CHECK: In 4 hours, Susan does $4\left(\frac{1}{6}\right) = \frac{4}{6} = \frac{2}{3}$ of the job; in 4 hours, John does $4\left(\frac{1}{12}\right) = \frac{4}{12} = \frac{1}{3}$ of the job. $\frac{2}{3} + \frac{1}{3} = 1$ (the entire job).

31. Let $w =$ width of the rectangle. Then $2w - 5 =$ length of the rectangle.

$$\frac{2w - 5}{w} = \frac{5}{3}$$

$$3\cancel{w}\left(\frac{2w - 5}{\cancel{w}}\right) = \cancel{3}w\left(\frac{5}{\cancel{3}}\right)$$

$$3(2w - 5) = 5w$$

$$6w - 15 = 5w$$

$$-15 = -w$$

$$15 = w$$

Then $2w - 5 = 2(15) - 5$
$= 30 - 5 = 25$.

So the dimensions are 15 by 25.

CHECK: $\frac{25}{15} = \frac{5}{3}$
and $25 = 2(15) - 5$.

Chapter 9 Practice Test

1. $\dfrac{4x^2}{x^2 - 4x} = \dfrac{4\overset{x}{\cancel{x^2}}}{\cancel{x}(x - 4)} = \dfrac{4x}{x - 4}$

3. $\dfrac{3}{x + 3} + \dfrac{2}{x + 2}$

$= \dfrac{3(x + 2)}{(x + 3)(x + 2)} + \dfrac{2(x + 3)}{(x + 2)(x + 3)}$

$= \dfrac{3x + 6}{(x + 3)(x + 2)} + \dfrac{2x + 6}{(x + 3)(x + 2)}$

$= \dfrac{3x + 6 + 2x + 6}{(x + 3)(x + 2)} = \dfrac{5x + 12}{(x + 3)(x + 2)}$

5. $\dfrac{5}{2x} - \dfrac{10}{x^2 + 4x}$

$= \dfrac{5}{2x} - \dfrac{10}{x(x + 4)}$

$= \dfrac{5(x + 4)}{2x(x + 4)} - \dfrac{10(2)}{x(x + 4)(2)}$

$= \dfrac{5x + 20}{2x(x + 4)} - \dfrac{20}{2x(x + 4)}$

$= \dfrac{5x + 20 - 20}{2x(x + 4)} = \dfrac{5\cancel{x}}{2\cancel{x}(x + 4)} = \dfrac{5}{2(x + 4)}$

7. $\dfrac{2x - 5}{x^2 - 3x} - \dfrac{3x + 7}{x^2 - 3x} + \dfrac{6x - 3}{x^2 - 3x} = \dfrac{2x - 5 - (3x + 7) + 6x - 3}{x^2 - 3x} = \dfrac{2x - 5 - 3x - 7 + 6x - 3}{x^2 - 3x}$

$= \dfrac{5x - 15}{x^2 - 3x} = \dfrac{5\cancel{(x - 3)}}{x\cancel{(x - 3)}} = \dfrac{5}{x}$

9.
$$at + b = \frac{3t}{2} + 7$$
$$2(at + b) = 2\left(\frac{3t}{2} + 7\right)$$
$$2at + 2b = 3t + 14$$
$$\underline{-3t-3t}$$
$$2at - 3t + 2b = 14$$
$$\underline{-2b-2b}$$
$$2at - 3t = 14 - 2b$$
$$(2a - 3)t = 14 - 2b$$
$$t = \frac{14 - 2b}{2a - 3}$$

11. Let t = number of hours needed to go from A to B.
Then $14 - t$ = number of hours needed to go from B to A.
$$45t = 60(14 - t)$$
$$45t = 840 - 60t$$
$$105t = 840$$
$$t = 8$$
Thus, the distance between the towns is $45(8) = 360$ kilometers.
CHECK: Time going $= \frac{360}{45} = 8$. Time returning $= \frac{360}{60} = 6$ hours. $8 + 6 \stackrel{\checkmark}{=} 14$ hours.

13. Let x = number of hours Haleema works.
Then $9 - x$ = number of hours Andrea works.
$$x\left(\frac{1}{8}\right) + (9 - x)\left(\frac{1}{12}\right) = 1$$
$$\frac{x}{8} + \frac{9 - x}{12} = 1$$
$$24\left(\frac{x}{8} + \frac{9 - x}{12}\right) = 24 \cdot 1$$
$$\frac{\overset{3}{\cancel{24}}}{1} \cdot \frac{x}{\cancel{8}} + \frac{\overset{2}{\cancel{24}}}{1} \cdot \frac{9 - x}{\cancel{12}} = 24$$
$$3x + 2(9 - x) = 24$$
$$3x + 18 - 2x = 24$$
$$x + 18 = 24$$
$$x = 6$$
Then $9 - x = 9 - (6) = 3$. So Haleema works on the puzzle for 6 hours and Andrea works on the puzzle for 3 hours.
CHECK: In 6 hours, Haleema completes $6\left(\frac{1}{8}\right) = \frac{6}{8} = \frac{3}{4}$ of the puzzle; in 3 hours, Andrea completes $3\left(\frac{1}{12}\right) = \frac{3}{12} = \frac{1}{4}$ of the puzzle. $\frac{3}{4} + \frac{1}{4} \stackrel{\checkmark}{=} 1$.

Chapters 7-9
Cumulative Review

1. $(x+8)(x-5)$
   $= x^2 - 5x + 8x - 40$
   $= x^2 + 3x - 40$

3. $(a+b+c)(a+b-c)$
   $= a(a+b-c) + b(a+b-c) + c(a+b-c)$
   $= a^2 + ab - ac + ab + b^2 - bc + ac + bc - c^2$
   $= a^2 + 2ab + b^2 - c^2$

5. $(5a-3c)(4a+3c)$
   $= 20a^2 + 15ac - 12ac - 9c^2$
   $= 20a^2 + 3ac - 9c^2$

7. $(x+3)(x-12) + (x+6)^2$
   $= (x+3)(x-12) + (x+6)(x+6)$
   $= x^2 - 12x + 3x - 36 + x^2 + 6x + 6x + 36$
   $= 2x^2 + 3x$

9. $(a-3)(2a+3)(2a-3)$
   $= (a-3)(4a^2 - 6a + 6a - 9)$
   $= (a-3)(4a^2 - 9)$
   $= 4a^3 - 9a - 12a^2 + 27$
   $= 4a^3 - 12a^2 - 9a + 27$

11. $(x^2 - xy + 3y^2) + (5x^2 - 8y^2) + (y^2 - 6x^2)$
    $= x^2 + 5x^2 - 6x^2 - xy + 3y^2 - 8y^2 + y^2$
    $= -xy - 4y^2$

13. (a) degree: 4
    (b) coefficient: -3.

15. $\dfrac{x^2 + 8x}{2x} = \dfrac{\cancel{x^2}}{2\cancel{x}} + \dfrac{\overset{4}{\cancel{8x}}}{\cancel{2x}} = \dfrac{x}{2} + 4$ or $\dfrac{x+8}{2}$

17. $\begin{array}{r} y - 1 \\ y-2 \overline{\smash{\big)}\, y^2 - 3y + 4} \\ \underline{-(y^2 - 2y)} \\ -y + 4 \\ \underline{-(-y + 2)} \\ 2 \end{array}$

19. $\begin{array}{r} 6x^2 + 8x + 9 \\ 3x-4 \overline{\smash{\big)}\, 18x^3 + 0x^2 - 5x - 28} \\ \underline{-(18x^3 - 24x^2)} \\ 24x^2 - 5x \\ \underline{-(24x^2 - 32x)} \\ 27x - 28 \\ \underline{-(27x - 36)} \\ 8 \end{array}$

21. $5^0 + 2^{-3} + 2^{-4}$
    $= 1 + \dfrac{1}{2^3} + \dfrac{1}{2^4}$
    $= 1 + \dfrac{1}{8} + \dfrac{1}{16} = \dfrac{16}{16} + \dfrac{2}{16} + \dfrac{1}{16}$
    $= \dfrac{16 + 2 + 1}{16} = \dfrac{19}{16}$

23. $\dfrac{(x^2)^3}{x^2 x^3} = \dfrac{x^6}{x^5} = x^{6-5} = x$

25. $\dfrac{(2x^3)^4}{4(x^5)^3} = \dfrac{2^4(x^3)^4}{4(x^5)^3} = \dfrac{16x^{12}}{4x^{15}}$

$= \dfrac{16}{4} \cdot \dfrac{x^{12}}{x^{15}} = \dfrac{4}{1} \cdot x^{-3}$

$= \dfrac{4}{1} \cdot \dfrac{1}{x^3} = \dfrac{4}{x^3}$

27. $\dfrac{(3a^{-3}t^2)^{-3}}{(a^{-1}t^{-2})^2} = \dfrac{3^{-3}(a^{-3})^{-3}(t^2)^{-3}}{(a^{-1})^2(t^{-2})^2} = \dfrac{3^{-3}a^9t^{-6}}{a^{-2}t^{-4}}$

$= \dfrac{1}{3^3} \cdot \dfrac{a^9}{a^{-2}} \cdot \dfrac{t^{-6}}{t^{-4}} = \dfrac{1}{27}a^{11}t^{-2}$

$= \dfrac{1}{27} \cdot \dfrac{a^{11}}{1} \cdot \dfrac{1}{t^2} = \dfrac{a^{11}}{27t^2}$

29. $0.000439 = 4.39 \times 10^{-4}$

31. $\dfrac{(4 \times 10^{-3})(5 \times 10^4)}{2 \times 10^{-3}} = \dfrac{\cancel{(4)}^2(5)}{\cancel{2}_1} \times \dfrac{10^{-3}10^4}{10^{-3}} = 10 \times \dfrac{10^1}{10^{-3}} = 10 \times 10^4 = 1 \times 10^5 \text{ or } 10^5$

33. $x^2 + 6x + 5 = (x+1)(x+5)$
35. $x^2 - 5x + 6 = (x-2)(x-3)$
37. $6x^3y - 12xy^2 - 9x^2y = 3xy(2x^2 - 4y - 3x)$
39. $u^2 - 49 = (u-7)(u+7)$
41. $2r^2 + r - 15 = (2r-5)(r+3)$
43. $5t^2 + 10t + 15 = 5(t^2 + 2t + 3)$
45. $6x^2 - 17xy + 12y^2 = (3x-4y)(2x-3y)$
47. $x^2 + 16x = x(x+16)$

49. $x^2 + ax + xy + ay$
$= (x^2 + ax) + (xy + ay)$
$= x(x+a) + y(x+a)$
$= (x+a)(x+y)$

51. $x^2 - 4x - ax + 4a$
$= (x^2 - 4x) + (-ax + 4a)$
$= x(x-4) - a(x-4)$
$= (x-4)(x-a)$

53. $\dfrac{x^2 - 4x}{x^2 - 16} = \dfrac{x\cancel{(x-4)}}{(x+4)\cancel{(x-4)}} = \dfrac{x}{x+4}$

55. $\dfrac{3}{4x} + \dfrac{5}{x+4} = \dfrac{3(x+4)}{4x(x+4)} + \dfrac{5(4x)}{(x+4)(4x)} = \dfrac{3x+12}{4x(x+4)} + \dfrac{20x}{4x(x+4)}$

$= \dfrac{3x+12+20x}{4x(x+4)} = \dfrac{23x+12}{4x(x+4)}$

57. $\dfrac{x^2-5x}{10x} \cdot \dfrac{x^2}{x^2-25} = \dfrac{x\cancel{(x-5)}}{10\cancel{x}} \cdot \dfrac{\cancel{x^2}^{\,x}}{\cancel{(x-5)}(x+5)} = \dfrac{x^2}{10(x+5)}$

59. $\dfrac{6}{x^2+2x} - \dfrac{4}{x^2-2x} = \dfrac{6}{x(x+2)} - \dfrac{4}{x(x-2)} = \dfrac{6(x-2)}{x(x+2)(x-2)} - \dfrac{4(x+2)}{x(x-2)(x+2)}$

$= \dfrac{6x-12}{x(x+2)(x-2)} - \dfrac{4x+8}{x(x+2)(x-2)}$

$= \dfrac{6x-12-(4x+8)}{x(x+2)(x-2)} = \dfrac{6x-12-4x-8}{x(x+2)(x-2)}$

$= \dfrac{2x-20}{x(x+2)(x-2)} = \dfrac{2(x-10)}{x(x+2)(x-2)}$

61. $\dfrac{\dfrac{a}{2}-\dfrac{8}{a}}{\dfrac{a^2-8a+16}{4}} = \dfrac{4a\left(\dfrac{a}{2}-\dfrac{8}{a}\right)}{4a\left(\dfrac{a^2-8a+16}{4}\right)}$

$= \dfrac{\dfrac{\overset{2}{\cancel{4a}}}{1}\cdot\dfrac{a}{\cancel{2}} - \dfrac{\cancel{4a}}{1}\cdot\dfrac{8}{\cancel{a}}}{\dfrac{\cancel{4a}}{1}\cdot\dfrac{a^2-8a+16}{\cancel{4}}}$

$= \dfrac{2a^2-32}{a(a^2-8a+16)}$

$= \dfrac{2(a^2-16)}{a(a^2-8a+16)}$

$= \dfrac{2\cancel{(a-4)}(a+4)}{a\cancel{(a-4)}(a-4)} = \dfrac{2(a+4)}{a(a-4)}$

63. $\dfrac{2z+9}{4z+12} - \dfrac{5z+8}{4z+12} + \dfrac{3z+1}{4z+12}$

$= \dfrac{2z+9-(5z+8)+3z+1}{4z+12}$

$= \dfrac{2z+9-5z-8+3z+1}{4z+12}$

$= \dfrac{2}{4z+12} = \dfrac{\overset{1}{\cancel{2}}}{\underset{2}{\cancel{4}}(z+3)} = \dfrac{1}{2(z+3)}$

65. $\dfrac{4}{3a+6} - \dfrac{3}{2} = \dfrac{5}{6a+12}$

$\dfrac{4}{3(a+2)} - \dfrac{3}{2} = \dfrac{5}{6(a+2)}$

$6(a+2)\left(\dfrac{4}{3(a+2)} - \dfrac{3}{2}\right) = 6(a+2)\left(\dfrac{5}{6(a+2)}\right)$

$\dfrac{\overset{2}{\cancel{6(a+2)}}}{1}\cdot\dfrac{4}{\cancel{3(a+2)}} - \dfrac{\overset{3}{\cancel{6}}(a+2)}{1}\cdot\dfrac{3}{\cancel{2}} = \dfrac{\cancel{6(a+2)}}{1}\cdot\dfrac{5}{\cancel{6(a+2)}}$

$8 - 9(a+2) = 5$
$8 - 9a - 18 = 5$
$-9a - 10 = 5$
$-9a = 15$
$a = \dfrac{15}{-9} = -\dfrac{5}{3}$

67. $\dfrac{3y}{5} - a = 4y + 3a - 6$

$5\left(\dfrac{3y}{5} - a\right) = 5(4y + 3a - 6)$

$\begin{array}{rl} 3y - 5a = & 20y + 15a - 30 \\ -20y & -20y \\ \hline -17y - 5a = & 15a - 30 \\ +5a & +5a \\ \hline -17y = & 20a - 30 \end{array}$

$y = \dfrac{20a - 30}{-17}$

$y = \dfrac{30 - 20a}{17}$

69. $\dfrac{x+8}{8} - \dfrac{x+6}{6} = \dfrac{x+3}{3} - \dfrac{x+4}{4}$

$24\left(\dfrac{x+8}{8} - \dfrac{x+6}{6}\right) = 24\left(\dfrac{x+3}{3} - \dfrac{x+4}{4}\right)$

$\dfrac{\overset{3}{\cancel{24}}}{1}\cdot\dfrac{x+8}{\cancel{8}} - \dfrac{\overset{4}{\cancel{24}}}{1}\cdot\dfrac{x+6}{\cancel{6}} = \dfrac{\overset{8}{\cancel{24}}}{1}\cdot\dfrac{x+3}{\cancel{3}} - \dfrac{\overset{6}{\cancel{24}}}{1}\cdot\dfrac{x+4}{\cancel{4}}$

$3(x+8) - 4(x+6) = 8(x+3) - 6(x+4)$
$3x + 24 - 4x - 24 = 8x + 24 - 6x - 24$
$-x = 2x$
$0 = 3x$
$0 = x$

71.
$$\frac{3c+1}{3c-2} = \frac{3c}{3c-1}$$
$$\frac{(3c-2)(3c-1)}{1}\left(\frac{3c+1}{3c-2}\right) = \frac{(3c-2)(3c-1)}{1}\left(\frac{3c}{3c-1}\right)$$
$$(3c-1)(3c+1) = (3c-2)(3c)$$
$$9c^2 + 3c - 3c - 1 = 9c^2 - 6c$$
$$9c^2 - 1 = 9c^2 - 6c$$
$$\underline{-9c^2 \qquad\quad -9c^2}$$
$$-1 = -6c$$
$$\frac{1}{6} = c$$

73.
$$3x^2 - 5x - 7 = x^2 - 6x + 8$$
$$\underline{-x^2 \qquad\qquad -x^2}$$
$$2x^2 - 5x - 7 = -6x + 8$$
$$\underline{+6x \qquad\qquad +6x}$$
$$2x^2 + x - 7 = 8$$
$$\underline{-8 \qquad -8}$$
$$2x^2 + x - 15 = 0$$
$$(2x - 5)(x + 3) = 0$$
$$2x - 5 = 0 \quad \text{or} \quad x + 3 = 0$$
$$2x = 5 \quad \text{or} \quad x = -3$$
$$x = \frac{5}{2}$$

75.
$$(x+4)(x-6) = (x-2)(x+5)$$
$$x^2 - 6x + 4x - 24 = x^2 + 5x - 2x - 10$$
$$x^2 - 2x - 24 = x^2 + 3x - 10$$
$$\underline{-x^2 \qquad\qquad -x^2}$$
$$-2x - 24 = 3x - 10$$
$$\underline{+2x \qquad\qquad +2x}$$
$$-24 = 5x - 10$$
$$\underline{+10 \qquad\qquad +10}$$
$$-14 = 5x$$
$$-\frac{14}{5} = x$$

77. Let x = number of hours Pat needs working alone.
$$5\left(\frac{1}{8}\right) + 5\left(\frac{1}{x}\right) = 1$$
$$\frac{5}{8} + \frac{5}{x} = 1$$
$$8x\left(\frac{5}{8} + \frac{5}{x}\right) = 8x(1)$$
$$8x\left(\frac{5}{8}\right) + 8x\left(\frac{5}{x}\right) = 8x$$
$$5x + 40 = 8x$$
$$40 = 3x$$
$$\frac{40}{3} = x$$

So Pat needs $\frac{40}{3}$ or $13\frac{1}{3}$ hours to overhaul the engine working alone.

Chapters 7-9
Cumulative Practice Test

1. $(x - 2y)(x^2 - 3xy - y^2)$
    $= x(x^2 - 3xy - y^2) - 2y(x^2 - 3xy - y^2)$
    $= x^3 - 3x^2y - xy^2 - 2x^2y + 6xy^2 + 2y^3$
    $= x^3 - 5x^2y + 5xy^2 + 2y^3$

3. $2a(3a - 5) + (a - 6)(a - 4)$
    $= 6a^2 - 10a + a^2 - 4a - 6a + 24$
    $= 7a^2 - 20a + 24$

5. $\dfrac{(2x^3)^{-4}}{(x^{-3})^3} = \dfrac{2^{-4}(x^3)^{-4}}{(x^{-3})^3} = \dfrac{x^{-12}}{2^4 \cdot x^{-9}}$

$= \dfrac{x^{-3}}{16} = \dfrac{1}{16x^3}$

7.
$$\begin{array}{r}
4x^2 + 5x + 15 \\
x-2\overline{\smash{\big)}\,4x^3 - 3x^2 + 5x - 20} \\
\underline{-(4x^3 - 8x^2)} \\
5x^2 + 5x \\
\underline{-(5x^2 - 10x)} \\
15x - 20 \\
\underline{-(15x - 30)} \\
10
\end{array}$$

9. (a) $0.000916 = 9.16 \times 10^{-4}$

   (b) $916,000 = 9.16 \times 10^5$

11. $x^3 - x^2 - x + 1 - (x^3 - 3x + x^2 - 5x) = x^3 - x^2 - x + 1 - (x^3 + x^2 - 8x)$

$= x^3 - x^2 - x + 1 - x^3 - x^2 + 8x$

$= -2x^2 + 7x + 1$

13. $6a^2b^5 - 3ab^3 = 3ab^3(2ab^2 - 1)$

15. $6x^2 - 36x + 72 = 6(x^2 - 6x + 12)$

17. $a(a+5) - 7(a+5) = (a+5)(a-7)$

19. $2u^4 - 32 = 2(u^4 - 16)$

$= 2(u^2 + 4)(u^2 - 4)$

$= 2(u^2 + 4)(u + 2)(u - 2)$

21. $\dfrac{t^2 - t - 6}{t^2 + t - 6} = \dfrac{(t-3)(t+2)}{(t+3)(t-2)}$

(cannot be reduced)

23. $\dfrac{w^2 - 3w - 10}{4w^2 + 8w} \cdot \dfrac{w^2}{w^2 - 10w + 25} = \dfrac{\cancel{(w-5)}\cancel{(w+2)}}{4\cancel{w}\cancel{(w+2)}} \cdot \dfrac{\overset{w}{\cancel{w^2}}}{\cancel{(w-5)}(w-5)}$

$= \dfrac{w}{4(w-5)}$

25. $\dfrac{u^2 - 9u}{u^2} \div (u^2 - 81) = \dfrac{u^2 - 9u}{u^2} \cdot \dfrac{1}{u^2 - 81}$

$= \dfrac{\overset{1}{\cancel{u}}\cancel{(u-9)}}{\underset{u}{\cancel{u^2}}} \cdot \dfrac{1}{(u+9)\cancel{(u-9)}}$

$= \dfrac{1}{u(u+9)}$

27.
$$\frac{9}{4t-12} - \frac{2}{3} = \frac{11}{12t-36}$$
$$\frac{9}{4(t-3)} - \frac{2}{3} = \frac{11}{12(t-3)}$$
$$12(t-3)\left(\frac{9}{4(t-3)} - \frac{2}{3}\right) = 12(t-3)\left(\frac{11}{12(t-3)}\right)$$
$$\frac{\cancel{12(t-3)}^3}{1} \cdot \frac{9}{\cancel{4(t-3)}_1} - \frac{\cancel{12(t-3)}^4}{1} \cdot \frac{2}{\cancel{3}_1} = \frac{\cancel{12(t-3)}^1}{1} \cdot \frac{11}{\cancel{12(t-3)}_1}$$
$$27 - 8(t-3) = 11$$
$$27 - 8t + 24 = 11$$
$$-8t + 51 = 11$$
$$-8t = -40$$
$$t = 5$$

29.
$$\frac{10}{x+4} + \frac{3}{5} = \frac{6-x}{x+4}$$
$$5(x+4)\left(\frac{10}{x+4} + \frac{3}{5}\right) = 5(x+4)\left(\frac{6-x}{x+4}\right)$$
$$\frac{5\cancel{(x+4)}}{1} \cdot \frac{10}{\cancel{(x+4)}_1} + \frac{\cancel{5}(x+4)}{1} \cdot \frac{3}{\cancel{5}_1} = \frac{5\cancel{(x+4)}}{1} \cdot \frac{6-x}{\cancel{(x+4)}_1}$$
$$50 + 3(x+4) = 5(6-x)$$
$$50 + 3x + 12 = 30 - 5x$$
$$3x + 62 = 30 - 5x$$
$$8x + 62 = 30$$
$$8x = -32$$
$$x = -4$$

When this value of x is checked in the original equation, it produces fractions with zero denominators. Therefore, the equation has no solution.

31. Let $x =$ Roni's speed (in kph).

Then $x + 5 =$ Lamar's speed (in kph).
$$\frac{140}{x} = \frac{160}{x+5}$$
$$\frac{\cancel{x}(x+5)}{1} \cdot \frac{140}{\cancel{x}} = \frac{x\cancel{(x+5)}}{1} \cdot \frac{160}{\cancel{x+5}}$$
$$140(x+5) = 160x$$
$$140x + 700 = 160x$$
$$700 = 20x$$
$$35 = x$$

Since Roni drives 140 km at a speed of 35 kph, her driving time is 4 hours.

Chapter 10
Radical Expressions

Exercises 10.1

1. $\sqrt{4} = 2$ because $2 \cdot 2 = 4$
3. $-\sqrt{4} = -2$
5. $\sqrt{-4}$ is undefined
7. $\sqrt{25} = 5$ because $5 \cdot 5 = 25$
9. $-\sqrt{100} = -10$
11. $\sqrt{64} = 8$ because $8 \cdot 8 = 64$
13. $\sqrt{121} = 11$ because $11 \cdot 11 = 121$
15. $\sqrt{169} = 13$ because $13 \cdot 13 = 169$
17. $\sqrt{225} = 15$ because $15 \cdot 15 = 225$
19. $\sqrt{289} = 17$ because $17 \cdot 17 = 289$
21. $\sqrt{361} = 19$ because $19 \cdot 19 = 361$
23. $\sqrt{3}\sqrt{3} = 3$
25. $\sqrt{29}\sqrt{29} = 29$
27. $\left(\sqrt{11}\right)^2 = 11$
29. $\left(\sqrt{33}\right)^2 = 33$
31. $\sqrt{x}\sqrt{x} = x$
33. $\left(\sqrt{7}\right)^4 = \underbrace{\sqrt{7} \cdot \sqrt{7}}_{7} \cdot \underbrace{\sqrt{7} \cdot \sqrt{7}}_{7}$
    $= 7 \cdot 7$
    $= 49$
35. $\left(\sqrt{7}\right)^5 = \underbrace{\sqrt{7} \cdot \sqrt{7}}_{7} \cdot \underbrace{\sqrt{7} \cdot \sqrt{7}}_{7} \cdot \sqrt{7}$
    $= 7 \cdot 7 \cdot \sqrt{7}$
    $= 49\sqrt{7}$
37. $\sqrt{25 - 9} = \sqrt{16} = 4$
39. $\sqrt{25} - \sqrt{9} = 5 - 3 = 2$
41. $\left(\sqrt{25} - \sqrt{9}\right)^2 = (5-3)^2 = 2^2 = 4$
43. $\left(\sqrt{25-9}\right)^2 = \left(\sqrt{16}\right)^2 = 16$
45. 20.62
47. 25.24
49. 8.58
51. 1.45
53. 0.19
55. $\sqrt{17} = 4.1$ correct to 1 place
    $= 4.12$ correct to 2 places
    $= 4.123$ correct to 3 places
57. $\sqrt{110} = 10.5$ correct to 1 place
    $= 10.49$ correct to 2 places
    $= 10.488$ correct to 3 places
59. Since $64 < 73 < 81$,
    $\sqrt{64} < \sqrt{73} < \sqrt{81}$. That is,
    $8 < \sqrt{73} < 9$. $\sqrt{73} = 8.54$,
    rounded to two decimal places.
61. Since $196 < 217 < 225$,
    $\sqrt{196} < \sqrt{217} < \sqrt{225}$. That is,
    $14 < \sqrt{217} < 15$. $\sqrt{217} = 14.73$,
    rounded to two decimal places.
63. Since $625 < 673 < 676$,
    $\sqrt{625} < \sqrt{673} < \sqrt{676}$. That is,
    $25 < \sqrt{673} < 26$. $\sqrt{673} = 25.94$,
    rounded to two decimal places.
65. Since $256 < 285 < 289$,
    $\sqrt{256} < \sqrt{285} < \sqrt{289}$. That is,
    $16 < \sqrt{285} < 17$. $\sqrt{285} = 16.88$,
    rounded to two decimal places.

67. Since $20^2 = 400$, $30^2 = 900$, and 648 is between 400 and 900, $\sqrt{648}$ must be between 20 and 30. Consider the following table:
$0^2 = 0 \quad 1^2 = 1 \quad 2^2 = 4 \quad 3^2 = 9 \quad 4^2 = 16$
$5^2 = 25 \quad 6^2 = 36 \quad 7^2 = 49 \quad 8^2 = 64 \quad 9^2 = 81$
This implies that any perfect square must end in one of the digits underlined: 0, 1, 4, 9, 6, or 5. Therefore, any number that ends in either 2, 3, 7, or 8 cannot be a perfect square. So 648 cannot be a perfect square.

68. Using the same argument as in (67), $\sqrt{841}$ must be between 20 and 30. Since 841 ends in the digit 1, there are only two possibilities, if 841 is a perfect square: $(21)^2 = 841$ or $(29)^2 = 841$. Since 841 is closer to 900 than it is to 400, we try $(29)^2$ first, since 29 is closer to 30 than it is to 20. Since $29 \cdot 29 = 841$, we conclude that 841 is a perfect square, and that $\sqrt{841} = 29$.

69. It is incorrect to claim that $\sqrt{1+1} = \sqrt{1} + \sqrt{1}$. In fact, if a and b are positive, it is never true that $\sqrt{a+b} = \sqrt{a} + \sqrt{b}$. That is, the square root of the sum of two positive numbers is never equal to the sum of the square roots of those numbers.

70. For any non-negative number a, $\sqrt{a}\sqrt{a} = a$ tells us that $\sqrt{a}$ is the non-negative quantity whose square is equal to a.

71. $\dfrac{x^2 - 6x}{x^2} \cdot \dfrac{2x}{x^2 - 12x + 36}$

$= \dfrac{\cancel{x}(x-6)}{\cancel{x^2}} \cdot \dfrac{2\cancel{x}}{(x-6)(x-6)}$

$= \dfrac{2}{x-6}$

73. $\dfrac{3}{2x} + \dfrac{5}{x+2}$

$= \dfrac{3(x+2)}{2x(x+2)} + \dfrac{5(2x)}{(x+2)(2x)}$

$= \dfrac{3(x+2) + 5(2x)}{2x(x+2)} = \dfrac{3x + 6 + 10x}{2x(x+2)}$

$= \dfrac{13x + 6}{2x(x+2)}$

75. Let $x = $ number of regular bulbs.
Then $40 - x = $ number of long-life bulbs.

$$0.75x + 0.89(40 - x) = 31.68$$
$$100[0.75x + 0.89(40 - x)] = 100(31.68)$$
$$75x + 89(40 - x) = 3168$$
$$75x + 3560 - 89x = 3168$$
$$3560 - 14x = 3168$$
$$-14x = -392$$
$$x = 28$$

Then $40 - x = 40 - 28 = 12$, so 28 regular bulbs and 12 long-life bulbs are sold.
CHECK: $28(0.75) + 12(0.89) = 21.00 + 10.68 = 31.68$ and $28 + 12 \stackrel{\checkmark}{=} 40$.

Exercises 10.2

1. $\sqrt{64} = 8$

3. $\sqrt{18} = \sqrt{9 \cdot 2} = \sqrt{9}\sqrt{2} = 3\sqrt{2}$

5. $\sqrt{32} = \sqrt{16 \cdot 2} = \sqrt{16}\sqrt{2} = 4\sqrt{2}$

7. $\sqrt{50} = \sqrt{25 \cdot 2} = \sqrt{25}\sqrt{2} = 5\sqrt{2}$

9. $\sqrt{400} = 20$

11. $\sqrt{x^6} = x^3$, because $x^3 x^3 = x^6$

13. $\sqrt{x^7} = \sqrt{x^6 \cdot x} = \sqrt{x^6}\sqrt{x} = x^3\sqrt{x}$

15. $\sqrt{16x^{16}} = \sqrt{16}\sqrt{x^{16}} = 4x^8$

17. $\sqrt{9x^9} = \sqrt{9}\sqrt{x^8}\sqrt{x} = 3x^4\sqrt{x}$

19. $\sqrt{40x^8} = \sqrt{4x^8}\sqrt{10} = \sqrt{4}\sqrt{x^8}\sqrt{10}$
    $= 2x^4\sqrt{10}$

21. $\sqrt{25x^7} = \sqrt{25}\sqrt{x^6}\sqrt{x} = 5x^3\sqrt{x}$

23. $\sqrt{12x^5} = \sqrt{4x^4}\sqrt{3x} = \sqrt{4}\sqrt{x^4}\sqrt{3x}$
    $= 2x^2\sqrt{3x}$

25. $\sqrt{a^2 b^4} = \sqrt{a^2}\sqrt{b^4} = ab^2$

27. $\sqrt{x^6 + y^8}$ cannot be simplified

29. $\sqrt{49a^8 b^{12}} = \sqrt{49}\sqrt{a^8}\sqrt{b^{12}} = 7a^4 b^6$

31. $\sqrt{28x^9 y^6} = \sqrt{4x^8 y^6}\sqrt{7x}$
    $= \sqrt{4}\sqrt{x^8}\sqrt{y^6}\sqrt{7x}$
    $= 2x^4 y^3 \sqrt{7x}$

33. $\sqrt{50m^7 n^{11}} = \sqrt{25m^6 n^{10}}\sqrt{2mn}$
    $= \sqrt{25}\sqrt{m^6}\sqrt{n^{10}}\sqrt{2mn}$
    $= 5m^3 n^5 \sqrt{2mn}$

35. $\sqrt{81 + x^6 + y^{10}}$ cannot be simplified

37. $\sqrt{48x^6 y^8 z^9} = \sqrt{16x^6 y^8 z^8}\sqrt{3z}$
    $= \sqrt{16}\sqrt{x^6}\sqrt{y^8}\sqrt{z^8}\sqrt{3z}$
    $= 4x^3 y^4 z^4 \sqrt{3z}$

39. $\sqrt{\dfrac{4}{9}} = \dfrac{\sqrt{4}}{\sqrt{9}} = \dfrac{2}{3}$

41. $\sqrt{\dfrac{7}{25}} = \dfrac{\sqrt{7}}{\sqrt{25}} = \dfrac{\sqrt{7}}{5}$

43. $\sqrt{\dfrac{5}{6}} = \dfrac{\sqrt{5}}{\sqrt{6}} = \dfrac{\sqrt{5} \cdot \sqrt{6}}{\sqrt{6} \cdot \sqrt{6}} = \dfrac{\sqrt{30}}{6}$

45. $\sqrt{\dfrac{1}{2}} = \dfrac{\sqrt{1}}{\sqrt{2}} = \dfrac{1 \cdot \sqrt{2}}{\sqrt{2} \cdot \sqrt{2}} = \dfrac{\sqrt{2}}{2}$

47. $\dfrac{1}{\sqrt{2}} = \dfrac{1 \cdot \sqrt{2}}{\sqrt{2} \cdot \sqrt{2}} = \dfrac{\sqrt{2}}{2}$

49. $\dfrac{18}{\sqrt{10}} = \dfrac{18 \cdot \sqrt{10}}{\sqrt{10} \cdot \sqrt{10}} = \dfrac{\cancel{18}^{9}\sqrt{10}}{\cancel{10}_{5}} = \dfrac{9\sqrt{10}}{5}$

51. $\dfrac{3}{\sqrt{x}} = \dfrac{3 \cdot \sqrt{x}}{\sqrt{x} \cdot \sqrt{x}} = \dfrac{3\sqrt{x}}{x}$

53. $\dfrac{15}{2\sqrt{7}} = \dfrac{15 \cdot \sqrt{7}}{2\sqrt{7} \cdot \sqrt{7}} = \dfrac{15\sqrt{7}}{2 \cdot 7} = \dfrac{15\sqrt{7}}{14}$

55. $\dfrac{12}{5\sqrt{6}} = \dfrac{12 \cdot \sqrt{6}}{5\sqrt{6} \cdot \sqrt{6}} = \dfrac{\cancel{12}^{2}\sqrt{6}}{5 \cdot \cancel{6}} = \dfrac{2\sqrt{6}}{5}$

57. $\dfrac{8x}{\sqrt{2x}} = \dfrac{8x\sqrt{2x}}{\sqrt{2x}\sqrt{2x}} = \dfrac{\overset{4}{\cancel{8x}}\sqrt{2x}}{\cancel{2x}} = 4\sqrt{2x}$

59. $\dfrac{5}{2\sqrt{x}} = \dfrac{5\cdot\sqrt{x}}{2\sqrt{x}\cdot\sqrt{x}} = \dfrac{5\sqrt{x}}{2x}$

61. $\dfrac{x^2}{\sqrt{xy}} = \dfrac{x^2\sqrt{xy}}{\sqrt{xy}\sqrt{xy}} = \dfrac{\overset{x}{\cancel{x^2}}\sqrt{xy}}{\cancel{xy}} = \dfrac{x\sqrt{xy}}{y}$

63. $\dfrac{\sqrt{8}}{\sqrt{6}} = \dfrac{\sqrt{8}\cdot\sqrt{6}}{\sqrt{6}\cdot\sqrt{6}} = \dfrac{\sqrt{48}}{6} = \dfrac{\sqrt{16}\sqrt{3}}{6}$
$= \dfrac{\overset{2}{\cancel{4}}\sqrt{3}}{\underset{3}{\cancel{6}}} = \dfrac{2\sqrt{3}}{3}$

65. $r = \sqrt{\dfrac{A}{P}} - 1;\ A = 10{,}000\text{ and }P = 8{,}000$
$r = \sqrt{\dfrac{10{,}000}{8{,}000}} - 1 = \sqrt{\dfrac{5}{4}} - 1$
$= 1.1180 - 1 = 0.1180\text{ or }11.80\%$

67. $s = 8.3\sqrt{L};\ L = 50$
$s = 8.3\sqrt{50} = 8.3(7.1)$
$= 59\text{ mph (rounded to the nearest mph)}$

69. $t = \sqrt{\dfrac{h}{16}};\ h = 1{,}250$
$t = \sqrt{\dfrac{1250}{16}}$
$= 8.8\text{ seconds (rounded to the nearest tenth)}$

71. $d = \sqrt{8{,}000m};\ m = 6$
$d = \sqrt{8{,}000\cdot 6} = \sqrt{48{,}000}$
$= 219\text{ miles (rounded to the nearest mile)}$

73. $I = \sqrt{\dfrac{W}{R}};\ W = 200,\ R = 70$
$I = \sqrt{\dfrac{200}{70}} = \sqrt{\dfrac{20}{7}}$
$= 1.7\text{ amps (rounded to the nearest tenth)}$

75.

x	y	$\sqrt{xy}$	$\sqrt{x}\sqrt{y}$
0	1	$\sqrt{0(1)} = \sqrt{0} = 0$	$\sqrt{0}\sqrt{1} = 0(1) = 0$
4	9	$\sqrt{4(9)} = \sqrt{36} = 6$	$\sqrt{4}\sqrt{9} = 2(3) = 6$
9	16	$\sqrt{9(16)} = \sqrt{144} = 12$	$\sqrt{9}\sqrt{16} = 3(4) = 12$
4	5	$\sqrt{4(5)} = \sqrt{4}\sqrt{5} = 2\sqrt{5}$	$\sqrt{4}\sqrt{5} = 2\sqrt{5}$
9	7	$\sqrt{9(7)} = \sqrt{9}\sqrt{7} = 3\sqrt{7}$	$\sqrt{9}\sqrt{7} = 3\sqrt{7}$

It appears from the table that in general $\sqrt{xy} = \sqrt{x}\sqrt{y}$.

x	y	$\sqrt{x+y}$	$\sqrt{x}+\sqrt{y}$
0	1	$\sqrt{0+1} = \sqrt{1} = 1$	$\sqrt{0}+\sqrt{1} = 0+1 = 1$
4	9	$\sqrt{4+9} = \sqrt{13}$	$\sqrt{4}+\sqrt{9} = 2+3 = 5$
9	16	$\sqrt{9+16} = \sqrt{25} = 5$	$\sqrt{9}+\sqrt{16} = 3+4 = 7$
4	5	$\sqrt{4+5} = \sqrt{9} = 3$	$\sqrt{4}+\sqrt{5} = 2+\sqrt{5}$
9	7	$\sqrt{9+7} = \sqrt{16} = 4$	$\sqrt{9}+\sqrt{7} = 3+\sqrt{7}$

However, it is <u>not</u> true that in general $\sqrt{x+y} = \sqrt{x}+\sqrt{y}$.

76. When we square a real number other than 0 or 1, we get an answer that is different from the original number. So it is incorrect to say that $\dfrac{2}{\sqrt{5}} = \dfrac{2^2}{(\sqrt{5})^2}$. When we rationalize the denominator properly, we multiply $\dfrac{2}{\sqrt{5}}$ by $\dfrac{\sqrt{5}}{\sqrt{5}}$. This means that we multiply by 1, which does <u>not</u> change the value of the original number.

77. $\dfrac{1}{\sqrt{5}} = \dfrac{1 \cdot \sqrt{5}}{\sqrt{5} \cdot \sqrt{5}} = \dfrac{\sqrt{5}}{5}$

$\dfrac{1}{\sqrt{5}} = \dfrac{1}{2.2360} = 0.4472$, correct to 3 decimal places

$\dfrac{\sqrt{5}}{5} = \dfrac{2.2360}{5} = 0.4472$, correct to 3 decimal places

It is easier to compute $\dfrac{\sqrt{5}}{5}$ than to compute $\dfrac{1}{\sqrt{5}}$, since $\dfrac{1}{\sqrt{5}}$ involves division by a decimal quantity, whereas $\dfrac{\sqrt{5}}{5}$ does not.

78. $\dfrac{\sqrt{3}}{\sqrt{7}} = \dfrac{1.7321}{2.6458} = 0.6547$,

  correct to 3 decimal places

$\dfrac{\sqrt{3}}{\sqrt{7}} = \dfrac{\sqrt{3} \cdot \sqrt{7}}{\sqrt{7} \cdot \sqrt{7}} = \dfrac{\sqrt{21}}{7}$

$= \dfrac{4.5828}{7} = 0.6547$,

correct to 3 decimal places

79. $4x + 3y = 8$

$\dfrac{-3y}{4x} \dfrac{-3y}{= 8 - 3y}$

$\dfrac{4x}{4} = \dfrac{8 - 3y}{4}$

$x = \dfrac{8 - 3y}{4}$

81. $4(a + 3b) - 2(a - 5b) = 6 - a + b$
$4a + 12b - 2a + 10b = 6 - a + b$
$2a + 22b = 6 - a + b$
$+a +a$
$\overline{3a + 22b = 6 + b}$
$-22b -22b$
$\overline{3a = 6 - 21b}$
$\dfrac{3a}{3} = \dfrac{6 - 21b}{3}$
$a = \dfrac{6 - 21b}{3}$
$a = \dfrac{3(2 - 7b)}{3}$
$a = 2 - 7b$

83. Let $t =$ number of hours the slower train travels until the trains meet. Then $t - 2 =$ number of hours the faster train travels until the trains meet.

$110t + 140(t - 2) = 595$
$110t + 140t - 280 = 595$
$250t - 280 = 595$
$250t = 875$
$t = \dfrac{875}{250} = \dfrac{7}{2}$ or $3\dfrac{1}{2}$

Since the slower train left at 11:00 A.M., the trains will meet $3\dfrac{1}{2}$ hours later, or at 2:30 P.M.

CHECK: $110\left(\dfrac{7}{2}\right) = 385$ and $140\left(\dfrac{3}{2}\right) = 210$.

$385 + 210 \stackrel{\checkmark}{=} 595$.

Exercises 10.3

1. $x + 2x + 3x = 6x$

3. $\sqrt{5} + 2\sqrt{5} + 3\sqrt{5} = (1 + 2 + 3)\sqrt{5}$
   $= 6\sqrt{5}$

5. $4x - x = 3x$

7. $4\sqrt{6} - \sqrt{6} = (4 - 1)\sqrt{6} = 3\sqrt{6}$

9. $3x + 5y$ cannot be simplified.

11. $3\sqrt{5} + 5\sqrt{3}$ cannot be simplified.

13. $x \cdot x = x^2$

15. $\sqrt{5}\sqrt{5} = 5$

17. $x + x = 2x$

19. $\sqrt{5} + \sqrt{5} = 2\sqrt{5}$

21. $\sqrt{5} + 3\sqrt{7} - 4\sqrt{5} - 5\sqrt{7}$
    $= \sqrt{5} - 4\sqrt{5} + 3\sqrt{7} - 5\sqrt{7}$
    $= -3\sqrt{5} - 2\sqrt{7}$

23. $3\sqrt{3} + 4\sqrt{3} - \sqrt{2}$
    $= 7\sqrt{3} - \sqrt{2}$

25. $2(x - y) + 3(y - x)$
    $= 2x - 2y + 3y - 3x$
    $= -x + y$

27. $2(\sqrt{5} - \sqrt{3}) + 3(\sqrt{3} - \sqrt{5})$
    $= 2\sqrt{5} - 2\sqrt{3} + 3\sqrt{3} - 3\sqrt{5}$
    $= 2\sqrt{5} - 3\sqrt{5} - 2\sqrt{3} + 3\sqrt{3}$
    $= -\sqrt{5} + \sqrt{3}$

29. $5(3\sqrt{6} + 2\sqrt{7}) - 4(\sqrt{7} - 2\sqrt{6})$
    $= 15\sqrt{6} + 10\sqrt{7} - 4\sqrt{7} + 8\sqrt{6}$
    $= 23\sqrt{6} + 6\sqrt{7}$

31. $6(\sqrt{m} - \sqrt{n}) - (3\sqrt{m} + 6\sqrt{n})$
    $= 6\sqrt{m} - 6\sqrt{n} - 3\sqrt{m} - 6\sqrt{n}$
    $= 6\sqrt{m} - 3\sqrt{m} - 6\sqrt{n} - 6\sqrt{n}$
    $= 3\sqrt{m} - 12\sqrt{n}$

33. $\sqrt{8} + \sqrt{18} = \sqrt{4}\sqrt{2} + \sqrt{9}\sqrt{2}$
    $= 2\sqrt{2} + 3\sqrt{2} = 5\sqrt{2}$

35. $\sqrt{25} + \sqrt{24} = 5 + \sqrt{4}\sqrt{6} = 5 + 2\sqrt{6}$

37. $\sqrt{20} - \sqrt{5} = \sqrt{4}\sqrt{5} - \sqrt{5}$
    $= 2\sqrt{5} - \sqrt{5}$
    $= \sqrt{5}$

39. $4\sqrt{12} - \sqrt{75} = 4\sqrt{4}\sqrt{3} - \sqrt{25}\sqrt{3}$
    $= 4(2\sqrt{3}) - 5\sqrt{3}$
    $= 8\sqrt{3} - 5\sqrt{3} = 3\sqrt{3}$

41. $\sqrt{20} + \sqrt{40} + \sqrt{60}$
    $= \sqrt{4}\sqrt{5} + \sqrt{4}\sqrt{10} + \sqrt{4}\sqrt{15}$
    $= 2\sqrt{5} + 2\sqrt{10} + 2\sqrt{15}$

43. $3\sqrt{72} - 5\sqrt{32}$
    $= 3(\sqrt{36}\sqrt{2}) - 5(\sqrt{16}\sqrt{2})$
    $= 3(6\sqrt{2}) - 5(4\sqrt{2})$
    $= 18\sqrt{2} - 20\sqrt{2} = -2\sqrt{2}$

45. $5\sqrt{36} + 4\sqrt{30} = 5 \cdot 6 + 4\sqrt{30}$
$\phantom{5\sqrt{36} + 4\sqrt{30}} = 30 + 4\sqrt{30}$

47. $\sqrt{25x} + \sqrt{36x} = \sqrt{25}\sqrt{x} + \sqrt{36}\sqrt{x}$
$\phantom{\sqrt{25x} + \sqrt{36x}} = 5\sqrt{x} + 6\sqrt{x}$
$\phantom{\sqrt{25x} + \sqrt{36x}} = 11\sqrt{x}$

49. $\sqrt{12w} + \sqrt{27w}$
$= \sqrt{4}\sqrt{3w} + \sqrt{9}\sqrt{3w}$
$= 2\sqrt{3w} + 3\sqrt{3w}$
$= 5\sqrt{3w}$

51. $\sqrt{45x} - \sqrt{20x}$
$= \sqrt{9}\sqrt{5x} - \sqrt{4}\sqrt{5x}$
$= 3\sqrt{5x} - 2\sqrt{5x}$
$= \sqrt{5x}$

53. $\sqrt{20y^3} - \sqrt{45y^3}$
$= \sqrt{4y^2}\sqrt{5y} - \sqrt{9y^2}\sqrt{5y}$
$= 2y\sqrt{5y} - 3y\sqrt{5y} = -y\sqrt{5y}$

55. $x\sqrt{28xy^3} + y\sqrt{63x^3y}$
$= x\sqrt{4y^2}\sqrt{7xy} + y\sqrt{9x^2}\sqrt{7xy}$
$= x(2y)\sqrt{7xy} + y(3x)\sqrt{7xy}$
$= 2xy\sqrt{7xy} + 3xy\sqrt{7xy} = 5xy\sqrt{7xy}$

57. $\dfrac{\sqrt{32x^3y^2}}{2xy} - \sqrt{8x}$
$= \dfrac{\sqrt{16x^2y^2}\sqrt{2x}}{2xy} - \sqrt{4}\sqrt{2x}$
$= \dfrac{\overset{2}{\cancel{4xy}}\sqrt{2x}}{\underset{1}{\cancel{2xy}}} - 2\sqrt{2x}$
$= 2\sqrt{2x} - 2\sqrt{2x} = 0$

59. $\sqrt{2} + \sqrt{\dfrac{1}{2}} = \sqrt{2} + \dfrac{\sqrt{1}}{\sqrt{2}}$
$\phantom{\sqrt{2} + \sqrt{\dfrac{1}{2}}} = \sqrt{2} + \dfrac{1 \cdot \sqrt{2}}{\sqrt{2} \cdot \sqrt{2}}$
$\phantom{\sqrt{2} + \sqrt{\dfrac{1}{2}}} = \sqrt{2} + \dfrac{1}{2}\sqrt{2}$
$\phantom{\sqrt{2} + \sqrt{\dfrac{1}{2}}} = \left(1 + \dfrac{1}{2}\right)\sqrt{2} = \dfrac{3}{2}\sqrt{2}$

61. $\sqrt{27} + \dfrac{4}{\sqrt{3}} = \sqrt{9}\sqrt{3} + \dfrac{4 \cdot \sqrt{3}}{\sqrt{3} \cdot \sqrt{3}}$
$\phantom{\sqrt{27} + \dfrac{4}{\sqrt{3}}} = 3\sqrt{3} + \dfrac{4}{3}\sqrt{3}$
$\phantom{\sqrt{27} + \dfrac{4}{\sqrt{3}}} = \left(3 + \dfrac{4}{3}\right)\sqrt{3} = \dfrac{13}{3}\sqrt{3}$

63. $\dfrac{\sqrt{8}}{7} + \dfrac{7}{\sqrt{2}} = \dfrac{\sqrt{4}\sqrt{2}}{7} + \dfrac{7 \cdot \sqrt{2}}{\sqrt{2} \cdot \sqrt{2}}$
$\phantom{\dfrac{\sqrt{8}}{7} + \dfrac{7}{\sqrt{2}}} = \dfrac{2\sqrt{2}}{7} + \dfrac{7\sqrt{2}}{2}$
$\phantom{\dfrac{\sqrt{8}}{7} + \dfrac{7}{\sqrt{2}}} = \left(\dfrac{2}{7} + \dfrac{7}{2}\right)\sqrt{2} = \dfrac{53}{14}\sqrt{2}$

65. $\sqrt{\dfrac{2}{7}} + \sqrt{\dfrac{7}{2}} = \dfrac{\sqrt{2}}{\sqrt{7}} + \dfrac{\sqrt{7}}{\sqrt{2}}$
$\phantom{\sqrt{\dfrac{2}{7}} + \sqrt{\dfrac{7}{2}}} = \dfrac{\sqrt{2} \cdot \sqrt{7}}{\sqrt{7} \cdot \sqrt{7}} + \dfrac{\sqrt{7} \cdot \sqrt{2}}{\sqrt{2} \cdot \sqrt{2}}$
$\phantom{\sqrt{\dfrac{2}{7}} + \sqrt{\dfrac{7}{2}}} = \dfrac{\sqrt{14}}{7} + \dfrac{\sqrt{14}}{2}$
$\phantom{\sqrt{\dfrac{2}{7}} + \sqrt{\dfrac{7}{2}}} = \dfrac{1}{7}\sqrt{14} + \dfrac{1}{2}\sqrt{14}$
$\phantom{\sqrt{\dfrac{2}{7}} + \sqrt{\dfrac{7}{2}}} = \left(\dfrac{1}{7} + \dfrac{1}{2}\right)\sqrt{14} = \dfrac{9}{14}\sqrt{14}$

67. $\sqrt{80} = 8.9442719$
$\sqrt{80} = \sqrt{16 \cdot 5}$
$\phantom{\sqrt{80}} = \sqrt{16}\sqrt{5}$
$\phantom{\sqrt{80}} = 4\sqrt{5}$
$\phantom{\sqrt{80}} = 4(2.360679)$
$\phantom{\sqrt{80}} = 8.9442716$

The two results are the same, to 6 places. These results should be equal, and only appear to differ because of rounding off.

68. $\sqrt{150} = 12.247448$
$\sqrt{150} = \sqrt{25 \cdot 6} = \sqrt{25}\sqrt{6} = 5\sqrt{6} = 5(2.4494897) = 12.247448$

69. $4x - 3y = 12$

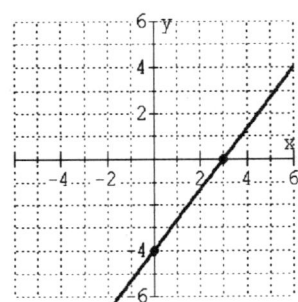

71. $x = 2$

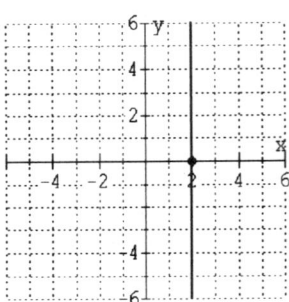

73. Let x = number of hours that the associate worked.
Then $14 - x$ = number of hours that the law clerk worked.
$$120x + 40(14 - x) = 1200$$
$$120x + 560 - 40x = 1200$$
$$80x + 560 = 1200$$
$$80x = 640$$
$$x = 8$$
Then $14 - x = 14 - 8 = 6$, so the associate worked 8 hours and the law clerk worked 6 hours.

Exercises 10.4

1. $\sqrt{3}\sqrt{11} = \sqrt{3 \cdot 11} = \sqrt{33}$

3. $\sqrt{3}\sqrt{5}\sqrt{13} = \sqrt{3 \cdot 5 \cdot 13} = \sqrt{195}$

5. $\sqrt{6} + \sqrt{24} = \sqrt{6} + \sqrt{4}\sqrt{6}$
$= \sqrt{6} + 2\sqrt{6} = 3\sqrt{6}$

7. $\sqrt{6}\sqrt{24} = \sqrt{6}(\sqrt{4}\sqrt{6}) = \sqrt{6}(2\sqrt{6})$
$= 2(\sqrt{6}\sqrt{6}) = 2 \cdot 6 = 12$

9. $\sqrt{3}\sqrt{5}\sqrt{6} = \sqrt{3}\sqrt{5}\sqrt{2}\sqrt{3}$
$= \left(\sqrt{3}\sqrt{3}\right)\sqrt{5}\sqrt{2}$
$= 3\sqrt{10}$

11. $\sqrt{3}(\sqrt{5} + \sqrt{6}) = \sqrt{3}\sqrt{5} + \sqrt{3}\sqrt{6}$
$= \sqrt{15} + \sqrt{18}$
$= \sqrt{15} + \sqrt{9}\sqrt{2}$
$= \sqrt{15} + 3\sqrt{2}$

13. $\sqrt{18}\sqrt{32} = \sqrt{9}\sqrt{2}\sqrt{16}\sqrt{2}$
$= 3 \cdot 4\left(\sqrt{2}\sqrt{2}\right)$
$= 12 \cdot 2 = 24$

15. $\sqrt{3}(2\sqrt{3} - 3\sqrt{2}) = \sqrt{3}(2\sqrt{3}) - \sqrt{3}(3\sqrt{2})$
$= 2(\sqrt{3}\sqrt{3}) - 3(\sqrt{3}\sqrt{2})$
$= 2 \cdot 3 - 3\sqrt{6} = 6 - 3\sqrt{6}$

17. $5\sqrt{x}(\sqrt{x} - 2\sqrt{5}) = 5\sqrt{x}\sqrt{x} - 5\sqrt{x} \cdot 2\sqrt{5}$
$= 5x - 10\sqrt{5x}$

19. $3\sqrt{2}(\sqrt{2}-4)+\sqrt{2}(5-\sqrt{2}) = 3\sqrt{2}(\sqrt{2})-3\sqrt{2}(4)+\sqrt{2}(5)-\sqrt{2}\sqrt{2}$
$= 3(\sqrt{2}\sqrt{2})-3\cdot 4\sqrt{2}+5\sqrt{2}-2$
$= 3\cdot 2-12\sqrt{2}+5\sqrt{2}-2$
$= 6-12\sqrt{2}+5\sqrt{2}-2$
$= 4-7\sqrt{2}$

21. $4\sqrt{x}(\sqrt{x}-\sqrt{2})-\sqrt{x}(3\sqrt{x}-2\sqrt{2}) = 4\sqrt{x}\sqrt{x}-4\sqrt{x}\sqrt{2}-3\sqrt{x}\sqrt{x}+2\sqrt{x}\sqrt{2}$
$= 4x-4\sqrt{2x}-3x+2\sqrt{2x}$
$= x-2\sqrt{2x}$

23. $(\sqrt{11}+3)(\sqrt{11}-6)$
$= \sqrt{11}\sqrt{11}-6\sqrt{11}+3\sqrt{11}-18$
$= 11-3\sqrt{11}-18$
$= -7-3\sqrt{11}$

25. $(\sqrt{x}+\sqrt{3})^2$
$= (\sqrt{x}+\sqrt{3})(\sqrt{x}+\sqrt{3})$
$= \sqrt{x}\sqrt{x}+\sqrt{x}\sqrt{3}+\sqrt{3}\sqrt{x}+\sqrt{3}\sqrt{3}$
$= x+\sqrt{3x}+\sqrt{3x}+3$
$= x+2\sqrt{3x}+3$

27. $(\sqrt{x}+\sqrt{3})(\sqrt{x}-\sqrt{3}) = \sqrt{x}\sqrt{x}-\sqrt{x}\sqrt{3}+\sqrt{3}\sqrt{x}-\sqrt{3}\sqrt{3}$
$= x-\sqrt{3x}+\sqrt{3x}-3 = x-3$

29. $(3\sqrt{2}-2\sqrt{5})^2 = (3\sqrt{2}-2\sqrt{5})(3\sqrt{2}-2\sqrt{5})$
$= (3\sqrt{2})(3\sqrt{2})-(3\sqrt{2})(2\sqrt{5})-(2\sqrt{5})(3\sqrt{2})+(2\sqrt{5})(2\sqrt{5})$
$= 3\cdot 3\sqrt{2}\sqrt{2}-3\cdot 2\sqrt{2}\sqrt{5}-2\cdot 3\sqrt{5}\sqrt{2}+2\cdot 2\sqrt{5}\sqrt{5}$
$= 9\cdot 2-6\sqrt{10}-6\sqrt{10}+4\cdot 5$
$= 18-6\sqrt{10}-6\sqrt{10}+20$
$= 38-12\sqrt{10}$

31. $(3\sqrt{x}-\sqrt{7})(3\sqrt{x}+\sqrt{7})$
$= (3\sqrt{x})(3\sqrt{x})+(3\sqrt{x})(\sqrt{7})-(\sqrt{7})(3\sqrt{x})-(\sqrt{7})(\sqrt{7})$
$= 3\cdot 3\sqrt{x}\sqrt{x}+3\sqrt{7x}-3\sqrt{7x}-7$
$= 9x-7$

33. $(2\sqrt{x}-3)(3\sqrt{x}+4) = (2\sqrt{x})(3\sqrt{x})+(2\sqrt{x})(4)-3(3\sqrt{x})-12$
$= 2\cdot 3\sqrt{x}\sqrt{x}+8\sqrt{x}-9\sqrt{x}-12$
$= 6x-\sqrt{x}-12$

35. $(\sqrt{28} - \sqrt{24})(\sqrt{7} - \sqrt{6}) = (\sqrt{4}\sqrt{7} - \sqrt{4}\sqrt{6})(\sqrt{7} - \sqrt{6})$
$= (2\sqrt{7} - 2\sqrt{6})(\sqrt{7} - \sqrt{6})$
$= 2\sqrt{7}\sqrt{7} - 2\sqrt{7}\sqrt{6} - 2\sqrt{6}\sqrt{7} + 2\sqrt{6}\sqrt{6}$
$= 2 \cdot 7 - 2\sqrt{42} - 2\sqrt{42} + 2 \cdot 6$
$= 14 - 2\sqrt{42} - 2\sqrt{42} + 12$
$= 26 - 4\sqrt{42}$

37. $(\sqrt{t+9})^2 + (\sqrt{t}+9)^2$
$= (t+9) + (\sqrt{t}+9)(\sqrt{t}+9)$
$= t + 9 + \sqrt{t}\sqrt{t} + 9\sqrt{t} + 9\sqrt{t} + 81$
$= t + 9 + t + 18\sqrt{t} + 81$
$= 2t + 18\sqrt{t} + 90$

39. $(\sqrt{x}+2)^2 - (\sqrt{x+2})^2$
$= (\sqrt{x}+2)(\sqrt{x}+2) - (x+2)$
$= \sqrt{x}\sqrt{x} + 2\sqrt{x} + 2\sqrt{x} + 4 - (x+2)$
$= x + 2\sqrt{x} + 2\sqrt{x} + 4 - x - 2$
$= 4\sqrt{x} + 2$

41. $\dfrac{10}{\sqrt{11}} = \dfrac{10 \cdot \sqrt{11}}{\sqrt{11} \cdot \sqrt{11}} = \dfrac{10\sqrt{11}}{11}$

43. $\dfrac{\sqrt{54}}{\sqrt{3}} = \sqrt{\dfrac{54}{3}} = \sqrt{18} = \sqrt{9}\sqrt{2} = 3\sqrt{2}$

45. $\dfrac{\sqrt{xy^3}}{\sqrt{x^3y}} = \sqrt{\dfrac{\cancel{x}y^{\cancel{3}}}{\cancel{x^{\cancel{3}}}\cancel{y}}} = \sqrt{\dfrac{y^2}{x^2}} = \dfrac{\sqrt{y^2}}{\sqrt{x^2}} = \dfrac{y}{x}$

47. $\dfrac{\sqrt{a^2b^5}}{\sqrt{ab^8}} = \sqrt{\dfrac{\cancel{a^2}b^{\cancel{5}}}{\cancel{a}b^{\cancel{8}}}} = \sqrt{\dfrac{a}{b^3}} = \dfrac{\sqrt{a}}{\sqrt{b^3}} = \dfrac{\sqrt{a}}{\sqrt{b^2}\sqrt{b}} = \dfrac{\sqrt{a}}{b\sqrt{b}} = \dfrac{\sqrt{a}\sqrt{b}}{b\sqrt{b}\sqrt{b}} = \dfrac{\sqrt{ab}}{b \cdot b} = \dfrac{\sqrt{ab}}{b^2}$

49. $\dfrac{10}{4-\sqrt{11}} = \dfrac{10(4+\sqrt{11})}{(4-\sqrt{11})(4+\sqrt{11})} = \dfrac{10(4+\sqrt{11})}{16-11} = \dfrac{\cancel{10}^{2}(4+\sqrt{11})}{\cancel{5}}$
$= 2(4+\sqrt{11}) \text{ or } 8 + 2\sqrt{11}$

51. $\dfrac{6}{\sqrt{x}+\sqrt{3}} = \dfrac{6(\sqrt{x}-\sqrt{3})}{(\sqrt{x}+\sqrt{3})(\sqrt{x}-\sqrt{3})} = \dfrac{6(\sqrt{x}-\sqrt{3})}{x-3} \text{ or } \dfrac{6\sqrt{x}-6\sqrt{3}}{x-3}$

53. $\dfrac{\sqrt{3}}{2+\sqrt{3}} = \dfrac{\sqrt{3}(2-\sqrt{3})}{(2+\sqrt{3})(2-\sqrt{3})} = \dfrac{2\sqrt{3}-\sqrt{3}\sqrt{3}}{4-3} = \dfrac{2\sqrt{3}-3}{1} = 2\sqrt{3}-3$

55. $\dfrac{\sqrt{5}+\sqrt{3}}{\sqrt{5}-\sqrt{3}} = \dfrac{(\sqrt{5}+\sqrt{3})(\sqrt{5}+\sqrt{3})}{(\sqrt{5}-\sqrt{3})(\sqrt{5}+\sqrt{3})} = \dfrac{\sqrt{5}\sqrt{5}+\sqrt{5}\sqrt{3}+\sqrt{3}\sqrt{5}+\sqrt{3}\sqrt{3}}{5-3}$
$= \dfrac{5+\sqrt{15}+\sqrt{15}+3}{5-3} = \dfrac{8+2\sqrt{15}}{2}$
$= \dfrac{\cancel{2}(4+\sqrt{15})}{\cancel{2}} = 4+\sqrt{15}$

57. $\dfrac{8}{\sqrt{5}-\sqrt{3}} - \dfrac{12}{\sqrt{3}}$

$= \dfrac{8(\sqrt{5}+\sqrt{3})}{(\sqrt{5}-\sqrt{3})(\sqrt{5}+\sqrt{3})} - \dfrac{12\sqrt{3}}{\sqrt{3}\sqrt{3}}$

$= \dfrac{8(\sqrt{5}+\sqrt{3})}{5-3} - \dfrac{12\sqrt{3}}{3}$

$= \dfrac{\overset{4}{\cancel{8}}(\sqrt{5}+\sqrt{3})}{\cancel{2}} - \dfrac{\overset{4}{\cancel{12}}\sqrt{3}}{\cancel{3}}$

$= 4\sqrt{5} + 4\sqrt{3} - 4\sqrt{3} = 4\sqrt{5}$

59. $\dfrac{6}{3-\sqrt{7}} - \dfrac{21}{\sqrt{7}}$

$= \dfrac{6(3+\sqrt{7})}{(3-\sqrt{7})(3+\sqrt{7})} - \dfrac{21\sqrt{7}}{\sqrt{7}\sqrt{7}}$

$= \dfrac{6(3+\sqrt{7})}{9-7} - \dfrac{21\sqrt{7}}{7}$

$= \dfrac{\overset{3}{\cancel{6}}(3+\sqrt{7})}{\cancel{2}} - \dfrac{\overset{3}{\cancel{21}}\sqrt{7}}{\cancel{7}}$

$= 9 + 3\sqrt{7} - 3\sqrt{7} = 9$

61. $(3+\sqrt{5})^2 + \dfrac{8}{3-\sqrt{5}} = (3+\sqrt{5})(3+\sqrt{5}) + \dfrac{8(3+\sqrt{5})}{(3-\sqrt{5})(3+\sqrt{5})}$

$= 9 + 3\sqrt{5} + 3\sqrt{5} + 5 + \dfrac{24 + 8\sqrt{5}}{9-5}$

$= 14 + 6\sqrt{5} + \dfrac{\overset{2}{\cancel{8}}(3+\sqrt{5})}{\cancel{4}}$

$= 14 + 6\sqrt{5} + 6 + 2\sqrt{5} = 20 + 8\sqrt{5}$

63. $\dfrac{4+\sqrt{8}}{6} = \dfrac{4+\sqrt{4}\sqrt{2}}{6}$

$= \dfrac{4+2\sqrt{2}}{6} = \dfrac{\cancel{2}(2+\sqrt{2})}{\underset{3}{\cancel{6}}}$

$= \dfrac{2+\sqrt{2}}{3}$

65. $\dfrac{12-\sqrt{20}}{10} = \dfrac{12-\sqrt{4}\sqrt{5}}{10}$

$= \dfrac{12-2\sqrt{5}}{10} = \dfrac{\cancel{2}(6-\sqrt{5})}{\underset{5}{\cancel{10}}}$

$= \dfrac{6-\sqrt{5}}{5}$

67. $(2+\sqrt{10})^2 - 4(2+\sqrt{10}) - 6 = (2+\sqrt{10})(2+\sqrt{10}) - 4(2+\sqrt{10}) - 6$

$= 4 + 2\sqrt{10} + 2\sqrt{10} + 10 - 8 - 4\sqrt{10} - 6$

$= 0$

69. (a) The "2" in the numerator is under the square root and is thus $\sqrt{2}$. This cannot be canceled with the "2" in the denominator.

(b) The cancellation is not valid since 2 is not a common factor of the numerator. Remember that terms cannot be canceled.

71. $x^2 - 6x - 7 = (x-7)(x+1)$

73. $x^2 + 10x - 16$ is not factorable.

75. $w^3 - 8w^2 + 16w = w(w^2 - 8w + 16)$
$= w(w-4)(w-4)$

77. $6u^2 - 5u - 14 = (6u+7)(u-2)$

79. Let x = price of a ticket sold at the door (in dollars).

Then $x - 1.50$ = price of a ticket sold in advance (in dollars).

$$200(x - 1.50) + 75x = 1075$$
$$200x - 300 + 75x = 1075$$
$$275x - 300 = 1075$$
$$275x = 1375$$
$$x = 5$$

So a ticket sold at the door cost $5.

Exercises 10.5

1. $\sqrt{x} = 5$ CHECK $x = 25$:
    $(\sqrt{x})^2 = (5)^2$ $\sqrt{x} = 5$
    $x = 25$ $\sqrt{25} \stackrel{?}{=} 5$
                       $5 \stackrel{\checkmark}{=} 5$

3. $9 = \sqrt{a}$ CHECK $a = 81$:
    $(9)^2 = (\sqrt{a})^2$ $9 = \sqrt{a}$
    $81 = a$ $9 \stackrel{?}{=} \sqrt{81}$
                       $9 \stackrel{\checkmark}{=} 9$

5. $\sqrt{u} = 6$ CHECK $u = 36$:
    $(\sqrt{u})^2 = (6)^2$ $\sqrt{u} = 6$
    $u = 36$ $\sqrt{36} \stackrel{?}{=} 6$
                       $6 \stackrel{\checkmark}{=} 6$

7. $\sqrt{u} = -8$ CHECK $u = 64$:
    $(\sqrt{u})^2 = (-8)^2$ $\sqrt{u} = -8$
    $u = 64$ $\sqrt{64} \stackrel{?}{=} -8$
                       $8 \neq -8$

    So the given equation has no solution.

9. $\sqrt{x + 3} = 10$ CHECK $x = 97$:
    $(\sqrt{x + 3})^2 = (10)^2$ $\sqrt{x + 3} = 10$
    $x + 3 = 100$ $\sqrt{97 + 3} \stackrel{?}{=} 10$
    $x = 97$ $\sqrt{100} \stackrel{?}{=} 10$
                       $10 \stackrel{\checkmark}{=} 10$

11. $\sqrt{x} + 3 = 10$ CHECK $x = 49$:
    $\phantom{\sqrt{x}} -3 -3$ $\sqrt{x} + 3 = 10$
    $\sqrt{x} = 7$ $\sqrt{49} + 3 \stackrel{?}{=} 10$
    $(\sqrt{x})^2 = (7)^2$ $7 + 3 \stackrel{?}{=} 10$
    $x = 49$ $10 \stackrel{\checkmark}{=} 10$

13. $\sqrt{2t - 1} = 5$ CHECK $t = 13$:
    $(\sqrt{2t - 1})^2 = (5)^2$ $\sqrt{2t - 1} = 5$
    $2t - 1 = 25$ $\sqrt{2(13) - 1} \stackrel{?}{=} 5$
    $2t = 26$ $\sqrt{26 - 1} \stackrel{?}{=} 5$
    $t = 13$ $\sqrt{25} \stackrel{?}{=} 5$
                       $5 \stackrel{\checkmark}{=} 5$

15. $\sqrt{2t} - 1 = 5$ CHECK $t = 18$:
    $\phantom{\sqrt{2t}} +1 +1$ $\sqrt{2t} - 1 = 5$
    $\sqrt{2t} = 6$ $\sqrt{2(18)} - 1 \stackrel{?}{=} 5$
    $(\sqrt{2t})^2 = (6)^2$ $\sqrt{36} - 1 \stackrel{?}{=} 5$
    $2t = 36$ $6 - 1 \stackrel{?}{=} 5$
    $t = 18$ $5 \stackrel{\checkmark}{=} 5$

17. $\sqrt{5-4x} = 7$ CHECK $x = -11$:
    $\left(\sqrt{5-4x}\right)^2 = (7)^2$ $\sqrt{5-4x} = 7$
    $5 - 4x = 49$ $\sqrt{5-4(-11)} \stackrel{?}{=} 7$
    $-4x = 44$ $\sqrt{5+44} \stackrel{?}{=} 7$
    $x = -11$ $\sqrt{49} \stackrel{?}{=} 7$
    $7 \stackrel{\checkmark}{=} 7$

19. $8 = \sqrt{8-14u}$ CHECK $u = -4$:
    $(8)^2 = \left(\sqrt{8-14u}\right)^2$ $8 = \sqrt{8-14u}$
    $64 = 8 - 14u$ $8 \stackrel{?}{=} \sqrt{8-14(-4)}$
    $56 = -14u$ $8 \stackrel{?}{=} \sqrt{8+56}$
    $-4 = u$ $8 \stackrel{?}{=} \sqrt{64}$
    $8 \stackrel{\checkmark}{=} 8$

21. $\sqrt{1+x} + 8 = 4$ CHECK $x = 15$:
    $\quad\quad\quad -8 \ -8$ $\sqrt{1+x} + 8 = 4$
    $\sqrt{1+x} = -4$ $\sqrt{1+15} + 8 \stackrel{?}{=} 4$
    $\left(\sqrt{1+x}\right)^2 = (-4)^2$ $\sqrt{16} + 8 \stackrel{?}{=} 4$
    $1 + x = 16$ $4 + 8 \stackrel{?}{=} 4$
    $x = 15$ $12 \neq 4$
    So the given equation has no solution.

23. $\sqrt{3+x} + 8 = 3$ CHECK $x = 22$:
    $\quad\quad\quad -8 \ -8$ $\sqrt{3+x} + 8 = 3$
    $\sqrt{3+x} = -5$ $\sqrt{3+22} + 8 \stackrel{?}{=} 3$
    $\left(\sqrt{3+x}\right)^2 = (-5)^2$ $\sqrt{25} + 8 \stackrel{?}{=} 3$
    $3 + x = 25$ $5 + 8 \stackrel{?}{=} 3$
    $x = 22$ $13 \neq 3$
    So the given equation has no solution.

25. $4 - \sqrt{x} = 0$ CHECK $x = 16$:
    $4 = \sqrt{x}$ $4 - \sqrt{x} = 0$
    $(4)^2 = (\sqrt{x})^2$ $4 - \sqrt{16} \stackrel{?}{=} 0$
    $16 = x$ $4 - 4 \stackrel{?}{=} 0$
    $0 \stackrel{\checkmark}{=} 0$

27. $4 - \sqrt{x} = 3$ CHECK $x = 1$:
    $-\sqrt{x} = -1$ $4 - \sqrt{x} = 3$
    $(-\sqrt{x})^2 = (-1)^2$ $4 - \sqrt{1} \stackrel{?}{=} 3$
    $x = 1$ $4 - 1 \stackrel{?}{=} 3$
    $3 \stackrel{\checkmark}{=} 3$

29. $10 - 2\sqrt{y} = 4$ CHECK $y = 9$:
    $-10 \quad\quad\quad -10$ $10 - 2\sqrt{y} = 4$
    $-2\sqrt{y} = -6$ $10 - 2\sqrt{9} \stackrel{?}{=} 4$
    $\sqrt{y} = 3$ $10 - 2(3) \stackrel{?}{=} 4$
    $\left(\sqrt{y}\right)^2 = (3)^2$ $10 - 6 \stackrel{?}{=} 4$
    $y = 9$ $4 \stackrel{\checkmark}{=} 4$

31. $10 - \sqrt{2y} = 2$ CHECK $y = 32$:
    $-10 \quad\quad\quad -10$ $10 - \sqrt{2y} = 2$
    $-\sqrt{2y} = -8$ $10 - \sqrt{2(32)} \stackrel{?}{=} 2$
    $\sqrt{2y} = 8$ $10 - \sqrt{64} \stackrel{?}{=} 2$
    $\left(\sqrt{2y}\right)^2 = (8)^2$ $10 - 8 \stackrel{?}{=} 2$
    $2y = 64$ $2 \stackrel{\checkmark}{=} 2$
    $y = 32$

33. $6 - \sqrt{a} = 11$ CHECK $a = 25$:
    $-\sqrt{a} = 5$ $6 - \sqrt{a} = 11$
    $\left(-\sqrt{a}\right)^2 = (5)^2$ $6 - \sqrt{25} \stackrel{?}{=} 11$
    $a = 25$ $6 - 5 \stackrel{?}{=} 11$
    $1 \neq 11$
    So the given equation has no solution.

35. $7 - \sqrt{a} = 9$ CHECK $a = 4$:
    $-\sqrt{a} = 2$ $7 - \sqrt{a} = 9$
    $\left(-\sqrt{a}\right)^2 = (2)^2$ $7 - \sqrt{4} \stackrel{?}{=} 9$
    $a = 4$ $7 - 2 \stackrel{?}{=} 9$
    $5 \neq 9$
    So the given equation has no solution.

37. $6\sqrt{c} - 3 = 3\sqrt{c}$
$\phantom{6\sqrt{c}}\underline{-6\sqrt{c} -6\sqrt{c}}$
$\phantom{6\sqrt{c} -\ } -3 = -3\sqrt{c}$
$\phantom{6\sqrt{c} - 3 =\ \ } 1 = \sqrt{c}$
$\phantom{6\sqrt{c} - 3\ } (1)^2 = \left(\sqrt{c}\right)^2$
$\phantom{6\sqrt{c} - 3 =\ \ } 1 = c$

CHECK $c = 1$:
$6\sqrt{c} - 3 = 3\sqrt{c}$
$6\sqrt{1} - 3 \stackrel{?}{=} 3\sqrt{1}$
$6(1) - 3 \stackrel{?}{=} 3(1)$
$6 - 3 \stackrel{?}{=} 3$
$3 \stackrel{\checkmark}{=} 3$

39. $\sqrt{2x+1} - 1 = 4$
$\phantom{\sqrt{2x+1}\ \ }\underline{+1 +1}$
$\sqrt{2x+1} = 5$
$\left(\sqrt{2x+1}\right)^2 = (5)^2$
$2x + 1 = 25$
$2x = 24$
$x = 12$

CHECK $x = 12$:
$\sqrt{2x+1} - 1 = 4$
$\sqrt{2(12)+1} - 1 \stackrel{?}{=} 4$
$\sqrt{25} - 1 \stackrel{?}{=} 4$
$5 - 1 \stackrel{?}{=} 4$
$4 \stackrel{\checkmark}{=} 4$

41. $6 + \sqrt{5x+1} = 12$
$\underline{-6 \phantom{+\sqrt{5x+1}=}-6}$
$ \sqrt{5x+1} = 6$
$\left(\sqrt{5x+1}\right)^2 = (6)^2$
$ 5x + 1 = 36$
$ 5x = 35$
$ x = 7$

CHECK $x = 7$:
$6 + \sqrt{5x+1} = 12$
$6 + \sqrt{5(7)+1} \stackrel{?}{=} 12$
$6 + \sqrt{36} \stackrel{?}{=} 12$
$6 + 6 \stackrel{?}{=} 12$
$12 \stackrel{\checkmark}{=} 12$

43. $5\sqrt{w} + 3 = 4\sqrt{w} + 11$
$\underline{-4\sqrt{w} -4\sqrt{w}}$
$\sqrt{w} + 3 = \phantom{-4\sqrt{w} +\ } 11$
$\phantom{-\sqrt{w} +\ } \underline{-3 -3}$
$\sqrt{w} = \phantom{-4\sqrt{w} + 1} 8$
$(\sqrt{w})^2 = (8)^2$
$ w = 64$

CHECK $w = 64$:
$5\sqrt{w} + 3 = 4\sqrt{w} + 11$
$5\sqrt{64} + 3 \stackrel{?}{=} 4\sqrt{64} + 11$
$5(8) + 3 \stackrel{?}{=} 4(8) + 11$
$40 + 3 \stackrel{?}{=} 32 + 11$
$43 \stackrel{\checkmark}{=} 43$

45. $2\sqrt{w} + 5 = 7\sqrt{w} - 10$
$\underline{-2\sqrt{w} -2\sqrt{w}}$
$\phantom{-2\sqrt{w} +\ } 5 = 5\sqrt{w} - 10$
$\phantom{-2\sqrt{w}\ \ } \underline{+10 \phantom{=\ 5\sqrt{w}} +10}$
$\phantom{-2\sqrt{w} +\ } 15 = 5\sqrt{w}$
$\phantom{-2\sqrt{w} +\ } 3 = \sqrt{w}$
$\phantom{-2\sqrt{w} +\ } (3)^2 = \left(\sqrt{w}\right)^2$
$\phantom{-2\sqrt{w} +\ } 9 = w$

CHECK $w = 9$:
$2\sqrt{w} + 5 = 7\sqrt{w} - 10$
$2\sqrt{9} + 5 \stackrel{?}{=} 7\sqrt{9} - 10$
$2(3) + 5 \stackrel{?}{=} 7(3) - 10$
$6 + 5 \stackrel{?}{=} 21 - 10$
$11 \stackrel{\checkmark}{=} 11$

47. $$\begin{aligned} 2\sqrt{2x-1}+7 &= 10 \\ -7 & -7 \\ \hline 2\sqrt{2x-1} &= 3 \\ \left(2\sqrt{2x-1}\right)^2 &= (3)^2 \\ 4(2x-1) &= 9 \\ 8x-4 &= 9 \\ 8x &= 13 \\ x &= \frac{13}{8} \end{aligned}$$

CHECK $x = \frac{13}{8}$:
$$\begin{aligned} 2\sqrt{2x-1}+7 &= 10 \\ 2\sqrt{2\left(\frac{13}{8}\right)-1}+7 &\stackrel{?}{=} 10 \\ 2\sqrt{\frac{13}{4}-1}+7 &\stackrel{?}{=} 10 \\ 2\sqrt{\frac{9}{4}}+7 &\stackrel{?}{=} 10 \\ 2\left(\frac{3}{2}\right)+7 &\stackrel{?}{=} 10 \\ 3+7 &\stackrel{?}{=} 10 \\ 10 &\stackrel{\checkmark}{=} 10 \end{aligned}$$

49. $$\begin{aligned} 4\sqrt{7-8x}+3 &= 15 \\ -3 & -3 \\ \hline 4\sqrt{7-8x} &= 12 \\ \sqrt{7-8x} &= 3 \\ \left(\sqrt{7-8x}\right)^2 &= (3)^2 \\ 7-8x &= 9 \\ -8x &= 2 \\ x &= -\frac{1}{4} \end{aligned}$$

CHECK $x = -\frac{1}{4}$:
$$\begin{aligned} 4\sqrt{7-8x}+3 &= 15 \\ 4\sqrt{7-8\left(-\frac{1}{4}\right)}+3 &\stackrel{?}{=} 15 \\ 4\sqrt{7+2}+3 &\stackrel{?}{=} 15 \\ 4\sqrt{9}+3 &\stackrel{?}{=} 15 \\ 4(3)+3 &\stackrel{?}{=} 15 \\ 12+3 &\stackrel{?}{=} 15 \\ 15 &\stackrel{\checkmark}{=} 15 \end{aligned}$$

51. $$\begin{aligned} \sqrt{2x-1} &= \sqrt{x+5} \\ \left(\sqrt{2x-1}\right)^2 &= \left(\sqrt{x+5}\right)^2 \\ 2x-1 &= x+5 \\ -x & -x \\ \hline x-1 &= 5 \\ x &= 6 \end{aligned}$$

CHECK $x = 6$:
$$\begin{aligned} \sqrt{2x-1} &= \sqrt{x+5} \\ \sqrt{2(6)-1} &\stackrel{?}{=} \sqrt{6+5} \\ \sqrt{12-1} &\stackrel{?}{=} \sqrt{11} \\ \sqrt{11} &\stackrel{\checkmark}{=} \sqrt{11} \end{aligned}$$

53. $$\begin{aligned} \sqrt{5x-1} &= \sqrt{3x+9} \\ \left(\sqrt{5x-1}\right)^2 &= \left(\sqrt{3x+9}\right)^2 \\ 5x-1 &= 3x+9 \\ -3x & -3x \\ \hline 2x-1 &= 9 \\ 2x &= 10 \\ x &= 5 \end{aligned}$$

CHECK $x = 5$:
$$\begin{aligned} \sqrt{5x-1} &= \sqrt{3x+9} \\ \sqrt{5(5)-1} &\stackrel{?}{=} \sqrt{3(5)+9} \\ \sqrt{25-1} &\stackrel{?}{=} \sqrt{15+9} \\ \sqrt{24} &\stackrel{\checkmark}{=} \sqrt{24} \end{aligned}$$

55. $I = \sqrt{\dfrac{W}{R}}$; $I = 12$, $R = 9$

$12 = \sqrt{\dfrac{W}{9}}$

$(12)^2 = \left(\sqrt{\dfrac{W}{9}}\right)^2$

$144 = \dfrac{W}{9}$

$1296 = W$

57. $t = \sqrt{\dfrac{h}{16}}$; $t = 3.4$

$3.4 = \sqrt{\dfrac{h}{16}}$

$(3.4)^2 = \left(\sqrt{\dfrac{h}{16}}\right)^2$

$11.56 = \dfrac{h}{16}$

$184.96 = h$

The height of the building is 185 ft (rounded to the nearest foot).

59. $d = \sqrt{8000a}$; $d = 150$

$150 = \sqrt{8000a}$

$(150)^2 = \left(\sqrt{8000a}\right)^2$

$22500 = 8000a$

$2.8125 = a$

The altitude of the plane is 2.8 miles (rounded to the nearest tenth).

61. $r = \sqrt{\dfrac{A}{P}} - 1$; $A = 5000$, $r = 0.06$

$0.06 = \sqrt{\dfrac{5000}{P}} - 1$

$1.06 = \sqrt{\dfrac{5000}{P}}$

$(1.06)^2 = \left(\sqrt{\dfrac{5000}{P}}\right)^2$

$1.1236 = \dfrac{5000}{P}$

$1.1236P = 5000$

$P = 4449.98$

The original investment must be \$4,450 (rounded to the nearest dollar).

63. The equations $x = 5$ and $x^2 = 25$ are not equivalent, since their solution sets are not the same. The solution set of the second equation includes -5, which is not in the solution set of the first equation. In general, squaring both sides of an equation does not yield an equivalent equation, since the "squared" equation might have a solution that fails to satisfy the original equation.

64. The equation $\sqrt{x} + 3 = 2$ implies that $\sqrt{x} = -1$. We know this is impossible because by definition, the $\sqrt{}$ symbol stands for the non-negative square root. Therefore, the given equation has no solution.

65. $\dfrac{8x^3 + 12x - 4x^2}{2x} = \dfrac{\overset{4x^2}{\cancel{8x^3}}}{\cancel{2x}} + \dfrac{\overset{6}{\cancel{12x}}}{\cancel{2x}} - \dfrac{\overset{2x}{\cancel{4x^2}}}{\cancel{2x}}$

$= 4x^2 + 6 - 2x$

67.
$$\begin{array}{r}
3x - 2\\
x-2{\overline{\smash{\big)}\,3x^2 - 8x + 4}}\\
\underline{-(3x^2 - 6x)}\\
-2x + 4\\
\underline{-(-2x + 4)}\\
0
\end{array}$$

Chapter 10 Review Exercises

1. $\sqrt{49} = 7$

3. $\sqrt{-16}$ is undefined

5. $\sqrt{96} = \sqrt{16 \cdot 6} = \sqrt{16}\sqrt{6} = 4\sqrt{6}$

7. $\sqrt{9x^9} = \sqrt{9x^8}\sqrt{x} = \sqrt{9}\sqrt{x^8}\sqrt{x} = 3x^4\sqrt{x}$

9. $\sqrt{\dfrac{4}{9}} = \dfrac{\sqrt{4}}{\sqrt{9}} = \dfrac{2}{3}$

11. $\sqrt{\dfrac{3}{5}} = \dfrac{\sqrt{3}}{\sqrt{5}} = \dfrac{\sqrt{3} \cdot \sqrt{5}}{\sqrt{5} \cdot \sqrt{5}} = \dfrac{\sqrt{15}}{5}$

13. $8\sqrt{7} - 5\sqrt{7} - \sqrt{7} = (8 - 5 - 1)\sqrt{7}$
$\phantom{8\sqrt{7} - 5\sqrt{7} - \sqrt{7}} = 2\sqrt{7}$

15. $\sqrt{45} - \sqrt{20} = \sqrt{9}\sqrt{5} - \sqrt{4}\sqrt{5}$
$\phantom{\sqrt{45} - \sqrt{20}} = 3\sqrt{5} - 2\sqrt{5} = \sqrt{5}$

17. $\sqrt{75x} + \sqrt{12x}$
$= \sqrt{25}\sqrt{3x} + \sqrt{4}\sqrt{3x}$
$= 5\sqrt{3x} + 2\sqrt{3x} = 7\sqrt{3x}$

19. $\dfrac{\sqrt{12x^3y^2}}{xy} + \sqrt{27x}$
$= \dfrac{\sqrt{4x^2y^2}\sqrt{3x}}{xy} + \sqrt{9}\sqrt{3x}$
$= \dfrac{2\cancel{xy}\sqrt{3x}}{\cancel{xy}} + 3\sqrt{3x}$
$= 2\sqrt{3x} + 3\sqrt{3x} = 5\sqrt{3x}$

21. $\sqrt{5}\left(3\sqrt{5} + \sqrt{2}\right) = 3\sqrt{5}\sqrt{5} + \sqrt{5}\sqrt{2} = 3 \cdot 5 + \sqrt{10} = 15 + \sqrt{10}$

23. $(3\sqrt{7} - 2\sqrt{3})(2\sqrt{7} + 5\sqrt{3})$
$= (3\sqrt{7})(2\sqrt{7}) + (3\sqrt{7})(5\sqrt{3}) - (2\sqrt{3})(2\sqrt{7}) - (2\sqrt{3})(5\sqrt{3})$
$= 3 \cdot 2\sqrt{7}\sqrt{7} + 3 \cdot 5\sqrt{7}\sqrt{3} - 2 \cdot 2\sqrt{3}\sqrt{7} - 2 \cdot 5\sqrt{3}\sqrt{3}$
$= 6 \cdot 7 + 15\sqrt{21} - 4\sqrt{21} - 10 \cdot 3$
$= 42 + 15\sqrt{21} - 4\sqrt{21} - 30$
$= 12 + 11\sqrt{21}$

25. $\left(\sqrt{x} - 3\right)^2$
$= \left(\sqrt{x} - 3\right)\left(\sqrt{x} - 3\right)$
$= \sqrt{x}\sqrt{x} - 3\sqrt{x} - 3\sqrt{x} + 9$
$= x - 6\sqrt{x} + 9$

27. $\dfrac{7}{\sqrt{3}} = \dfrac{7\sqrt{3}}{\sqrt{3}\sqrt{3}} = \dfrac{7\sqrt{3}}{3}$

29. $\dfrac{x^2}{\sqrt{x}} = \dfrac{x^2\sqrt{x}}{\sqrt{x}\sqrt{x}} = \dfrac{\overset{x}{\cancel{x^2}}\sqrt{x}}{\cancel{x}} = x\sqrt{x}$

31. $\dfrac{18}{\sqrt{12}} = \dfrac{18}{\sqrt{4}\sqrt{3}} = \dfrac{\overset{9}{\cancel{18}}}{\cancel{2}\sqrt{3}} = \dfrac{9}{\sqrt{3}}$
$= \dfrac{9\sqrt{3}}{\sqrt{3}\sqrt{3}} = \dfrac{\overset{3}{\cancel{9}}\sqrt{3}}{\cancel{3}_1} = 3\sqrt{3}$

33. $\dfrac{14}{3-\sqrt{2}} = \dfrac{14(3+\sqrt{2})}{(3-\sqrt{2})(3+\sqrt{2})}$

$= \dfrac{14(3+\sqrt{2})}{9-2} = \dfrac{\cancel{14}^{2}(3+\sqrt{2})}{\cancel{7}}$

$= 2(3+\sqrt{2})$ or $6+2\sqrt{2}$

35. $\dfrac{2+\sqrt{5}}{6+\sqrt{5}} = \dfrac{(2+\sqrt{5})(6-\sqrt{5})}{(6+\sqrt{5})(6-\sqrt{5})}$

$= \dfrac{12-2\sqrt{5}+6\sqrt{5}-5}{36-5}$

$= \dfrac{7+4\sqrt{5}}{31}$

37. $\left(\sqrt{x+7}\right)^2 - \left(\sqrt{x}+\sqrt{7}\right)^2 = x+7 - \left(\sqrt{x}+\sqrt{7}\right)\left(\sqrt{x}+\sqrt{7}\right)$

$= x+7 - \left(x + \sqrt{x}\sqrt{7} + \sqrt{7}\sqrt{x} + 7\right)$

$= x+7 - \left(x + \sqrt{7x} + \sqrt{7x} + 7\right)$

$= x+7 - x - \sqrt{7x} - \sqrt{7x} - 7$

$= -2\sqrt{7x}$

39. $\sqrt{x+5} = 7$ CHECK $x=44$:

$(\sqrt{x+5})^2 = (7)^2$ $\sqrt{x+5} = 7$

$x+5 = 49$ $\sqrt{44+5} \stackrel{?}{=} 7$

$x = 44$ $\sqrt{49} \stackrel{?}{=} 7$

$7 \stackrel{\checkmark}{=} 7$

41. $10 - \sqrt{4u} = 2$ CHECK $u=16$:

$\underline{-10 \qquad\qquad -10}$ $10 - \sqrt{4u} = 2$

$-\sqrt{4u} = -8$ $10 - \sqrt{4(16)} \stackrel{?}{=} 2$

$\sqrt{4u} = 8$ $10 - \sqrt{64} \stackrel{?}{=} 2$

$(\sqrt{4u})^2 = (8)^2$ $10 - 8 \stackrel{?}{=} 2$

$4u = 64$ $2 \stackrel{\checkmark}{=} 2$

$u = 16$

43. $8 - \sqrt{y-5} = 11$ CHECK $y=14$:

$\underline{-8 \qquad\qquad -8}$ $8 - \sqrt{y-5} = 11$

$-\sqrt{y-5} = 3$ $8 - \sqrt{14-5} \stackrel{?}{=} 11$

$\sqrt{y-5} = -3$ $8 - \sqrt{9} \stackrel{?}{=} 11$

$(\sqrt{y-5})^2 = (-3)^2$ $8 - 3 \stackrel{?}{=} 11$

$y-5 = 9$ $5 \neq 11$

$y = 14$

So the given equation has no solution.

45. $T = 2\pi\sqrt{\dfrac{L}{980}}$; $T = 1.5$

$1.5 = 2(3.14)\sqrt{\dfrac{L}{980}}$

$1.5 = 6.28\sqrt{\dfrac{L}{980}}$

$0.2388 = \sqrt{\dfrac{L}{980}}$

$(0.2388)^2 = \left(\sqrt{\dfrac{L}{980}}\right)^2$

$0.0570 = \dfrac{L}{980}$

$55.88 = L$

The length of the pendulum is 55.9 cm. (rounded to the nearest tenth).

Chapter 10 Practice Test

1. $\sqrt{25x^{16}y^6} = \sqrt{25}\sqrt{x^{16}}\sqrt{y^6} = 5x^8y^3$

3. $\sqrt{50x^3} - x\sqrt{32x} = \sqrt{25x^2}\sqrt{2x} - x(\sqrt{16}\sqrt{2x})$
$= 5x\sqrt{2x} - x(4\sqrt{2x})$
$= 5x\sqrt{2x} - 4x\sqrt{2x} = x\sqrt{2x}$

5. $\sqrt{20x^8y^9} + 3x^4y^4\sqrt{5y} = \sqrt{4x^8y^8}\sqrt{5y} + 3x^4y^4\sqrt{5y}$
$= 2x^4y^4\sqrt{5y} + 3x^4y^4\sqrt{5y}$
$= 5x^4y^4\sqrt{5y}$

7. $(2\sqrt{x} - \sqrt{5})(\sqrt{x} + 3\sqrt{5}) = 2\sqrt{x}\sqrt{x} + (2\sqrt{x})(3\sqrt{5}) - \sqrt{5}\sqrt{x} - 3\sqrt{5}\sqrt{5}$
$= 2x + 6\sqrt{5x} - \sqrt{5x} - 3 \cdot 5$
$= 2x + 5\sqrt{5x} - 15$

9. $(\sqrt{x} - 4)^2 - (\sqrt{x-4})^2 = (\sqrt{x} - 4)(\sqrt{x} - 4) - (x - 4)$
$= \sqrt{x}\sqrt{x} - 4\sqrt{x} - 4\sqrt{x} + 16 - x + 4$
$= x - 8\sqrt{x} + 16 - x + 4$
$= -8\sqrt{x} + 20$

11. $x^2 - 2x = 1$
$(1 - \sqrt{2})^2 - 2(1 - \sqrt{2}) \stackrel{?}{=} 1$
$(1 - \sqrt{2})(1 - \sqrt{2}) - 2(1 - \sqrt{2}) \stackrel{?}{=} 1$
$1 - \sqrt{2} - \sqrt{2} + 2 - 2 + 2\sqrt{2} \stackrel{?}{=} 1$
$1 \stackrel{\checkmark}{=} 1$

So $1 - \sqrt{2}$ is a solution to the equation.

13. $C = 1.8\sqrt{1000 - n};\ C = 18$
$18 = 1.8\sqrt{1000 - n}$
$10 = \sqrt{1000 - n}$
$(10)^2 = (\sqrt{1000 - n})^2$
$100 = 1000 - n$
$-900 = -n$
$900 = n$

So 900 items are produced.

Chapter 11
Quadratic Equations

Exercises 11.1

1. $$x^2 + 4x = 32$$
    $$x^2 + 4x - 32 = 0$$
    $$(x+8)(x-4) = 0$$
    $$x + 8 = 0 \quad \text{or} \quad x - 4 = 0$$
    $$x = -8 \quad \text{or} \quad x = 4$$

3. $$(x+3)(x+5) = 80$$
    $$x^2 + 5x + 3x + 15 = 80$$
    $$x^2 + 8x + 15 = 80$$
    $$x^2 + 8x - 65 = 0$$
    $$(x+13)(x-5) = 0$$
    $$x + 13 = 0 \quad \text{or} \quad x - 5 = 0$$
    $$x = -13 \quad \text{or} \quad x = 5$$

5. $$x^2 = 25$$
    $$x = \pm\sqrt{25} = \pm 5$$

7. $$b^2 - 16 = 0$$
    $$b^2 = 16$$
    $$b = \pm\sqrt{16} = \pm 4$$

9. $$9b^2 - 16 = 0$$
    $$9b^2 = 16$$
    $$b^2 = \frac{16}{9}$$
    $$b = \pm\sqrt{\frac{16}{9}}$$
    $$b = \pm\frac{\sqrt{16}}{\sqrt{9}} = \pm\frac{4}{3}$$

11. $$b^2 + 16 = 0$$
    $$b^2 = -16$$
    $$b = \pm\sqrt{-16}$$
    Since $\sqrt{-16}$ is not a real number, the equation has no real solutions.

13. $$25x^2 = 4$$
    $$x^2 = \frac{4}{25}$$
    $$x = \pm\sqrt{\frac{4}{25}}$$
    $$x = \pm\frac{\sqrt{4}}{\sqrt{25}} = \pm\frac{2}{5}$$

15. $$36x^2 - 15 = 0$$
    $$36x^2 = 15$$
    $$x^2 = \frac{15}{36}$$
    $$x = \pm\sqrt{\frac{15}{36}}$$
    $$x = \pm\frac{\sqrt{15}}{\sqrt{36}} = \pm\frac{\sqrt{15}}{6}$$

17. $$3b^2 = 11$$
    $$b^2 = \frac{11}{3}$$
    $$b = \pm\sqrt{\frac{11}{3}} = \pm\frac{\sqrt{11}}{\sqrt{3}}$$
    $$b = \pm\frac{\sqrt{11}\sqrt{3}}{\sqrt{3}\sqrt{3}} = \pm\frac{\sqrt{33}}{3}$$

19. $$3b^2 = 12$$
    $$b^2 = 4$$
    $$b = \pm\sqrt{4} = \pm 2$$

21. $9a^2 = 20$

$a^2 = \dfrac{20}{9}$

$a = \pm\sqrt{\dfrac{20}{9}} = \pm\dfrac{\sqrt{20}}{\sqrt{9}} = \pm\dfrac{\sqrt{20}}{3}$

$a = \pm\dfrac{\sqrt{4}\sqrt{5}}{3} = \pm\dfrac{2\sqrt{5}}{3}$

23. $3y^2 = 32$

$y^2 = \dfrac{32}{3}$

$y = \pm\sqrt{\dfrac{32}{3}} = \pm\dfrac{\sqrt{32}}{\sqrt{3}} = \dfrac{\sqrt{16}\sqrt{2}}{\sqrt{3}}$

$y = \pm\dfrac{4\sqrt{2}\sqrt{3}}{\sqrt{3}\sqrt{3}} = \pm\dfrac{4\sqrt{6}}{3}$

25. $7y^2 - 4 = 5y^2 + 6$

$2y^2 - 4 = 6$

$2y^2 = 10$

$y^2 = 5$

$y = \pm\sqrt{5}$

27. $5a^2 - 3a + 4 = 2a^2 - 3a + 13$

$3a^2 - 3a + 4 = -3a + 13$

$3a^2 + 4 = 13$

$3a^2 = 9$

$a^2 = 3$

$a = \pm\sqrt{3}$

29. $3a^2 - 18 = 5a^2 - 10$

$-2a^2 - 18 = -10$

$-2a^2 = 8$

$a^2 = -4$

$a = \pm\sqrt{-4}$

Since $\sqrt{-4}$ is not a real number, the equation has no real solutions.

31. $(y-3)(y+4) = y$

$y^2 + y - 12 = y$

$y^2 - 12 = 0$

$y^2 = 12$

$y = \pm\sqrt{12}$

$y = \pm\sqrt{4}\sqrt{3} = \pm 2\sqrt{3}$

33. $(2v-1)(v+2) = 3(v+4)$

$2v^2 + 3v - 2 = 3v + 12$

$2v^2 - 2 = 12$

$2v^2 = 14$

$v^2 = 7$

$v = \pm\sqrt{7}$

35. $(x+2)^2 = 4(x+7)$

$x^2 + 4x + 4 = 4x + 28$

$x^2 + 4 = 28$

$x^2 = 24$

$x = \pm\sqrt{24}$

$x = \pm\sqrt{4}\sqrt{6} = \pm 2\sqrt{6}$

37. $(t-2)^2 = 9$

$t - 2 = \pm\sqrt{9}$

$t - 2 = \pm 3$

$t - 2 = 3$ or $t - 2 = -3$

$t = 5$ or $t = -1$

39. $(a+5)^2 = 7$

$a + 5 = \pm\sqrt{7}$

$a = -5 \pm \sqrt{7}$

41. $(x-6)^2 = 12$

$x - 6 = \pm\sqrt{12}$

$x = 6 \pm \sqrt{4}\sqrt{3}$

$x = 6 \pm 2\sqrt{3}$

43. $(x+5)^2 = 10$

$x + 5 = \pm\sqrt{10}$

$x = -5 \pm \sqrt{10}$

45.
$$(x-6)^2 = 3x$$
$$x^2 - 12x + 36 = 3x$$
$$x^2 - 15x + 36 = 0$$
$$(x-12)(x-3) = 0$$
$$x - 12 = 0 \quad \text{or} \quad x - 3 = 0$$
$$x = 12 \quad \text{or} \quad x = 3$$

47.
$$(2x-3)(x+4) = x(x+9)$$
$$2x^2 + 5x - 12 = x^2 + 9x$$
$$x^2 - 4x - 12 = 0$$
$$(x-6)(x+2) = 0$$
$$x - 6 = 0 \quad \text{or} \quad x + 2 = 0$$
$$x = 6 \quad \text{or} \quad x = -2$$

49.
$$\left(m - \frac{2}{3}\right)^2 = \frac{4}{9}$$
$$m - \frac{2}{3} = \pm\sqrt{\frac{4}{9}}$$
$$m - \frac{2}{3} = \pm\frac{\sqrt{4}}{\sqrt{9}}$$
$$m - \frac{2}{3} = \pm\frac{2}{3}$$
$$m - \frac{2}{3} = \frac{2}{3} \quad \text{or} \quad m - \frac{2}{3} = -\frac{2}{3}$$
$$m = \frac{4}{3} \quad \text{or} \quad m = 0$$

51.
$$\left(x + \frac{2}{5}\right)^2 = \frac{3}{25}$$
$$x + \frac{2}{5} = \pm\sqrt{\frac{3}{25}}$$
$$x + \frac{2}{5} = \pm\frac{\sqrt{3}}{\sqrt{25}}$$
$$x + \frac{2}{5} = \pm\frac{\sqrt{3}}{5}$$
$$x = -\frac{2}{5} \pm \frac{\sqrt{3}}{5} = \frac{-2 \pm \sqrt{3}}{5}$$

53.
$$\left(y - \frac{1}{2}\right)^2 = \frac{2}{3}$$
$$y - \frac{1}{2} = \pm\sqrt{\frac{2}{3}}$$
$$y - \frac{1}{2} = \pm\frac{\sqrt{2}}{\sqrt{3}}$$
$$y - \frac{1}{2} = \pm\frac{\sqrt{2}\sqrt{3}}{\sqrt{3}\sqrt{3}}$$
$$y = \frac{1}{2} \pm \frac{\sqrt{6}}{3} = \frac{3 \pm 2\sqrt{6}}{6}$$

55.
$$x + \frac{1}{x} = 2$$
$$x\left(x + \frac{1}{x}\right) = x(2)$$
$$x^2 + 1 = 2x$$
$$x^2 - 2x + 1 = 0$$
$$(x-1)^2 = 0$$
$$x - 1 = 0$$
$$x = 1$$

57.
$$\frac{x-1}{x+1} = \frac{x}{x+3}$$
$$(x+1)(x+3)\left(\frac{x-1}{x+1}\right) = (x+1)(x+3)\left(\frac{x}{x+3}\right)$$
$$(x+3)(x-1) = (x+1)(x)$$
$$x^2 + 2x - 3 = x^2 + x$$
$$2x - 3 = x$$
$$-3 = -x$$
$$3 = x$$

59.
$$\frac{2x}{x+2}+1=x$$
$$(x+2)\left(\frac{2x}{x+2}+1\right)=(x+2)x$$
$$\cancel{(x+2)}\left(\frac{2x}{\cancel{x+2}}\right)+(x+2)(1)=(x+2)x$$

$$2x+x+2=x^2+2x$$
$$3x+2=x^2+2x$$
$$0=x^2-x-2$$
$$0=(x-2)(x+1)$$
$$0=x-2 \quad \text{or} \quad 0=x+1$$
$$2=x \quad \text{or} \quad -1=x$$

61.
$$a-\frac{5a}{a+1}=\frac{5}{a+1}$$
$$(a+1)\left(a-\frac{5a}{a+1}\right)=(a+1)\left(\frac{5}{a+1}\right)$$
$$(a+1)(a)-\cancel{(a+1)}\left(\frac{5a}{\cancel{a+1}}\right)=\cancel{(a+1)}\left(\frac{5}{\cancel{a+1}}\right)$$
$$(a+1)(a)-5a=5$$
$$a^2+a-5a=5$$
$$a^2-4a=5$$
$$a^2-4a-5=0$$
$$(a-5)(a+1)=0$$

$$a-5=0 \quad \text{or} \quad a+1=0$$
$$a=5 \quad \text{or} \quad a=-1$$

Reject $a=-1$, since this value of a produces denominators of 0 when it is checked in the original equation, and division by zero is undefined.

63.
$$\frac{3}{x-2}+\frac{7}{x+2}=\frac{x+1}{x-2}$$
$$(x-2)(x+2)\left(\frac{3}{x-2}+\frac{7}{x+2}\right)=(x-2)(x+2)\left(\frac{x+1}{x-2}\right)$$
$$\cancel{(x-2)}(x+2)\left(\frac{3}{\cancel{x-2}}\right)+(x-2)\cancel{(x+2)}\left(\frac{7}{\cancel{x+2}}\right)=\cancel{(x-2)}(x+2)\left(\frac{x+1}{\cancel{x-2}}\right)$$
$$3(x+2)+7(x-2)=(x+2)(x+1)$$
$$3x+6+7x-14=x^2+3x+2$$
$$10x-8=x^2+3x+2$$
$$0=x^2-7x+10$$
$$0=(x-5)(x-2)$$

$$0=x-5 \quad \text{or} \quad 0=x-2$$
$$5=x \quad \text{or} \quad 2=x$$

Reject $x=2$, since this value of x produces denominators of 0 when it is checked in the original equation, and division by zero is undefined.

65. $2x^2 + 7x - 5 = 3x^2 + 9x - 4$
$7x - 5 = x^2 + 9x - 4$
$-5 = x^2 + 2x - 4$
$0 = x^2 + 2x + 1$
$0 = (x+1)(x+1)$
$0 = x + 1$ or $0 = x + 1$
$-1 = x$ or $-1 = x$
So $x = -1$.

67. $(y-2)(y+3) = y + 10$
$y^2 + y - 6 = y + 10$
$y^2 - 6 = 10$
$y^2 = 16$
$y = \pm\sqrt{16} = \pm 4$

69. $(y-2)(y+3) = (2y-7)(y+4)$
$y^2 + y - 6 = 2y^2 + y - 28$
$y - 6 = y^2 + y - 28$
$22 = y^2$
$\pm\sqrt{22} = y$

71. $\dfrac{2x}{x+1} = \dfrac{x}{x-1}$

$(x+1)(x-1)\left(\dfrac{2x}{x+1}\right) = (x+1)(x-1)\left(\dfrac{x}{x-1}\right)$

$2x(x-1) = x(x+1)$
$2x^2 - 2x = x^2 + x$
$x^2 - 2x = x$
$x^2 - 3x = 0$
$x(x-3) = 0$
$x = 0$ or $x - 3 = 0$
or $x = 3$

73. $4(x+1) = \dfrac{9}{x+1}$

$(x+1)[4(x+1)] = \dfrac{x+1}{1} \cdot \dfrac{9}{x+1}$

$4(x+1)^2 = 9$

$(x+1)^2 = \dfrac{9}{4}$

$x + 1 = \pm\sqrt{\dfrac{9}{4}} = \pm\dfrac{\sqrt{9}}{\sqrt{4}}$

$x + 1 = \pm\dfrac{3}{2}$

$x + 1 = \dfrac{3}{2}$ or $x + 1 = -\dfrac{3}{2}$

$x = \dfrac{1}{2}$ or $x = -\dfrac{5}{2}$

75. $x = \pm 2.65$

77. $k = \pm 2.58$

79. $x = \pm 2.22$

81. $a = 1.52, a = -1.22$

83. $y = 0.23, y = -1.19$

85. Step 1: We can add the same quantity, 7, to both sides of an equation to obtain an equivalent equation.
Step 2: We can add the same quantity, 4, to both sides of an equation to obtain an equivalent equation.
Step 3: $7 + 4 = 11$
Step 4: $x^2 + 4x + 4$ factors as the square of the sum $x + 2$.
Step 5: $u^2 = d$ implies that $u = \pm\sqrt{d}$.
Step 6: We can subtract the same quantity, 2, from both sides of an equation to obtain an equivalent equation.

86. $(2+\sqrt{3})^2 - 4(2+\sqrt{3}) + 1 = (2+\sqrt{3})(2+\sqrt{3}) - 4(2+\sqrt{3}) + 1$
$$= 4 + 2\sqrt{3} + 2\sqrt{3} + 3 - 8 - 4\sqrt{3} + 1$$
$$= 0$$

So $2 + \sqrt{3}$ is a solution to the equation $x^2 - 4x + 1 = 0$.

$(2-\sqrt{3})^2 - 4(2-\sqrt{3}) + 1 = (2-\sqrt{3})(2-\sqrt{3}) - 4(2-\sqrt{3}) + 1$
$$= 4 - 2\sqrt{3} - 2\sqrt{3} + 3 - 8 + 4\sqrt{3} + 1$$
$$= 0$$

So $2 - \sqrt{3}$ is also a solution to the equation $x^2 - 4x + 1 = 0$. (This is no coincidence. In fact, whenever $p + \sqrt{d}$ satisfies the quadratic equation $ax^2 + bx + c = 0$, where a, b, and c are integers, $a \neq 0$, it must be true that $p - \sqrt{d}$ also satisfies the equation.)

87. $\begin{cases} 2x + y = 6 \\ 3x - 2y = 23 \end{cases}$ $\xrightarrow[\text{as is}]{\text{multiply by 2}}$ $\begin{aligned} 4x + 2y &= 12 \\ 3x - 2y &= 23 \quad \text{Add.} \\ \hline 7x &= 35 \\ x &= 5 \end{aligned}$ $\quad\begin{aligned} 2x + y &= 6 \\ 2(5) + y &= 6 \\ 10 + y &= 6 \\ y &= -4 \end{aligned}$

Solution: $(5, -4)$

89. $\begin{cases} 6x - 4y = 9 \\ 3x - 2y = 2 \end{cases}$ $\xrightarrow[\text{multiply by } -2]{\text{as is}}$ $\begin{aligned} 6x - 4y &= 9 \\ -6x + 4y &= -4 \quad \text{Add} \\ \hline 0 &= 5 \end{aligned}$

Contradiction. This system has no solution.

91. Let x = number of hours needed if they work together.
$$\frac{x}{4} + \frac{x}{3} = 1$$
$$12\left(\frac{x}{4} + \frac{x}{3}\right) = 12(1)$$
$$3x + 4x = 12$$
$$7x = 12$$
$$x = \frac{12}{7} \text{ or } 1\frac{5}{7}$$

To prepare the meal together, Lorraine and Renaldo need $1\frac{5}{7}$ hours.

Exercises 11.2

1. $\begin{aligned} x^2 + 8x + 6 &= 0 \\ x^2 + 8x &= -6 \quad \left[\tfrac{1}{2}(8)\right]^2 = (4)^2 = 16 \\ \underline{+16 \quad +16}& \\ x^2 + 8x + 16 &= 10 \\ (x+4)^2 &= 10 \\ x + 4 &= \pm\sqrt{10} \\ x &= -4 \pm \sqrt{10} \end{aligned}$

3. $\begin{aligned} x^2 - 4x - 3 &= 0 \\ x^2 - 4x &= 3 \quad \left[\tfrac{1}{2}(-4)\right]^2 = (-2)^2 = 4 \\ \underline{+4 \quad +4}& \\ x^2 - 4x + 4 &= 7 \\ (x-2)^2 &= 7 \\ x - 2 &= \pm\sqrt{7} \\ x &= 2 \pm \sqrt{7} \end{aligned}$

5. $\begin{aligned} x^2 - 10x &= 15 \quad \left[\tfrac{1}{2}(-10)\right]^2 = (-5)^2 = 25 \\ +25 & +25 \\ \hline x^2 - 10x + 25 &= 40 \\ (x-5)^2 &= 40 \\ x - 5 &= \pm\sqrt{40} \\ x - 5 &= \pm 2\sqrt{10} \\ x &= 5 \pm 2\sqrt{10} \end{aligned}$

7. $\begin{aligned} a^2 - 8a - 20 &= 0 \\ a^2 - 8a &= 20 \quad \left[\tfrac{1}{2}(-8)\right]^2 = (-4)^2 = 16 \\ +16 & +16 \\ \hline a^2 - 8a + 16 &= 36 \\ (a-4)^2 &= 36 \\ a - 4 &= \pm\sqrt{36} \\ a - 4 &= \pm 6 \\ a - 4 = 6 \quad &\text{or} \quad a - 4 = -6 \\ a = 10 \quad &\text{or} \quad a = -2 \end{aligned}$

9. $\begin{aligned} -x^2 - 12x &= 6 \\ x^2 + 12x &= -6 \quad \left[\tfrac{1}{2}(12)\right]^2 = (6)^2 = 36 \\ +36 & +36 \\ \hline x^2 + 12x + 36 &= 30 \\ (x+6)^2 &= 30 \\ x + 6 &= \pm\sqrt{30} \\ x &= -6 \pm \sqrt{30} \end{aligned}$

11. $\begin{aligned} 2z^2 - 12z + 4 &= 0 \\ 2z^2 - 12z &= -4 \\ z^2 - 6z &= -2 \quad \left[\tfrac{1}{2}(-6)\right]^2 = (-3)^2 = 9 \\ +9 & +9 \\ \hline z^2 - 6z + 9 &= 7 \\ (z-3)^2 &= 7 \\ z - 3 &= \pm\sqrt{7} \\ z &= 3 \pm \sqrt{7} \end{aligned}$

13. $\begin{aligned} 10 &= 5y^2 + 20y \\ 2 &= y^2 + 4y \quad \left[\tfrac{1}{2}(4)\right]^2 = (2)^2 = 4 \\ +4 & +4 \\ \hline 6 &= y^2 + 4y + 4 \\ 6 &= (y+2)^2 \\ \pm\sqrt{6} &= y + 2 \\ -2 \pm \sqrt{6} &= y \end{aligned}$

15. $\begin{aligned} u^2 + 5u - 2 &= 0 \\ u^2 + 5u &= 2 \quad \left[\tfrac{1}{2}(5)\right]^2 = \left(\tfrac{5}{2}\right)^2 = \tfrac{25}{4} \\ +\tfrac{25}{4} & +\tfrac{25}{4} \\ \hline u^2 + 5u + \tfrac{25}{4} &= 2 + \tfrac{25}{4} \end{aligned}$

$\left(u + \dfrac{5}{2}\right)^2 = \dfrac{33}{4}$

$u + \dfrac{5}{2} = \pm\sqrt{\dfrac{33}{4}} = \pm\dfrac{\sqrt{33}}{2}$

$u = -\dfrac{5}{2} \pm \dfrac{\sqrt{33}}{2} = \dfrac{-5 \pm \sqrt{33}}{2}$

17. $\begin{aligned} z^2 + 5 &= 7z \\ z^2 &= 7z - 5 \\ z^2 - 7z &= -5 \quad \left[\tfrac{1}{2}(-7)\right]^2 = \left(-\tfrac{7}{2}\right)^2 = \tfrac{49}{4} \\ +\tfrac{49}{4} & +\tfrac{49}{4} \\ \hline z^2 - 7z + \tfrac{49}{4} &= -5 + \tfrac{49}{4} \end{aligned}$

$\left(z - \dfrac{7}{2}\right)^2 = \dfrac{29}{4}$

$z - \dfrac{7}{2} = \pm\sqrt{\dfrac{29}{4}} = \pm\dfrac{\sqrt{29}}{2}$

$z = \dfrac{7}{2} \pm \dfrac{\sqrt{29}}{2} = \dfrac{7 \pm \sqrt{29}}{2}$

19.
$$w^2 - 3w = 2w^2 - 7w + 2$$
$$-w^2 - 3w = -7w + 2$$
$$-w^2 + 4w = 2$$
$$w^2 - 4w = -2 \quad \left[\tfrac{1}{2}(-4)\right]^2 = (-2)^2 = 4$$
$$\underline{+4 \quad +4}$$
$$w^2 - 4w + 4 = 2$$
$$(w-2)^2 = 2$$
$$w - 2 = \pm\sqrt{2}$$
$$w = 2 \pm \sqrt{2}$$

21.
$$x^2 + 4x + 5 = 2x - 3$$
$$x^2 + 2x + 5 = -3$$
$$x^2 + 2x = -8 \quad \left[\tfrac{1}{2}(2)\right]^2 = (1)^2 = 1$$
$$\underline{+1 \quad +1}$$
$$x^2 + 2x + 1 = -7$$
$$(x+1)^2 = -7$$
$$x + 1 = \pm\sqrt{-7}$$
Since $\sqrt{-7}$ is not a real number, the equation has no real solutions.

23.
$$(x-3)(x+2) = 9x - 1$$
$$x^2 - x - 6 = 9x - 1$$
$$x^2 - 10x - 6 = -1$$
$$x^2 - 10x = 5 \quad \left[\tfrac{1}{2}(-10)\right]^2 = (-5)^2 = 25$$
$$\underline{+25 \quad +25}$$
$$x^2 - 10x + 25 = 30$$
$$(x-5)^2 = 30$$
$$x - 5 = \pm\sqrt{30}$$
$$x = 5 \pm \sqrt{30}$$

25.
$$(a-2)(a+1) = 6$$
$$a^2 - a - 2 = 6$$
$$a^2 - a = 8 \quad \left[\tfrac{1}{2}(-1)\right]^2 = (-\tfrac{1}{2})^2 = \tfrac{1}{4}$$
$$\underline{+\tfrac{1}{4} \quad +\tfrac{1}{4}}$$
$$a^2 - a + \tfrac{1}{4} = 8 + \tfrac{1}{4}$$
$$\left(a - \tfrac{1}{2}\right)^2 = \tfrac{33}{4}$$
$$a - \tfrac{1}{2} = \pm\sqrt{\tfrac{33}{4}} = \pm\tfrac{\sqrt{33}}{2}$$
$$a = \tfrac{1}{2} \pm \tfrac{\sqrt{33}}{2}$$
$$a = \tfrac{1 \pm \sqrt{33}}{2}$$

27.
$$(x-4)(x+3) = 1 - x$$
$$x^2 - x - 12 = 1 - x$$
$$x^2 - 12 = 1$$
$$x^2 = 13$$
$$x = \pm\sqrt{13}$$

29.
$$2x^2 + 3 = 6x$$
$$2x^2 - 6x + 3 = 0$$
$$2x^2 - 6x = -3$$
$$x^2 - 3x = -\tfrac{3}{2} \quad \left[\tfrac{1}{2}(-3)\right]^2 = (-\tfrac{3}{2})^2 = \tfrac{9}{4}$$
$$\underline{+\tfrac{9}{4} \quad +\tfrac{9}{4}}$$
$$x^2 - 3x + \tfrac{9}{4} = -\tfrac{3}{2} + \tfrac{9}{4}$$
$$\left(x - \tfrac{3}{2}\right)^2 = \tfrac{3}{4}$$
$$x - \tfrac{3}{2} = \pm\sqrt{\tfrac{3}{4}} = \pm\tfrac{\sqrt{3}}{2}$$
$$x = \tfrac{3}{2} \pm \tfrac{\sqrt{3}}{2} = \tfrac{3 \pm \sqrt{3}}{2}$$

31.
$$6t^2 - 5t = t^2 + 3t - 1$$
$$5t^2 - 5t = 3t - 1$$
$$5t^2 - 8t = -1$$
$$t^2 - \frac{8}{5}t = -\frac{1}{5} \quad \left[\frac{1}{2}\left(-\frac{8}{5}\right)\right]^2 = \left(-\frac{4}{5}\right)^2 = \frac{16}{25}$$
$$\underline{\phantom{t^2 - \frac{8}{5}t} + \frac{16}{25} + \frac{16}{25}}$$
$$t^2 - \frac{8}{5}t + \frac{16}{25} = -\frac{1}{5} + \frac{16}{25}$$
$$\left(t - \frac{4}{5}\right)^2 = \frac{11}{25}$$
$$t - \frac{4}{5} = \pm\sqrt{\frac{11}{25}}$$
$$t = \frac{4}{5} \pm \frac{\sqrt{11}}{5} = \frac{4 \pm \sqrt{11}}{5}$$

33.
$$4z^2 + 20z + 19 = 0$$
$$4z^2 + 20z = -19$$
$$z^2 + 5z = -\frac{19}{4} \quad \left[\frac{1}{2}(5)\right]^2 = \left(\frac{5}{2}\right)^2 = \frac{25}{4}$$
$$\underline{ + \frac{25}{4} + \frac{25}{4}}$$
$$z^2 + 5z + \frac{25}{4} = \frac{-19}{4} + \frac{25}{4}$$
$$\left(z + \frac{5}{2}\right)^2 = \frac{6}{4}$$
$$z + \frac{5}{2} = \pm\sqrt{\frac{6}{4}} = \pm\frac{\sqrt{6}}{2}$$
$$z = -\frac{5}{2} \pm \frac{\sqrt{6}}{2}$$
$$z = \frac{-5 \pm \sqrt{6}}{2}$$

35.
$$(x + 3)^2 = 6$$
$$x + 3 = \pm\sqrt{6}$$
$$x = -3 \pm \sqrt{6}$$

37.
$$(x + 3)^2 = 6x$$
$$x^2 + 6x + 9 = 6x$$
$$x^2 + 9 = 0$$
$$x^2 = -9$$
This has no real solution.

39.
$$x^2 + 8x - 9 = 0$$
$$(x + 9)(x - 1) = 0$$
$$x + 9 = 0 \quad \text{or} \quad x - 1 = 0$$
$$x = -9 \quad \text{or} \quad x = 1$$

41.
$$3x^2 + 4 = 8x$$
$$3x^2 - 8x + 4 = 0$$
$$(3x - 2)(x - 2) = 0$$
$$3x - 2 = 0 \quad \text{or} \quad x - 2 = 0$$
$$x = \frac{2}{3} \quad \text{or} \quad x = 2$$

43. The constant of a perfect square is the square of one-half of the coefficient of the middle term, provided that the leading coefficient is equal to 1.

44.
$$\frac{x}{x - 1} = \frac{2}{x - 2}$$
$$\frac{(x-1)(x-2)}{1} \cdot \frac{x}{x-1} = \frac{(x-1)(x-2)}{1} \cdot \frac{2}{x-2}$$
$$x(x - 2) = 2(x - 1)$$
$$x^2 - 2x = 2x - 2$$
$$x^2 - 4x = -2$$
$$\underline{ + 4 + 4}$$
$$x^2 - 4x + 4 = 2$$
$$(x - 2)^2 = 2$$
$$x - 2 = \pm\sqrt{2}$$
$$x = 2 \pm \sqrt{2}$$

CHECK: $x = 2 + \sqrt{2}$:

$$\frac{x}{x-1} = \frac{2}{x-2}$$

$$\frac{2+\sqrt{2}}{2+\sqrt{2}-1} \stackrel{?}{=} \frac{2}{2+\sqrt{2}-2}$$

$$\frac{2+\sqrt{2}}{1+\sqrt{2}} \stackrel{?}{=} \frac{2}{\sqrt{2}}$$

$$\frac{2+\sqrt{2}}{1+\sqrt{2}} \cdot \frac{1-\sqrt{2}}{1-\sqrt{2}} \stackrel{?}{=} \frac{2}{\sqrt{2}} \cdot \frac{\sqrt{2}}{\sqrt{2}}$$

$$\frac{2-2\sqrt{2}+\sqrt{2}-2}{1-2} \stackrel{?}{=} \frac{2\sqrt{2}}{2}$$

$$\frac{-\sqrt{2}}{-1} \stackrel{?}{=} \sqrt{2}$$

$$\sqrt{2} \stackrel{\checkmark}{=} \sqrt{2}$$

CHECK: $x = 2 - \sqrt{2}$:

$$\frac{x}{x-1} = \frac{2}{x-2}$$

$$\frac{2-\sqrt{2}}{2-\sqrt{2}-1} \stackrel{?}{=} \frac{2}{2-\sqrt{2}-2}$$

$$\frac{2-\sqrt{2}}{1-\sqrt{2}} \stackrel{?}{=} \frac{2}{-\sqrt{2}}$$

$$\frac{2-\sqrt{2}}{1-\sqrt{2}} \cdot \frac{1+\sqrt{2}}{1+\sqrt{2}} \stackrel{?}{=} \frac{2}{-\sqrt{2}} \cdot \frac{\sqrt{2}}{\sqrt{2}}$$

$$\frac{2+2\sqrt{2}-\sqrt{2}-2}{1-2} \stackrel{?}{=} \frac{2\sqrt{2}}{-2}$$

$$\frac{\sqrt{2}}{-1} \stackrel{?}{=} -\sqrt{2}$$

$$-\sqrt{2} \stackrel{\checkmark}{=} -\sqrt{2}$$

45. $\sqrt{48} - \sqrt{75} = \sqrt{16 \cdot 3} - \sqrt{25 \cdot 3}$
$= \sqrt{16}\sqrt{3} - \sqrt{25}\sqrt{3}$
$= 4\sqrt{3} - 5\sqrt{3} = -\sqrt{3}$

47. $\sqrt{\frac{2}{7}} = \frac{\sqrt{2}}{\sqrt{7}} = \frac{\sqrt{2}\sqrt{7}}{\sqrt{7}\sqrt{7}} = \frac{\sqrt{14}}{7}$

49. $\frac{3}{3-\sqrt{2}} = \frac{3(3+\sqrt{2})}{(3-\sqrt{2})(3+\sqrt{2})}$
$= \frac{3(3+\sqrt{2})}{9-2} = \frac{3(3+\sqrt{2})}{7}$

51. $\frac{6+4\sqrt{3}}{2} = \frac{\overset{3}{\cancel{6}}}{\cancel{2}} + \frac{\overset{2}{\cancel{4}}\sqrt{3}}{\cancel{2}} = 3 + 2\sqrt{3}$

53. Let $x =$ number of hours Yvonne needs to finish the puzzle.

$$\frac{3}{10} + \frac{x}{8} = 1$$

$$40\left(\frac{3}{10} + \frac{x}{8}\right) = 40(1)$$

$$12 + 5x = 40$$

$$5x = 28$$

$$x = \frac{28}{5} \text{ or } 5\frac{3}{5}$$

It takes Yvonne $5\frac{3}{5}$ hours or 5 hours and 36 minutes to finish the puzzle.

Exercises 11.3

1. $x^2 + 3x - 5 = 0$
   $a = 1, b = 3, c = -5$

3. $t^2 - 7t = 6$
   $t^2 - 7t - 6 = 0$
   $a = 1, b = -7, c = -6$

5. $2u^2 = 8u$
    $2u^2 - 8u = 0$
    $a = 2, b = -8, c = 0$

7. $3x^2 - 11 = 0$
    $a = 3, b = 0, c = -11$

9. $x^2 + 3x - 5 = 0$
    $a = 1, b = 3, c = -5$
    $x = \dfrac{-b \pm \sqrt{b^2 - 4ac}}{2a}$
    $x = \dfrac{-(3) \pm \sqrt{(3)^2 - 4(1)(-5)}}{2(1)}$
    $x = \dfrac{-3 \pm \sqrt{9 + 20}}{2}$
    $x = \dfrac{-3 \pm \sqrt{29}}{2}$

11. $y^2 + 4y - 6 = 0$
    $a = 1, b = 4, c = -6$
    $y = \dfrac{-b \pm \sqrt{b^2 - 4ac}}{2a}$
    $y = \dfrac{-(4) \pm \sqrt{(4)^2 - 4(1)(-6)}}{2(1)}$
    $y = \dfrac{-4 \pm \sqrt{16 + 24}}{2} = \dfrac{-4 \pm \sqrt{40}}{2}$
    $y = \dfrac{-4 \pm 2\sqrt{10}}{2} = \dfrac{\cancel{2}(-2 \pm \sqrt{10})}{\cancel{2}}$
    $y = -2 \pm \sqrt{10}$

13. $u^2 - 2u + 3 = 0$
    $a = 1, b = -2, c = 3$
    $u = \dfrac{-b \pm \sqrt{b^2 - 4ac}}{2a}$
    $u = \dfrac{-(-2) \pm \sqrt{(-2)^2 - 4(1)(3)}}{2(1)}$
    $u = \dfrac{2 \pm \sqrt{4 - 12}}{2}$
    $u = \dfrac{2 \pm \sqrt{-8}}{2}$
    No real solutions, since the answer involves the square root of a negative number.

15. $t^2 - 7t = 6$
    $t^2 - 7t - 6 = 0$
    $a = 1, b = -7, c = -6$
    $t = \dfrac{-b \pm \sqrt{b^2 - 4ac}}{2a}$
    $t = \dfrac{-(-7) \pm \sqrt{(-7)^2 - 4(1)(-6)}}{2(1)}$
    $t = \dfrac{7 \pm \sqrt{49 + 24}}{2} =$
    $t = \dfrac{7 \pm \sqrt{73}}{2}$

17. $2x^2 - 3x - 1 = 0$
    $a = 2, b = -3, c = -1$
    $x = \dfrac{-b \pm \sqrt{b^2 - 4ac}}{2a}$
    $x = \dfrac{-(-3) \pm \sqrt{(-3)^2 - 4(2)(-1)}}{2(2)}$
    $x = \dfrac{3 \pm \sqrt{9 + 8}}{4}$
    $x = \dfrac{3 \pm \sqrt{17}}{4}$

19. $5x^2 - x = 2$
    $5x^2 - x - 2 = 0$
    $a = 5, b = -1, c = -2$
    $x = \dfrac{-b \pm \sqrt{b^2 - 4ac}}{2a}$
    $x = \dfrac{-(-1) \pm \sqrt{(-1)^2 - 4(5)(-2)}}{2(5)}$
    $x = \dfrac{1 \pm \sqrt{1 + 40}}{10}$
    $x = \dfrac{1 \pm \sqrt{41}}{10}$

21. $t^2 - 3t + 4 = 2t^2 + 4t - 3$
$-3t + 4 = t^2 + 4t - 3$
$4 = t^2 + 7t - 3$
$0 = t^2 + 7t - 7$
$a = 1, b = 7, c = -7$
$t = \dfrac{-b \pm \sqrt{b^2 - 4ac}}{2a}$
$t = \dfrac{-(7) \pm \sqrt{(7)^2 - 4(1)(-7)}}{2(1)}$
$t = \dfrac{-7 \pm \sqrt{49 + 28}}{2}$
$t = \dfrac{-7 \pm \sqrt{77}}{2}$

23. $(5w + 2)(w - 1) = 3w + 1$
$5w^2 - 3w - 2 = 3w + 1$
$5w^2 - 6w - 2 = 1$
$5w^2 - 6w - 3 = 0$
$a = 5, b = -6, c = -3$
$w = \dfrac{-b \pm \sqrt{b^2 - 4ac}}{2a}$
$w = \dfrac{-(-6) \pm \sqrt{(-6)^2 - 4(5)(-3)}}{2(5)}$
$w = \dfrac{6 \pm \sqrt{36 + 60}}{10} = \dfrac{6 \pm \sqrt{96}}{10}$
$w = \dfrac{6 \pm 4\sqrt{6}}{10} = \dfrac{\cancel{2}(3 \pm 2\sqrt{6})}{\underset{5}{\cancel{10}}}$
$w = \dfrac{3 \pm 2\sqrt{6}}{5}$

25. $(x - 1)^2 = x(x - 5)$
$x^2 - 2x + 1 = x^2 - 5x$
$-2x + 1 = -5x$
$1 = -3x$
$-\dfrac{1}{3} = x$

27. $3x^2 - 5x + 7 = 2x(x - 5) + 9x + 5$
$3x^2 - 5x + 7 = 2x^2 - 10x + 9x + 5$
$3x^2 - 5x + 7 = 2x^2 - x + 5$
$x^2 - 5x + 7 = -x + 5$
$x^2 - 4x + 7 = 5$
$x^2 - 4x + 2 = 0$
$a = 1, b = -4, c = 2$
$x = \dfrac{-b \pm \sqrt{b^2 - 4ac}}{2a}$
$x = \dfrac{-(-4) \pm \sqrt{(-4)^2 - 4(1)(2)}}{2(1)}$
$x = \dfrac{4 \pm \sqrt{16 - 8}}{2} = \dfrac{4 \pm \sqrt{8}}{2}$
$x = \dfrac{4 \pm 2\sqrt{2}}{2} = \dfrac{\cancel{2}(2 \pm \sqrt{2})}{\cancel{2}}$
$x = 2 \pm \sqrt{2}$

29.
$$2u^2 = 6u + 3$$
$$2u^2 - 6u = 3$$
$$2u^2 - 6u - 3 = 0$$
$$a = 2, b = -6, c = -3$$
$$u = \frac{-b \pm \sqrt{b^2 - 4ac}}{2a}$$
$$u = \frac{-(-6) \pm \sqrt{(-6)^2 - 4(2)(-3)}}{2(2)}$$
$$u = \frac{6 \pm \sqrt{36 + 24}}{4} = \frac{6 \pm \sqrt{60}}{4}$$
$$u = \frac{6 \pm 2\sqrt{15}}{4} = \frac{\cancel{2}(3 \pm \sqrt{15})}{\cancel{4}_2}$$
$$u = \frac{3 \pm \sqrt{15}}{2}$$

31.
$$x^2(x - 1) = (x - 1)^3$$
$$x^3 - x^2 = x^3 - 3x^2 + 3x - 1$$
$$-x^2 = -3x^2 + 3x - 1$$
$$0 = -2x^2 + 3x - 1$$
$$0 = 2x^2 - 3x + 1$$
$$0 = (2x - 1)(x - 1)$$
$$0 = 2x - 1 \quad \text{or} \quad 0 = x - 1$$
$$1 = 2x \quad \text{or} \quad 1 = x$$
$$\frac{1}{2} = x \quad \text{or} \quad 1 = x$$

33.
$$4x = 9x^2$$
$$0 = 9x^2 - 4x$$
$$0 = x(9x - 4)$$
$$0 = x \quad \text{or} \quad 0 = 9x - 4$$
$$0 = x \quad \text{or} \quad 4 = 9x$$
$$0 = x \quad \text{or} \quad \frac{4}{9} = x$$

35.
$$4 = 9x^2$$
$$\frac{4}{9} = x^2$$
$$\pm\sqrt{\frac{4}{9}} = x$$
$$\pm\frac{2}{3} = x$$

37.
$$\frac{w}{2} = \frac{3}{w + 2}$$
$$\frac{\cancel{2}(w + 2)}{1} \cdot \frac{w}{\cancel{2}} = \frac{2(\cancel{w + 2})}{1} \cdot \frac{3}{\cancel{w+2}}$$
$$w(w + 2) = 6$$
$$w^2 + 2w = 6$$
$$w^2 + 2w - 6 = 0$$
$$a = 1, b = 2, c = -6$$
$$w = \frac{-b \pm \sqrt{b^2 - 4ac}}{2a}$$
$$w = \frac{-(2) \pm \sqrt{(2)^2 - 4(1)(-6)}}{2(1)}$$
$$w = \frac{-2 \pm \sqrt{4 + 24}}{2} = \frac{-2 \pm \sqrt{28}}{2}$$
$$w = \frac{-2 \pm 2\sqrt{7}}{2} = \frac{\cancel{2}(-1 \pm \sqrt{7})}{\cancel{2}} = -1 \pm \sqrt{7}$$

39.
$$\frac{y}{y + 1} = \frac{y + 2}{3y}$$
$$\frac{3y(\cancel{y+1})}{1} \cdot \frac{y}{(\cancel{y+1})} = \frac{\cancel{3y}(y + 1)}{1} \cdot \frac{y + 2}{\cancel{3y}}$$
$$3y^2 = (y + 1)(y + 2)$$
$$3y^2 = y^2 + 3y + 2$$
$$0 = -2y^2 + 3y + 2$$
$$0 = 2y^2 - 3y - 2$$
$$0 = (2y + 1)(y - 2)$$
$$0 = 2y + 1 \quad \text{or} \quad 0 = y - 2$$
$$-1 = 2y \qquad\qquad 2 = y$$
$$-\frac{1}{2} = y$$

41. $x = 1.62, x = -0.62$

43. $t = 8.22, t = -1.22$

45. $w = 1.72, w = -4.06$

47. $x = 3.06, x = -1.17$

49.
$$R = s^2 - 250s + 600$$
$$50,000 = s^2 - 250s + 600$$
$$0 = s^2 - 250s - 49,400$$
$$0 = (s - 380)(s + 130)$$

$0 = s - 380$ or $0 = s + 130$
$380 = s$ or $-130 = s$
Reject -130, since it makes no sense to manufacture a negative number of square feet of plastic. So the firm should manufacture 380 sq. ft. of plastic.

51.
$$h = 120 + 40t - 16t^2$$
$$140 = 120 + 40t - 16t^2$$
$$16t^2 - 40t + 20 = 0$$
$$4t^2 - 10t + 5 = 0$$
$$a = 4, b = -10, c = 5$$
$$t = \frac{-b \pm \sqrt{b^2 - 4ac}}{2a}$$
$$t = \frac{-(-10) \pm \sqrt{(-10)^2 - 4(4)(5)}}{2(4)}$$
$$t = \frac{10 \pm \sqrt{100 - 80}}{8} = \frac{10 \pm \sqrt{20}}{8}$$
$$t = \frac{10 \pm 2\sqrt{5}}{8} = \frac{2(5 \pm \sqrt{5})}{8} = \frac{5 \pm \sqrt{5}}{4}$$

$$t = \frac{5 + \sqrt{5}}{4} \quad \text{or} \quad t = \frac{5 - \sqrt{5}}{4}$$
$$t = 1.8 \quad \text{or} \quad t = 0.7$$
(rounded to the nearest tenth)

The ball is 140 ft. above the ground after 0.7 seconds (when it is on the way up) and again after 1.8 seconds (when it is on the way down).

53.
$$2x^2 + 7x + 4 = 0$$
$$a = 2, b = 7, c = 4$$
$$x = \frac{-b \pm \sqrt{b^2 - 4ac}}{2a}$$
$$x = \frac{-(7) \pm \sqrt{(7)^2 - 4(2)(4)}}{2(2)}$$
$$x = \frac{-7 \pm \sqrt{49 - 32}}{4}$$
$$x = \frac{-7 \pm \sqrt{17}}{4}$$

Using the quadratic formula is easier than the method of completing the square.

54. The factoring method, when it works, is usually the easiest method to use. However, it does not always work. Completing the square and the quadratic formula work for any quadratic equation. Generally, completing the square is the most complicated of these two methods.

55. (a) Since $b = -3$ and $c = -1$, $x = \dfrac{-(-3) \pm \sqrt{9 - 4(-1)}}{2} = \dfrac{3 \pm \sqrt{9 + 4}}{2} = \dfrac{3 \pm \sqrt{13}}{2}$.

(b) The minus under the square root should be a plus sign.

(c) The "5" should be divided by 2 as well. That is,
$$x = \frac{5 \pm \sqrt{25 - 12}}{2} = \frac{5 \pm \sqrt{13}}{2}.$$

(d) This is correct up until the last step. Then
$$x = \frac{6 \pm 4\sqrt{3}}{2} = \frac{2(3 \pm 2\sqrt{3})}{2} = 3 \pm 2\sqrt{3}, \text{ not } 6 \pm 2\sqrt{3}.$$

57. $3\sqrt{2}(\sqrt{2}+\sqrt{7})+\sqrt{7}(5\sqrt{2}-\sqrt{7}) = 3\sqrt{2}\sqrt{2}+3\sqrt{2}\sqrt{7}+5\sqrt{2}\sqrt{7}-\sqrt{7}\sqrt{7}$
$= 3(2)+3\sqrt{14}+5\sqrt{14}-7$
$= 6+8\sqrt{14}-7$
$= 8\sqrt{14}-1$

59. $(2\sqrt{x}-\sqrt{5})(3\sqrt{x}-4\sqrt{5}) = (2\sqrt{x})(3\sqrt{x})-(2\sqrt{x})(4\sqrt{5})-(\sqrt{5})(3\sqrt{x})+(\sqrt{5})(4\sqrt{5})$
$= 6x-8\sqrt{5x}-3\sqrt{5x}+20$
$= 6x-11\sqrt{5x}+20$

61. Let p = price of a pastrami sandwich.
Let t = price of a tuna sandwich.

$\begin{cases} 5p+6t = 42.00 \\ 4p+9t = 47.25 \end{cases}$ $\xrightarrow{\text{multiply by 3}}$ $\quad 15p+18t = 126.00 \qquad\qquad 5p+6t = 42.00$
$\xrightarrow{\text{multiply by }-2}$ $\quad \underline{-8p-18t = -94.50}$ Add. $\quad 5(4.50)+6t = 42.00$
$\qquad\qquad\qquad\qquad\qquad\qquad\qquad 7p \quad = 31.50 \qquad\qquad\qquad 22.50+6t = 42.00$
$\qquad\qquad\qquad\qquad\qquad\qquad\qquad\quad p = 4.50 \qquad\qquad\qquad\qquad\quad 6t = 19.50$
$\qquad\qquad\qquad\qquad\qquad\qquad\qquad\qquad\qquad\qquad\qquad\qquad\qquad\qquad\quad t = 3.25$

So a pastrami sandwich costs \$4.50 and a tuna sandwich costs \$3.25.

Exercises 11.4

1. $x^2+6x+5 = 0$
$(x+1)(x+5) = 0$
$x+1 = 0 \quad\text{or}\quad x+5 = 0$
$\quad x = -1 \quad\text{or}\qquad x = -5$

$x^2+6x \quad = -5$
$\underline{\qquad +9 \quad +9}$
$x^2+6x+9 = 4$
$(x+3)^2 = 4$
$x+3 = \pm\sqrt{4} = \pm 2$
$x+3 = 2 \quad\text{or}\quad x+3 = -2$
$\quad x = -1 \quad\text{or}\qquad x = -5$

3. $x^2+6x-5 = 0$
$x^2+6x \quad = 5 \quad \left[\tfrac{1}{2}(6)\right]^2 = (3)^2 = 9$
$\underline{\qquad +9 \quad +9}$
$x^2+6x+9 = 14$
$(x+3)^2 = 14$
$x+3 = \pm\sqrt{14}$
$\quad x = -3\pm\sqrt{14}$

$a = 1, b = 6, c = -5$
$x = \dfrac{-b\pm\sqrt{b^2-4ac}}{2a}$
$x = \dfrac{-(6)\pm\sqrt{(6)^2-4(1)(-5)}}{2(1)}$
$x = \dfrac{-6\pm\sqrt{36+20}}{2} = \dfrac{-6\pm\sqrt{56}}{2}$
$x = \dfrac{-6\pm 2\sqrt{14}}{2} = \dfrac{\cancel{2}(-3\pm\sqrt{14})}{\cancel{2}}$
$x = -3\pm\sqrt{14}$

5.
$$2r^2 + 1 = 3r$$
$$2r^2 - 3r + 1 = 0$$
$$(2r - 1)(r - 1) = 0$$
$$2r - 1 = 0 \quad \text{or} \quad r - 1 = 0$$
$$2r = 1 \quad \text{or} \quad r = 1$$
$$r = \frac{1}{2} \quad \text{or} \quad r = 1$$

$$a = 2, b = -3, c = 1$$
$$r = \frac{-b \pm \sqrt{b^2 - 4ac}}{2a}$$
$$r = \frac{-(-3) \pm \sqrt{(-3)^2 - 4(2)(1)}}{2(2)}$$
$$r = \frac{3 \pm \sqrt{9 - 8}}{4} = \frac{3 \pm \sqrt{1}}{4} = \frac{3 \pm 1}{4}$$
$$r = \frac{3 + 1}{4} = \frac{4}{4} = 1 \quad \text{or} \quad t = \frac{3 - 1}{4} = \frac{2}{4} = \frac{1}{2}$$

7.
$$w^2 = 4w + 5$$
$$w^2 - 4w = 5$$
$$w^2 - 4w - 5 = 0$$
$$(w - 5)(w + 1) = 0$$
$$w - 5 = 0 \quad \text{or} \quad w + 1 = 0$$
$$w = 5 \quad \text{or} \quad w = -1$$

$$w^2 = 4w + 5$$
$$w^2 - 4w = 5$$
$$\underline{ + 4 +4}$$
$$w^2 - 4w + 4 = 9$$
$$(w - 2)^2 = 9$$
$$w - 2 = \pm\sqrt{9} = \pm 3$$
$$w - 2 = 3 \quad \text{or} \quad w - 2 = -3$$
$$w = 5 \quad \text{or} \quad w = -1$$

9.
$$(x + 1)(x + 2) = (x + 3)(x + 4)$$
$$x^2 + 3x + 2 = x^2 + 7x + 12$$
$$3x + 2 = 7x + 12$$
$$2 = 4x + 12$$
$$-10 = 4x$$
$$-\frac{5}{2} = x$$

11.
$$(a + 1)^2 - (a + 3)(a - 2) = 2a + 6$$
$$a^2 + 2a + 1 - (a^2 + a - 6) = 2a + 6$$
$$a^2 + 2a + 1 - a^2 - a + 6 = 2a + 6$$
$$a + 7 = 2a + 6$$
$$7 = a + 6$$
$$1 = a$$

13.
$$4x^2 = 16x - 28$$
$$4x^2 - 16x = -28$$
$$x^2 - 4x = -7$$
$$\underline{+4 +4}$$
$$x^2 - 4x + 4 = -3$$
$$(x - 2)^2 = -3$$
$$x - 2 = \pm\sqrt{-3}$$

Thus, there are no real solutions, since square roots of negative numbers are not real.

$$4x^2 = 16x - 28$$
$$4x^2 - 16x = -28$$
$$4x^2 - 16x + 28 = 0$$
$$x^2 - 4x + 7 = 0$$
$$a = 1, b = -4, c = 7$$
$$x = \frac{-b \pm \sqrt{b^2 - 4ac}}{2a}$$
$$x = \frac{-(-4) \pm \sqrt{(-4)^2 - 4(1)(7)}}{2(1)}$$
$$x = \frac{4 \pm \sqrt{16 - 28}}{2} = \frac{4 \pm \sqrt{-12}}{2}$$, which leads to the same conclusion for the same reason.

15. $(x-1)^2 = 5$ $\qquad$ $(x-1)^2 = 5$
$x - 1 = \pm\sqrt{5}$ $\qquad$ $x^2 - 2x + 1 = 5$
$x = 1 \pm \sqrt{5}$ $\qquad$ $x^2 - 2x - 4 = 0$
$\qquad\qquad\qquad$ $a = 1, b = -2, c = -4$

$$x = \frac{-b \pm \sqrt{b^2 - 4ac}}{2a} = \frac{-(-2) \pm \sqrt{(-2)^2 - 4(1)(-4)}}{2(1)} = \frac{2 \pm \sqrt{4 + 16}}{2}$$

$$= \frac{2 \pm \sqrt{20}}{2} = \frac{2 \pm 2\sqrt{5}}{2} = \frac{\cancel{2}(1 \pm \sqrt{5})}{\cancel{2}} = 1 \pm \sqrt{5}$$

17. $(x-1)^2 = 5x$ $\qquad\qquad\qquad\qquad$ $(x-1)^2 = 5x$
$x^2 - 2x + 1 = 5x$ $\qquad\qquad\qquad$ $x^2 - 2x + 1 = 5x$
$x^2 - 7x + 1 = 0$ $\qquad\qquad\qquad$ $x^2 - 7x + 1 = 0$
$a = 1, b = -7, c = 1$ $\qquad\qquad\qquad$ $x^2 - 7x = -1$

$x = \dfrac{-b \pm \sqrt{b^2 - 4ac}}{2a}$ $\qquad\qquad\qquad$ $+\dfrac{49}{4}\qquad +\dfrac{49}{4}$

$x = \dfrac{-(-7) \pm \sqrt{(-7)^2 - 4(1)(1)}}{2(1)}$ $\qquad$ $x^2 - 7x + \dfrac{49}{4} = -1 + \dfrac{49}{4}$

$x = \dfrac{7 \pm \sqrt{49 - 4}}{2} = \dfrac{7 \pm \sqrt{45}}{2}$ $\qquad$ $\left(x - \dfrac{7}{2}\right)^2 = \dfrac{45}{4}$

$x = \dfrac{7 \pm 3\sqrt{5}}{2}$ $\qquad\qquad\qquad\qquad$ $x - \dfrac{7}{2} = \pm\sqrt{\dfrac{45}{4}} = \pm\dfrac{\sqrt{45}}{2}$

$\qquad\qquad\qquad\qquad\qquad\qquad\qquad$ $x - \dfrac{7}{2} = \pm\dfrac{3\sqrt{5}}{2}$

$\qquad\qquad\qquad\qquad\qquad\qquad\qquad$ $x = \dfrac{7}{2} \pm \dfrac{3\sqrt{5}}{2} = \dfrac{7 \pm 3\sqrt{5}}{2}$

19. $(u + 2)^2 = 4(u + 5)$ $\qquad\qquad\qquad$ $u^2 + 4 = 20$
$u^2 + 4u + 4 = 4u + 20$ $\qquad\qquad\qquad$ $u^2 - 16 = 0$
$u^2 + 4 = 20$ $\qquad\qquad\qquad\qquad\qquad$ $(u + 4)(u - 4) = 0$
$u^2 = 16$ $\qquad\qquad\qquad\qquad\qquad\quad$ $u + 4 = 0 \quad\text{or}\quad u - 4 = 0$
$u = \pm\sqrt{16} = \pm 4$ $\qquad\qquad\qquad\qquad$ $u = -4 \quad\text{or}\quad u = 4$

21. $y^2 - 4y + 10 = 5(y + 2)$ $\qquad\qquad$ $y^2 - 9y = 0$
$y^2 - 4y + 10 = 5y + 10$ $\qquad\qquad\quad$ $a = 1, b = -9, c = 0$
$y^2 - 9x + 10 = 10$ $\qquad\qquad\qquad\quad$ $y = \dfrac{-b \pm \sqrt{b^2 - 4ac}}{2a}$
$y^2 - 9y = 0$
$y(y - 9) = 0$ $\qquad\qquad\qquad\qquad\quad$ $y = \dfrac{-(-9) \pm \sqrt{(-9)^2 - 4(1)(0)}}{2(1)}$
$y = 0 \quad\text{or}\quad y - 9 = 0$
$y = 0 \quad\text{or}\quad y = 9$ $\qquad\qquad\qquad$ $y = \dfrac{9 \pm \sqrt{81 - 0}}{2} = \dfrac{9 \pm \sqrt{81}}{2} = \dfrac{9 \pm 9}{2}$

$\qquad\qquad\qquad\qquad\qquad\qquad\qquad$ $y = \dfrac{9 + 9}{2} = \dfrac{18}{2} = 9 \text{ or } y = \dfrac{9 - 9}{2} = \dfrac{0}{2} = 0$

23.
$(t+4)(t-8) = 13$
$t^2 - 4t - 32 = 13$
$t^2 - 4t - 45 = 0$
$(t-9)(t+5) = 0$
$t - 9 = 0$ or $t + 5 = 0$
$t = 9$ or $t = -5$

$(t+4)(t-8) = 13$
$t^2 - 4t - 32 = 13$
$t^2 - 4t = 45$
$ +4 +4$
$\overline{t^2 - 4t + 4 = 49}$
$(t-2)^2 = 49$
$t - 2 = \pm\sqrt{49} = \pm 7$
$t - 2 = 7$ or $t - 2 = -7$
$t = 9$ or $t = -5$

25.
$(n+2)(n+1) = 3$
$n^2 + 3n + 2 = 3$
$n^2 + 3n - 1 = 0$
$a = 1, b = 3, c = -1$
$n = \frac{-b \pm \sqrt{b^2 - 4ac}}{2a}$
$n = \frac{-(3) \pm \sqrt{(3)^2 - 4(1)(-1)}}{2(1)}$
$n = \frac{-3 \pm \sqrt{9+4}}{2} = \frac{-3 \pm \sqrt{13}}{2}$

$n^2 + 3n + 2 = 3$
$n^2 + 3n = 1$
$ +\frac{9}{4} +\frac{9}{4}$
$\overline{n^2 + 3n + \frac{9}{4} = 1 + \frac{9}{4}}$
$\left(n + \frac{3}{2}\right)^2 = \frac{13}{4}$
$n + \frac{3}{2} = \pm\sqrt{\frac{13}{4}} = \pm\frac{\sqrt{13}}{2}$
$n = -\frac{3}{2} \pm \frac{\sqrt{13}}{2} = \frac{-3 \pm \sqrt{13}}{2}$

27.
$z^2 - 3z = 3z - 9$
$z^2 - 6z = -9$
$z^2 - 6z + 9 = 0$
$(z-3)^2 = 0$
$z - 3 = 0$
$z = 3$

$z^2 - 3z = 3z - 9$
$z^2 - 6z = -9$
$z^2 - 6z + 9 = 0$
$a = 1, b = -6, c = 9$
$z = \frac{-b \pm \sqrt{b^2 - 4ac}}{2a}$
$z = \frac{-(-6) \pm \sqrt{(-6)^2 - 4(1)(9)}}{2(1)} = \frac{6 \pm \sqrt{36-36}}{2}$
$z = \frac{6 \pm \sqrt{0}}{2} = \frac{6 \pm 0}{2} = \frac{6}{2} = 3$

29.
$16z + 12 = 3z^2$
$-3z^2 + 16z + 12 = 0$
$3z^2 - 16z - 12 = 0$
$(3z + 2)(z - 6) = 0$
$3z + 2 = 0$ or $z - 6 = 0$
$3z = -2$ $z = 6$
$z = -\frac{2}{3}$

$3z^2 - 16z - 12 = 0$
$a = 3, b = -16, c = -12$
$z = \frac{-b \pm \sqrt{b^2 - 4ac}}{2a} = \frac{-(-16) \pm \sqrt{(-16)^2 - 4(3)(-12)}}{2(3)}$
$z = \frac{16 \pm \sqrt{256 + 144}}{6} = \frac{16 \pm \sqrt{400}}{6} = \frac{16 \pm 20}{6}$
$z = \frac{16 + 20}{6} = \frac{36}{6} = 6$ or $x = \frac{16 - 20}{6} = \frac{-4}{6} = -\frac{2}{3}$

31.
$$x + \frac{1}{x} = 2$$
$$x\left(x + \frac{1}{x}\right) = 2x$$
$$x^2 + 1 = 2x$$
$$x^2 - 2x + 1 = 0$$
$$(x - 1)^2 = 0$$
$$x - 1 = 0$$
$$x = 1$$

$$x^2 - 2x + 1 = 0$$
$$a = 1, b = -2, c = 1$$
$$x = \frac{-b \pm \sqrt{b^2 - 4ac}}{2a} = \frac{-(-2) \pm \sqrt{(-2)^2 - 4(1)(1)}}{2(1)}$$
$$x = \frac{2 \pm \sqrt{4 - 4}}{2} = \frac{2 \pm \sqrt{0}}{2} = \frac{2 \pm 0}{2} = \frac{2}{2} = 1$$

33.
$$x^2 + 1 = \frac{5}{2}x$$
$$2(x^2 + 1) = 2\left(\frac{5}{2}x\right)$$
$$2x^2 + 2 = \frac{\cancel{2}}{1} \cdot \frac{5}{\cancel{2}}x$$
$$2x^2 + 2 = 5x$$
$$2x^2 - 5x + 2 = 0$$
$$(2x - 1)(x - 2) = 0$$
$$2x - 1 = 0 \quad \text{or} \quad x - 2 = 0$$
$$2x = 1 \quad \text{or} \quad x = 2$$
$$x = \frac{1}{2} \quad \text{or} \quad x = 2$$

$$2x^2 - 5x + 2 = 0$$
$$a = 2, b = -5, c = 2$$
$$x = \frac{-b \pm \sqrt{b^2 - 4ac}}{2a}$$
$$x = \frac{-(-5) \pm \sqrt{(-5)^2 - 4(2)(2)}}{2(2)}$$
$$x = \frac{5 \pm \sqrt{25 - 16}}{4}$$
$$x = \frac{5 \pm \sqrt{9}}{4} = \frac{5 \pm 3}{4}$$
$$x = \frac{5 + 3}{4} = \frac{8}{4} = 2 \quad \text{or} \quad x = \frac{5 - 3}{4} = \frac{2}{4} = \frac{1}{2}$$

35.
$$\frac{x}{x + 1} = \frac{4}{x + 4}$$
$$\frac{\cancel{(x + 1)}(x + 4)}{1} \cdot \frac{x}{\cancel{x + 1}} = \frac{(x + 1)\cancel{(x + 4)}}{1} \cdot \frac{4}{\cancel{x + 4}}$$
$$x(x + 4) = 4(x + 1)$$
$$x^2 + 4x = 4x + 4$$
$$x^2 = 4$$
$$x = \pm\sqrt{4} = \pm 2$$

$$x^2 = 4$$
$$x^2 - 4 = 0$$
$$(x - 2)(x + 2) = 0$$
$$x - 2 = 0 \quad \text{or} \quad x + 2 = 0$$
$$x = 2 \quad \text{or} \quad x = -2$$

37.
$$\frac{x}{x + 1} = \frac{x + 2}{x + 3}$$
$$\frac{\cancel{(x + 1)}(x + 3)}{1} \cdot \frac{x}{\cancel{x + 1}} = \frac{(x + 1)\cancel{(x + 3)}}{1} \cdot \frac{x + 2}{\cancel{x + 3}}$$
$$x(x + 3) = (x + 1)(x + 2)$$
$$x^2 + 3x = x^2 + 3x + 2$$
$$3x = 3x + 2$$
$$0 = 2$$

This is a contradiction, hence the original equation has no solution.

39.
$$\frac{3x}{x+1} + \frac{2}{x-1} = 4$$
$$\frac{(x+1)(x-1)}{1} \cdot \frac{3x}{x+1} + \frac{(x+1)(x-1)}{1} \cdot \frac{2}{x-1} = (x+1)(x-1)4$$
$$3x(x-1) + 2(x+1) = 4(x+1)(x-1)$$
$$3x^2 - 3x + 2x + 2 = 4(x^2-1)$$
$$3x^2 - x + 2 = 4x^2 - 4$$
$$-x + 2 = x^2 - 4$$
$$2 = x^2 + x - 4$$
$$0 = x^2 + x - 6$$
$$0 = (x+3)(x-2)$$
$$0 = x+3 \quad \text{or} \quad 0 = x-2$$
$$-3 = x \quad \text{or} \quad 2 = x$$

$x^2 + x - 6 = 0$
$a = 1, b = 1, c = -6$
$$x = \frac{-b \pm \sqrt{b^2 - 4ac}}{2a} = \frac{-(1) \pm \sqrt{(1)^2 - 4(1)(-6)}}{2(1)} = \frac{-1 \pm \sqrt{1+24}}{2}$$
$$x = \frac{-1 \pm \sqrt{25}}{2} = \frac{-1 \pm 5}{2}$$
$$x = \frac{-1+5}{2} = \frac{4}{2} = 2 \quad \text{or} \quad x = \frac{-1-5}{2} = \frac{-6}{2} = -3$$

41. $2x^2 + 3x = 20$

Method 1-factoring:
$2x^2 + 3x - 20 = 0$
$(2x - 5)(x + 4) = 0$
$2x - 5 = 0 \quad \text{or} \quad x + 4 = 0$
$\quad 2x = 5 \quad \text{or} \quad x = -4$
$\quad x = \frac{5}{2} \quad \text{or} \quad x = -4$

Method 2-completing the square:
$2x^2 + 3x = 20$
$x^2 + \frac{3}{2}x = 10$
$\qquad\quad + \frac{9}{16} \quad + \frac{9}{16}$
$x^2 + \frac{3}{2}x + \frac{9}{16} = 10 + \frac{9}{16}$
$\left(x + \frac{3}{4}\right)^2 = \frac{169}{16}$
$x + \frac{3}{4} = \pm\sqrt{\frac{169}{16}} = \pm\frac{\sqrt{169}}{\sqrt{16}}$
$x + \frac{3}{4} = \pm\frac{13}{4}$
$x = -\frac{3}{4} \pm \frac{13}{4} = \frac{-3 \pm 13}{4}$
$x = \frac{-3+13}{4} = \frac{10}{4} = \frac{5}{2}$
$\text{or } x = \frac{-3-13}{4} = \frac{-16}{4} = -4$

Method 3-quadratic formula:
$2x^2 + 3x - 20 = 0$
$a = 2, b = 3, c = -20$
$$x = \frac{-b \pm \sqrt{b^2 - 4ac}}{2a}$$
$$x = \frac{-(3) \pm \sqrt{(3)^2 - 4(2)(-20)}}{2(2)} = \frac{-3 \pm \sqrt{9 + 160}}{4} = \frac{-3 \pm \sqrt{169}}{4} = \frac{-3 \pm 13}{4}$$
$$x = \frac{-3 + 13}{4} = \frac{10}{4} = \frac{5}{2} \text{ or } x = \frac{-3 - 13}{4} = \frac{-16}{4} = -4$$

The easiest of these methods is the first, while the second one appears to be the most difficult.

42. $3x^2 - 5x - 1 = 0$
$a = 3, b = -5, c = -1$
$$x = \frac{-b \pm \sqrt{b^2 - 4ac}}{2a}$$
$$x = \frac{-(-5) \pm \sqrt{(-5)^2 - 4(3)(-1)}}{2(3)}$$
$$x = \frac{5 \pm \sqrt{25 + 12}}{6} = \frac{5 \pm \sqrt{37}}{6}$$

CHECK $x = \dfrac{5 + \sqrt{37}}{6}$:

$$3x^2 - 5x - 1 = 0$$
$$3\left(\frac{5 + \sqrt{37}}{6}\right)^2 - 5\left(\frac{5 + \sqrt{37}}{6}\right) - 1 \stackrel{?}{=} 0$$
$$3\left(\frac{5 + \sqrt{37}}{6}\right)\left(\frac{5 + \sqrt{37}}{6}\right) - 5\left(\frac{5 + \sqrt{37}}{6}\right) - 1 \stackrel{?}{=} 0$$
$$\frac{\cancel{3}}{1}\left(\frac{25 + 10\sqrt{37} + 37}{\underset{12}{\cancel{36}}}\right) - \frac{5}{1}\left(\frac{5 + \sqrt{37}}{6}\right) - 1 \stackrel{?}{=} 0$$
$$\frac{62 + 10\sqrt{37}}{12} - \frac{25 + 5\sqrt{37}}{6} - 1 \stackrel{?}{=} 0$$
$$\frac{\cancel{2}(31 + 5\sqrt{37})}{\underset{6}{\cancel{12}}} - \frac{25 + 5\sqrt{37}}{6} - 1 \stackrel{?}{=} 0$$
$$\frac{31 + 5\sqrt{37} - (25 + 5\sqrt{37})}{6} - 1 \stackrel{?}{=} 0$$
$$\frac{31 + 5\sqrt{37} - 25 - 5\sqrt{37}}{6} - 1 \stackrel{?}{=} 0$$
$$\frac{6}{6} - 1 \stackrel{?}{=} 0$$
$$1 - 1 \stackrel{?}{=} 0$$
$$0 \stackrel{\checkmark}{=} 0$$

$$3x^2 - 5x - 1 = 0$$
$$\underline{+1 \qquad +1}$$
$$3x^2 - 5x = 1$$
$$x^2 - \frac{5}{3}x = \frac{1}{3}$$
$$\underline{+\frac{25}{36} \qquad\qquad +\frac{25}{36}}$$
$$x^2 - \frac{5}{3}x + \frac{25}{36} = \frac{1}{3} + \frac{25}{36}$$
$$\left(x - \frac{5}{6}\right)^2 = \frac{37}{36}$$
$$x - \frac{5}{6} = \pm\sqrt{\frac{37}{36}} = \pm\frac{\sqrt{37}}{\sqrt{36}}$$
$$x - \frac{5}{6} = \pm\frac{\sqrt{37}}{6}$$
$$x = \frac{5}{6} \pm \frac{\sqrt{37}}{6} = \frac{5 \pm \sqrt{37}}{6}$$

This second method of checking was by far the easier in this example. The first check was more complicated than the actual solution. If the first check happened to fail, we really would not know whether an error was made in the solution to the problem or in the check of that solution. Further, the first check only verified that $x = \dfrac{5 + \sqrt{37}}{6}$ is a valid solution.

We would still have to examine $x = \dfrac{5 - \sqrt{37}}{6}$ and verify that it satisfied the equation as well.

43. (a) $m = \dfrac{y_2 - y_1}{x_2 - x_1} = \dfrac{-4 - 3}{3 - (-1)} = \dfrac{-7}{4}$

   (b) $m = 0$, since the points have the same y-coordinate.

   (c) m is undefined, since the points have the same x-coordinate.

45. $m = \dfrac{y_2 - y_1}{x_2 - x_1} = \dfrac{-1 - (-5)}{1 - (-2)} = \dfrac{-1 + 5}{1 + 2} = \dfrac{4}{3}$

$(x_1, y_1) = (-2, -5)$

$y - y_1 = m(x - x_1)$

$y - (-5) = \dfrac{4}{3}[x - (-2)]$

$y + 5 = \dfrac{4}{3}(x + 2)$ or $y = \dfrac{4}{3}x - \dfrac{7}{3}$

Exercises 11.5

1. Let x = the number.
$$x + \frac{1}{x} = \frac{13}{6}$$
$$6x\left(x + \frac{1}{x}\right) = 6x\left(\frac{13}{6}\right)$$
$$6x^2 + \frac{6\!\!\!/x}{1} \cdot \frac{1}{\!\!\!/x} = \frac{\!\!\!/6x}{1} \cdot \frac{13}{\!\!\!/6}$$
$$6x^2 + 6 = 13x$$
$$6x^2 - 13x + 6 = 0$$
$$(2x - 3)(3x - 2) = 0$$
$$2x - 3 = 0 \quad \text{or} \quad 3x - 2 = 0$$
$$2x = 3 \qquad\qquad 3x = 2$$
$$x = \frac{3}{2} \qquad\qquad x = \frac{2}{3}$$

The number can be either $\dfrac{3}{2}$ or $\dfrac{2}{3}$.

CHECK: If the number is $\dfrac{3}{2}$ then its reciprocal is $\dfrac{2}{3}$.

$\dfrac{3}{2} + \dfrac{2}{3} \stackrel{\checkmark}{=} \dfrac{13}{6}$; If the number is $\dfrac{2}{3}$ then its reciprocal is $\dfrac{3}{2}$. $\dfrac{2}{3} + \dfrac{3}{2} \stackrel{\checkmark}{=} \dfrac{13}{6}$.

3. Let $x =$ one of the numbers.
   Then $20 - x =$ the other number.
   $$x(20 - x) = 96$$
   $$20x - x^2 = 96$$
   $$x^2 - 20x + 96 = 0$$
   $$(x - 8)(x - 12) = 0$$
   $$x - 8 = 0 \quad \text{or} \quad x - 12 = 0$$
   $$x = 8 \quad \text{or} \quad x = 12$$
   If $x = 8$, then $20 - x = 20 - 8 = 12$.
   If $x = 12$, then $20 - x = 20 - 12 = 8$.
   In both cases, we conclude that the numbers are 8 and 12.
   CHECK: $8 + 12 \stackrel{\checkmark}{=} 20$; $8(12) \stackrel{\checkmark}{=} 96$

5. Let $W =$ width of the rectangle.
   Then $2W + 3 =$ length of the rectangle.
   $$W(2W + 3) = 90$$
   $$2W^2 + 3W = 90$$
   $$2W^2 + 3W - 90 = 0$$
   $$(2W + 15)(W - 6) = 0$$
   $$2W + 15 = 0 \quad \text{or} \quad W - 6 = 0$$
   $$W = -\frac{15}{2} \quad \text{or} \quad W = 6$$
   Reject $W = -\frac{15}{2}$, since width cannot be negative.
   So $W = 6$. Then $2W + 3 = 2(6) + 3 = 15$.
   Thus, the rectangle has a width of 6 meters and a length of 15 meters. CHECK: 15 is 3 more than twice 6, and $15(6) = 90$.

7. Let $x =$ length of the other leg.
   From the Pythagorean Theorem, we get
   $$x^2 + 7^2 = 15^2$$
   $$x^2 + 49 = 225$$
   $$x^2 = 176$$
   $$x = \pm\sqrt{176} = \pm 4\sqrt{11}$$
   Reject $x = -4\sqrt{11}$, since we cannot have a negative length. So $x = 4\sqrt{11}$. Thus, the length of the other leg is $4\sqrt{11}$ cm.
   CHECK: $\left(4\sqrt{11}\right)^2 + 7^2 \stackrel{?}{=} 15^2$
   $$176 + 49 \stackrel{?}{=} 225$$
   $$225 \stackrel{\checkmark}{=} 225$$

9. Let $d =$ length of the diagonal.
   From the Pythagorean Theorem, we get
   $$8^2 + 8^2 = d^2$$
   $$64 + 64 = d^2$$
   $$128 = d^2$$
   $$\pm\sqrt{128} = d$$
   $$\pm 8\sqrt{2} = d$$
   Reject $d = -8\sqrt{2}$, since we cannot have a negative length. So $d = 8\sqrt{2}$. Thus, the length of the diagonal is $8\sqrt{2}$ inches.
   CHECK: $8^2 + 8^2 \stackrel{?}{=} \left(8\sqrt{2}\right)^2$
   $$64 + 64 \stackrel{?}{=} 64 \cdot 2$$
   $$128 \stackrel{\checkmark}{=} 128$$

11. Let $x =$ height ladder reaches on building.
    From the Pythagorean Theorem, we get
    $$8^2 + x^2 = 30^2$$
    $$64 + x^2 = 900$$
    $$x^2 = 836$$
    $$x = \pm\sqrt{836} = \pm 2\sqrt{209}$$
    Reject $x = -2\sqrt{209}$, since we cannot have a negative length. So $x = 2\sqrt{209}$. Thus, the ladder reaches $2\sqrt{209}$ ft up the building.
    CHECK: $8^2 + \left(2\sqrt{209}\right)^2 + \stackrel{?}{=} 30^2$
    $$64 + 836 \stackrel{?}{=} 900$$
    $$900 \stackrel{\checkmark}{=} 900$$

13. Let $x =$ length of the shortest side.
    Then $x + 1 =$ length of the middle side
    and $x + 2 =$ length of the longest side.
    Using Pythagorean Theorem, we get
    $$x^2 + (x+1)^2 = (x+2)^2$$
    $$x^2 + x^2 + 2x + 1 = x^2 + 4x + 4$$
    $$2x^2 + 2x + 1 = x^2 + 4x + 4$$
    $$x^2 - 2x - 3 = 0$$
    $$(x-3)(x+1) = 0$$
    $$x - 3 = 0 \quad \text{or} \quad x + 1 = 0$$
    $$x = 3 \quad \text{or} \quad x = -1$$
    Reject $x = -1$, since the side of a triangle cannot have negative length. So $x = 3$.
    Then $x + 1 = 3 + 1 = 4$ and
    $x + 2 = 3 + 2 = 5$.
    Thus, the sides of the triangle have lengths of 3, 4, and 5.
    CHECK: 3, 4, and 5 are three consecutive integers and $3^2 + 4^2 = 9 + 16 = 25 = 5^2$.

15. Let $n =$ numerator of the original fraction.
    Then $n + 1 =$ denominator of the original fraction.
    $$\frac{n+3}{n+1} = \frac{n}{n+1} + 1$$
    $$(n+1)\left(\frac{n+3}{n+1}\right) = (n+1)\left(\frac{n}{n+1} + 1\right)$$
    $$\frac{\not{n+1}}{1} \cdot \frac{n+3}{\not{n+1}} = \frac{\not{n+1}}{1} \cdot \frac{n}{\not{n+1}} + (n+1) \cdot 1$$
    $$n + 3 = n + n + 1$$
    $$n + 3 = 2n + 1$$
    $$3 = n + 1$$
    $$2 = n$$
    Then $n + 1 = 2 + 1 = 3$, Thus, the original fraction is $\frac{2}{3}$.
    CHECK: $\frac{2+3}{3} = \frac{5}{3}$, which is one more than $\frac{2}{3}$.

17. Let $s =$ number of seats in each row.
    Then $s - 8 =$ number of rows of seats.
    $$s(s - 8) = 768$$
    $$s^2 - 8s = 768$$
    $$s^2 - 8s - 768 = 0$$
    $$(s - 32)(s + 24) = 0$$
    $$s - 32 = 0 \quad \text{or} \quad s + 24 = 0$$
    $$s = 32 \quad \text{or} \quad s = -24$$
    Reject $s = -24$, since the number of seats in a row cannot be negative. So $x = 32$.
    Then $s - 8 = 32 - 8 = 24$. Thus, there are 24 rows of seats in the concert hall, and each row has 32 seats in it.
    CHECK: 24 is eight less than 32, and $24 \cdot 32 = 768$.

19. Let $x =$ speed for the first part of the trip.
    Then $x - 20 =$ speed for the second part of the trip.
    $$\frac{120}{x} + \frac{30}{x - 20} = 2$$
    $$x(x-20)\left(\frac{120}{x} + \frac{30}{x-20}\right) = x(x-20)2$$
    $$\frac{\not{x}(x-20)}{1} \cdot \frac{120}{\not{x}} + \frac{x\not{(x-20)}}{1} \cdot \frac{30}{\not{x-20}} = 2x(x-20)$$
    $$120(x-20) + 30x = 2x(x-20)$$
    $$120x - 2400 + 30x = 2x^2 - 40x$$
    $$150x - 2400 = 2x^2 - 40x$$
    $$-2x^2 + 190x - 2400 = 0$$
    $$x^2 - 95x + 1200 = 0$$
    $$(x - 15)(x - 80) = 0$$
    $$x - 15 = 0 \quad \text{or} \quad x - 80 = 0$$
    $$x = 15 \quad \text{or} \quad x = 80$$
    We reject $x = 15$, since this would mean that $x - 20 = 15 - 20 = -5$, and a negative speed is impossible. So $x = 80$. Thus, the speed for the first part of the trip is 80 kph.
    CHECK: At 80 kph, the motorist covers 120 km in $\frac{120}{80} = \frac{3}{2}$ hr.
    At $(80 - 20) = 60$ kph, the motorist covers the remaining 30 km in
    $\frac{30}{60} = \frac{1}{2}$ hr. Then $\frac{3}{2} + \frac{1}{2} = \frac{4}{2} = 2$, as required.

21. $P = 1000(-x^2 + 15x - 35)$
When $x = 5$,
$P = 1000[-(5)^2 + 15(5) - 35]$
$= \$15,000$
When $x = 4$,
$P = 1000[-(4)^2 + 15(4) - 35]$
$= \$9000$

23. Let $n =$ distance from P to R (in meters).
By the Pythagorean Theorem,
$x^2 = (8.6)^2 + (4.9)^2$
$= 73.96 + 24.01$
$= 97.97$
So $x = 9.9$
The distance between P and R is 9.9 meters.

25. Let $x =$ Neva's speed.
Then $x - 12 =$ Yomar's speed.

Neva's time = Yomar's time $- \dfrac{1}{2}$

$$\frac{180}{x} = \frac{180}{x-12} - \frac{1}{2}$$

$$2x(x-12)\left(\frac{180}{x}\right) = 2x(x-12)\left(\frac{180}{x-12} - \frac{1}{2}\right)$$

$$\frac{2\cancel{x}(x-12)}{1} \cdot \frac{180}{\cancel{x}} = \frac{2x\cancel{(x-12)}}{1} \cdot \frac{180}{\cancel{(x-12)}} - \frac{\cancel{2}x(x-12)}{1} \cdot \frac{1}{\cancel{2}}$$

$360(x - 12) = 360x - x(x - 12)$
$360x - 4320 = 360x - x^2 + 12x$
$360x - 4320 = -x^2 + 372x$
$x^2 + 360x - 4320 = 372x$
$x^2 - 12x - 4320 = 0$
$(x - 72)(x + 60) = 0$
$x - 72 = 0$ or $x + 60 = 0$
$\quad x = 72$ or $\quad x = -60$

Reject $x = -60$, since speed cannot be negative. Hence Neva's speed was 72 mph.

27. (a) $h = 1000 - 16t^2$
$h = 1000 - 16(3.5)^2$
$h = 804$
Its height is 804 feet.

(b) $h = 1000 - 16t^2$
$700 = 1000 - 16t^2$
$-300 = -16t^2$
$\dfrac{75}{4} = t^2$
$18.75 = t^2$
$4.3 = t$
The object takes 4.3 seconds to reach a height of 700 feet.

(c) $h = 1000 - 16t^2$
$0 = 1000 - 16t^2$
$16t^2 = 1000$
$t^2 = \dfrac{125}{2}$
$t^2 = 62.5$
$t = 7.9$
The object hits the ground 7.9 seconds after it is dropped.

29. (a) $P = 0.82x^2 + 4.25x + 95$
    $350 = 3.5$ hundred, so $x = 3.5$
    $P = 0.82(3.5)^2 + 4.25(3.5) + 95$
    $P = 119.92$
    The profit is $119.92.

    (b) $P = 0.82x^2 + 4.25x + 95$
    $130 = 0.82x^2 + 4.25x + 95$
    $0 = 0.82x^2 + 4.25x - 35$
    $a = 0.82, b = 4.25, c = -35$
    $$x = \frac{-b \pm \sqrt{b^2 - 4ac}}{2a}$$
    $$x = \frac{-(4.25) \pm \sqrt{(4.25)^2 - 4(0.82)(-35)}}{2(0.82)}$$
    $x = 4.44$ or $x = -9.62$
    Reject -9.62 since the number of sets cannot be negative. Hence, $x = 4.44$. 4.44 hundred sets or 440 sets (rounded to the nearest ten) should be produced.

31. Let $d =$ distance from home plate to second base.
    By the Pythagorean Theorem.
    $90^2 + 90^2 = d^2$
    $8100 + 8100 = d^2$
    $16200 = d^2$
    $\sqrt{16200} = d$
    $127 = d$
    The distance from home plate to second base is 127 ft. (rounded to the nearest foot).

33. Let $w =$ width of the path.
    $(15 + 2w)(10 + w) - 150 = 100$
    $150 + 35w + 2w^2 - 150 = 100$
    $2w^2 + 35w = 100$
    $2w^2 + 35w - 100 = 0$
    $(2w - 5)(w + 20) = 0$
    $2w - 5 = 0$ or $w + 20 = 0$
    $2w = 5$ or $w = -20$
    $w = \frac{5}{2}$ or $w = -20$
    Reject $w = -20$, since the width of the path cannot be negative. So the path should be $\frac{5}{2}$ or $2\frac{1}{2}$ ft.

35. Let $x =$ number of days for father painting alone.
    Then $x + 9 =$ number of days for son painting alone.
    $$\frac{6}{x} + \frac{6}{x+9} = 1$$
    $$x(x+9)\left(\frac{6}{x} + \frac{6}{x+9}\right) = x(x+9)(1)$$
    $$\not{x}(x+9)\left(\frac{6}{\not{x}}\right) + x(\not{x+9})\left(\frac{6}{\not{x+9}}\right) = x(x+9)$$
    $6(x+9) + 6x = x(x+9)$
    $6x + 54 + 6x = x^2 + 9x$
    $0 = x^2 - 3x - 54$
    $0 = (x-9)(x+6)$
    $0 = x - 9$ or $0 = x + 6$
    $9 = x$ or $-6 = x$

    Reject $x = -6$, since the number of days needed cannot be negative. Thus the father needs 9 days to paint the house alone, while the son needs $x + 9 = 9 + 9 = 18$ days to paint the house alone.

37. (a)
$$\frac{1}{l} = \frac{l}{1+l}$$
$$\frac{l(1+l)}{1} \cdot \frac{1}{l} = \frac{l(1+l)}{1} \cdot \frac{l}{1+l}$$
$$1 + l = l^2$$
$$0 = l^2 - l - 1$$
$$a = 1, b = -1, c = -1$$
$$l = \frac{-b \pm \sqrt{b^2 - 4ac}}{2a}$$
$$l = \frac{-(-1) \pm \sqrt{(-1)^2 - 4(1)(-1)}}{2(1)}$$
$$l = \frac{1 \pm \sqrt{1+4}}{2} = \frac{1 \pm \sqrt{5}}{2}$$
So $l = \dfrac{1+\sqrt{5}}{2}$ or $l = \dfrac{1-\sqrt{5}}{2}$.

Since $\sqrt{5}$ is greater than 1, $\dfrac{1-\sqrt{5}}{2}$ is negative, and thus must be rejected.

So $l = \dfrac{1+\sqrt{5}}{2}$ inches.

(b)
$$\frac{w}{l} = \frac{l}{w+l}$$
$$\frac{l(w+l)}{1} \cdot \frac{w}{l} = \frac{l(w+l)}{1} \cdot \frac{l}{w+l}$$
$$w(w+l) = l^2$$
$$w^2 + wl = l^2$$
$$0 = l^2 - wl - w^2$$
$$a = 1, b = -w, c = -w^2$$
$$l = \frac{-b \pm \sqrt{b^2 - 4ac}}{2a}$$
$$l = \frac{-(-w) \pm \sqrt{(-w)^2 - 4(1)(-w^2)}}{2(1)}$$
$$l = \frac{w \pm \sqrt{w^2 + 4w^2}}{2} = \frac{w \pm \sqrt{5w^2}}{2}$$
So $l = \dfrac{w \pm \sqrt{5}w}{2} = \dfrac{(1 \pm \sqrt{5})w}{2}$.

As in part (a), we reject $\dfrac{(1-\sqrt{5})w}{2}$ because it is negative. Thus, $l = \dfrac{(1+\sqrt{5})w}{2}$.

(If $w = 1$, we get the golden ratio of part (a).)

38.
$$\frac{1}{x-1} = \frac{x}{1}$$
$$\frac{x-1}{1} \cdot \frac{1}{x-1} = \frac{x-1}{1} \cdot \frac{x}{1}$$
$$1 = x(x-1)$$
$$1 = x^2 - x$$
$$0 = x^2 - x - 1$$

Proceed with the quadratic formula to find that $x = \dfrac{1+\sqrt{5}}{2}$. (This quadratic equation is the same as the one encountered in part (a) of exercise (37), except for the letter used to represent the variable.)

39. Let $x = $ length of the shortest side. Then $x + 2 = $ length of the middle side and $x + 4 = $ length of the longest side. Use the Pythagorean Theorem to find that
$$x^2 + (x+2)^2 = (x+4)^2$$
$$x^2 + x^2 + 4x + 4 = x^2 + 8x + 16$$
$$2x^2 + 4x + 4 = x^2 + 8x + 16$$
$$x^2 - 4x - 12 = 0$$
$$(x-6)(x+2) = 0$$
$$x - 6 = 0 \quad \text{or} \quad x + 2 = 0$$
$$x = 6 \quad \text{or} \quad x = -2$$

Reject $x = 6$, since it is not odd. Reject $x = -2$, since we cannot have a negative length. Since both possible solutions to the problem have been rejected, we conclude that it is impossible to find three consecutive odd integers that are the sides of a right triangle.

Exercises 11.6

1. $y = x^2$

   Vertex: $x = \dfrac{-b}{2a} = \dfrac{-0}{2(1)} = 0$; $(0, 0)$

   $0 = x^2$
   $0 = x$
   The x-intercept is 0.
   0 is also the y-intercept.

   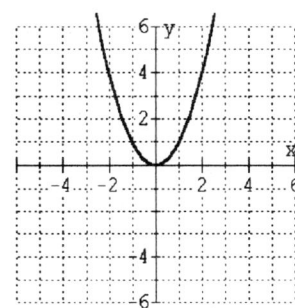

3. $y = x^2 + 2$

   Vertex: $x = \dfrac{-b}{2a} = \dfrac{-0}{2(1)} = 0$; $(0, 2)$

   There are no x-intercepts.
   The y-intercept is 2.

   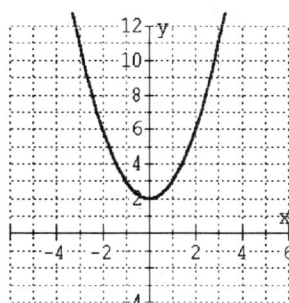

5. $y = x^2 - 9$

   Vertex: $x = \dfrac{-b}{2a} = \dfrac{-0}{2(1)} = 0$; $(0, -9)$

   $0 = x^2 - 9$
   $9 = x^2$
   $\pm 3 = x$
   The x-intercepts are 3 and -3.
   The y-intercept is -9.

   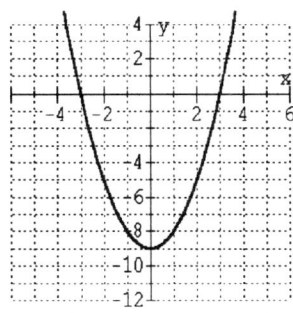

7. $y = -x^2 + 1$

   Vertex: $x = \dfrac{-b}{2a} = \dfrac{-0}{2(-1)} = 0$; $(0, 1)$

   $0 = -x^2 + 1$
   $x^2 = 1$
   $x = \pm 1$
   The x-intercepts are 1 and -1.
   The y-intercept is 1.

   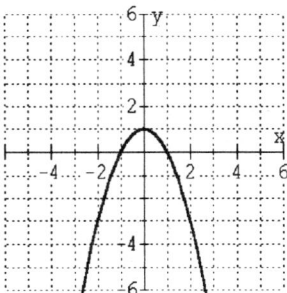

9. $y = x^2 - 2x$

   Vertex: $x = \dfrac{-b}{2a} = \dfrac{-(-2)}{2(1)} = 1$; $(1, -1)$

   $0 = x^2 - 2x$
   $0 = x(x - 2)$
   $0 = x$ or $0 = x - 2$
   $\qquad\qquad\quad 2 = x$
   The x-intercepts are 0 and 2.
   The y-intercept is 0.

   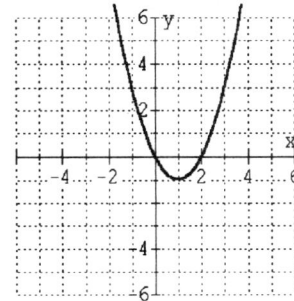

11. $y = 6x - 2x^2$

Vertex: $x = \dfrac{-b}{2a} = \dfrac{-(6)}{2(-2)} = \dfrac{3}{2}$; $\left(\dfrac{3}{2}, \dfrac{9}{2}\right)$

$0 = 6x - 2x^2$
$0 = 2x(3 - x)$
$0 = 2x$ or $0 = 3 - x$
$0 = x$ $\quad\quad x = 3$

The x-intercepts are 0 and 3.
The y-intercept is 0.

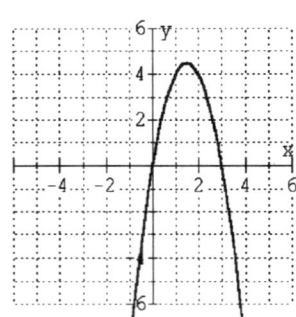

13. $y = x^2 - 2$

Vertex: $x = \dfrac{-b}{2a} = \dfrac{0}{2(1)} = 0$; $(0, -2)$

$0 = x^2 - 2$
$2 = x^2$
$\pm\sqrt{2} = x$

The x-intercepts are $\sqrt{2}$ and $-\sqrt{2}$.
The y-intercept is -2.

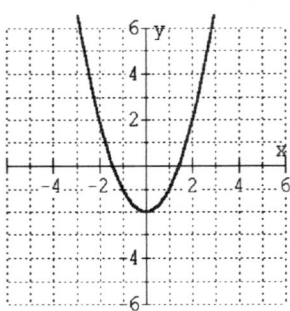

15. $y = 2x - 10$
$0 = 2x - 10$
$10 = 2x$
$5 = x$

The x-intercept is 5.
The y-intercept is -10.

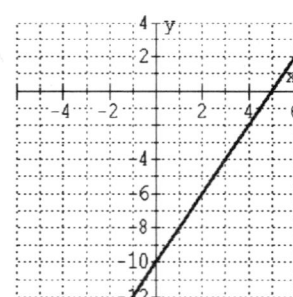

17. $y = x^2 - 4x + 3$

Vertex: $x = \dfrac{-b}{2a} = \dfrac{-(-4)}{2(1)} = 2$; $(2, -1)$

$0 = x^2 - 4x + 3$
$0 = (x - 1)(x - 3)$
$0 = x - 1$ or $0 = x - 3$
$1 = x$ $\quad$ or $\quad 3 = x$

The x-intercepts are 1 and 3.
The y-intercept is 3.

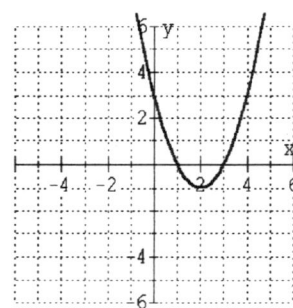

19. $y = x^2 + 2x - 8$

   Vertex: $x = \dfrac{-b}{2a} = \dfrac{-(2)}{2(1)} = -1$; $(-1, -9)$

   $0 = x^2 + 2x - 8$
   $0 = (x+4)(x-2)$
   $0 = x + 4$ or $0 = x - 2$
   ${-4} = x$ or $2 = x$

   The x-intercepts are -4 and 2.
   The y-intercept is -8.

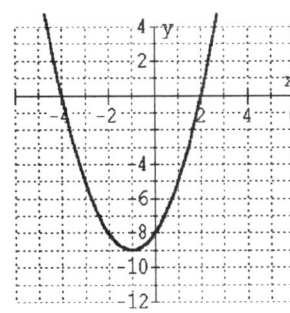

21. $y = -x^2 + 2x + 3$

   Vertex: $x = \dfrac{-b}{2a} = \dfrac{-(2)}{2(-1)} = 1$; $(1, 4)$

   $0 = -x^2 + 2x + 3$
   $0 = x^2 - 2x - 3$
   $0 = (x-3)(x+1)$
   $0 = x - 3$ or $0 = x + 1$
   $3 = x$ or $-1 = x$

   The x-intercepts are 3 and -1.
   The y-intercept is 3.

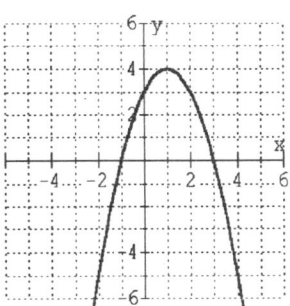

23. $y = x^2 + 4x + 4$

   Vertex: $x = \dfrac{-b}{2a} = \dfrac{-(4)}{2(1)} = -2$; $(-2, 0)$

   $0 = x^2 + 4x + 4$
   $0 = (x+2)(x+2) = (x+2)^2$
   $0 = x + 2$
   $-2 = x$

   The x-intercept is -2.
   The y-intercept is 4.

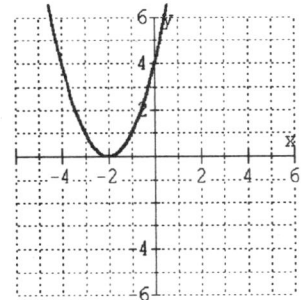

25. $y = (x-1)^2$
    $y = x^2 - 2x + 1$

   Vertex: $x = \dfrac{-b}{2a} = \dfrac{-(-2)}{2(1)} = 1$; $(1, 0)$

   $0 = (x-1)^2$
   $0 = x - 1$
   $1 = x$

   The x-intercept is 1.
   The y-intercept is 1.

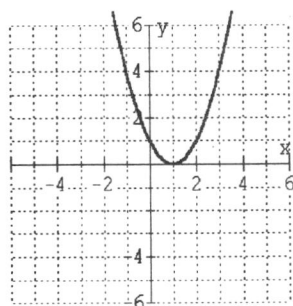

27. $y = x^2 + 3x + 2$

Vertex: $x = \dfrac{-b}{2a} = \dfrac{-(3)}{2(1)} = -\dfrac{3}{2}$;

$\left(-\dfrac{3}{2}, -\dfrac{1}{4}\right)$

$0 = x^2 + 3x + 2$
$0 = (x + 2)(x + 1)$
$0 = x + 2$ or $0 = x + 1$
$-2 = x$ $\quad\quad -1 = x$

The x-intercepts are -2 and -1.
The y-intercept is 2.

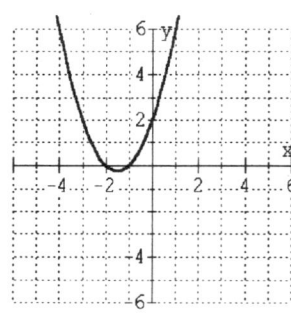

29. $y = -x^2 + x + 6$

Vertex: $x = \dfrac{-b}{2a} = \dfrac{-(1)}{2(-1)} = \dfrac{1}{2}$; $\left(\dfrac{1}{2}, \dfrac{25}{4}\right)$

$0 = -x^2 + x + 6$
$0 = x^2 - x - 6$
$0 = (x - 3)(x + 2)$
$0 = x - 3$ or $0 = x + 2$
$3 = x$ $\quad\quad$ or $\quad -2 = x$

The x-intercepts are 3 and -2.
The y-intercept is 6.

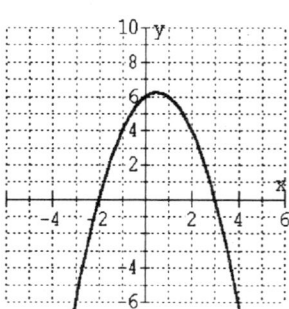

31. $y = x^2 - 4x + 2$

Vertex: $x = \dfrac{-b}{2a} = \dfrac{-(-4)}{2(1)} = 2$; $(2, -2)$

$0 = x^2 - 4x + 2$

$x = \dfrac{-b \pm \sqrt{b^2 - 4ac}}{2a}$

$= \dfrac{-(-4) \pm \sqrt{16 - 8}}{2} = \dfrac{4 \pm \sqrt{8}}{2}$

$= \dfrac{4 \pm 2\sqrt{2}}{2} = 2 \pm \sqrt{2}$

The x-intercepts are $2 + \sqrt{2} \approx 3.4$ and $2 - \sqrt{2} \approx 0.6$.
The y-intercept is 2.

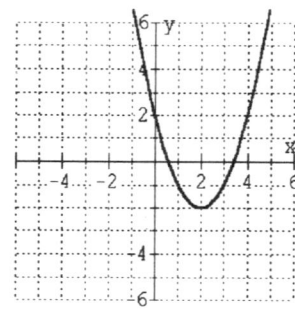

33. $y = -x^2 - 8x - 5$

Vertex: $x = \dfrac{-b}{2a} = \dfrac{-(-8)}{2(-1)} = -4$; $(-4, 11)$

$0 = -x^2 - 8x - 5$
$0 = x^2 + 8x + 5$

$x = \dfrac{-b \pm \sqrt{b^2 - 4ac}}{2a}$

$= \dfrac{-8 \pm \sqrt{64 - 20}}{2} = \dfrac{-8 \pm \sqrt{44}}{2}$

$= \dfrac{-8 \pm 2\sqrt{11}}{2} = -4 \pm \sqrt{11}$

The x-intercepts are $-4 + \sqrt{11} \approx -0.7$ and $-4 - \sqrt{11} \approx -7.3$.
The y-intercept is -5.

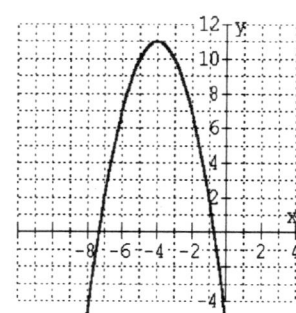

35. $y = x^2 - 4x + 5$
Vertex: $x = \dfrac{-b}{2a} = \dfrac{-(-4)}{2(1)} = 2$; $(2, 1)$
$0 = x^2 - 4x + 5$
$x = \dfrac{-b \pm \sqrt{b^2 - 4ac}}{2a}$
$= \dfrac{-(-4) \pm \sqrt{16 - 20}}{2} = \dfrac{4 \pm \sqrt{-4}}{2}$
There are no real solutions, hence there are no x-intercepts.
The y-intercept is 5.

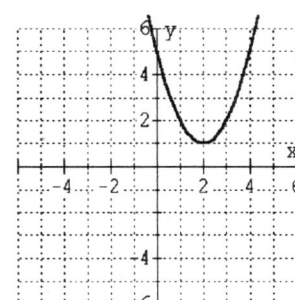

37. $y = 3x + 6$
$0 = 3x + 6$
$-6 = 3x$
$-2 = x$
The x-intercept is -2.
The y-intercept is 6.

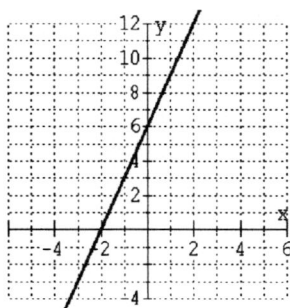

39. $y = 4x^2 - 8x - 5$
Vertex: $x = \dfrac{-b}{2a} = \dfrac{-(-8)}{2(4)} = 1$; $(1, -9)$
$0 = 4x^2 - 8x - 5$
$0 = (2x + 1)(2x - 5)$
$0 = 2x + 1$ or $0 = 2x - 5$
$-1 = 2x$ $\qquad$ $5 = 2x$
$-\dfrac{1}{2} = x$ $\qquad$ $\dfrac{5}{2} = x$
The x-intercepts are $-\dfrac{1}{2}$ and $\dfrac{5}{2}$.
The y-intercept is -5.

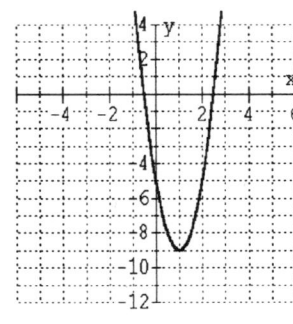

41. (a)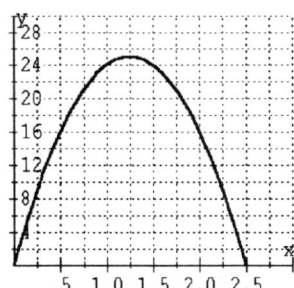

(b) At its highest point, the object is 25 ft above the ground.
(c) The object reaches its highest point 1.25 seconds after it is thrown.

43. (a) $(3, 6)$ $\qquad$ (b) $x = 3$
(c) $x \leq 3$ $\qquad$ (d) $x \geq 3$

45. (a) $(-4, 0)$ $\qquad$ (b) $x = -4$
(c) $x \geq -4$ $\qquad$ (d) $x \leq -4$

47.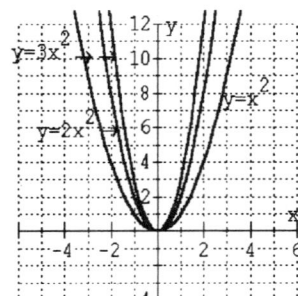

As the coefficient of x^2 increases from 1 to 2 to 3 the corresponding parabolas become narrower.

48.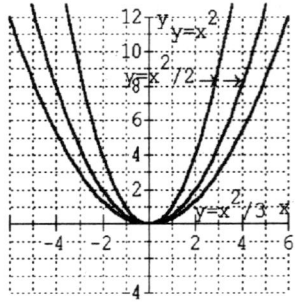

As the coefficient of x^2 decreases from 1 to $\dfrac{1}{2}$ to $\dfrac{1}{3}$ the corresponding parabolas become wider.

49.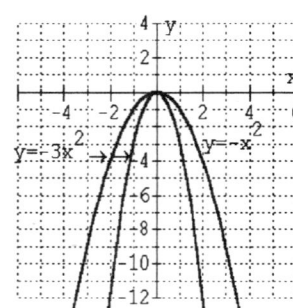

The coefficient of -3 narrows the graph of $y = x^2$ and then "flips" the resulting graph over the x-axis.

50.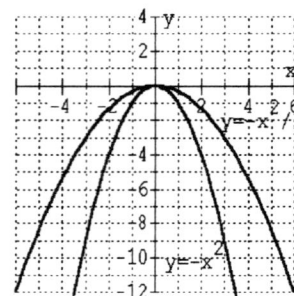

The coefficient of $-\dfrac{1}{3}$ widens the graph of $y = x^2$ and then "flips" the resulting graph over the x-axis.

51. $(x-5)^2$
$= (x-5)(x-5)$
$= x^2 - 10x + 25$

53. $(x+5)(x-5) = x^2 - 25$

55. $(x+p)^2$
$= (x+p)(x+p)$
$= x^2 + 2px + p^2$

57. Let $t =$ number of hours the assistant works.
Then $t - 1 =$ number of hours the handyman works.
$11t + 19(t-1) = 146$
$11t + 19t - 19 = 146$
$30t - 19 = 146$
$30t = 165$
$t = 5.5$

Since the assistant started at 10:00 A.M. and worked for 5.5 hours to finish the job, the job was completed at 3:30 P.M.

CHECK: The assistant worked for 5.5 hours and was paid $11 per hour, so he earned $5.5(\$11) = \$60.50$. The handyman worked for 4.5 hours and was paid $19 per hour, so he earned $4.5(\$19) = \$85.50$. $\$60.50 + \$85.50 \stackrel{\checkmark}{=} \$146$.

Chapter 11 Review Exercises

1. $x^2 - 7x - 6 = 0$
$a = 1, b = -7, c = -6$
$x = \dfrac{-b \pm \sqrt{b^2 - 4ac}}{2a}$
$x = \dfrac{-(-7) \pm \sqrt{(-7)^2 - 4(1)(-6)}}{2(1)}$
$x = \dfrac{7 \pm \sqrt{49 + 24}}{2} = \dfrac{7 \pm \sqrt{73}}{2}$

3. $x^2 + 5 = 4x$
$x^2 - 4x + 5 = 0$
$x^2 - 4x = -5$
$+4 +4$
$\overline{x^2 - 4x + 4 = -1}$
$(x - 2)^2 = -1$
$x - 2 = \pm\sqrt{-1}$
Therefore, there are no real solutions.

5. $(u - 6)^2 = 13$
$u - 6 = \pm\sqrt{13}$
$u = 6 \pm \sqrt{13}$

7. $2y^2 + 7y = 15$
$2y^2 + 7y - 15 = 0$
$(2y - 3)(y + 5) = 0$
$2y - 3 = 0 \quad \text{or} \quad y + 5 = 0$
$y = \dfrac{3}{2} \quad \text{or} \quad y = -5$

9. $18x^2 - 24x + 6 = 0$
$3x^2 - 4x + 1 = 0$
$(3x - 1)(x - 1) = 0$
$3x - 1 = 0 \quad \text{or} \quad x - 1 = 0$
$x = \dfrac{1}{3} \quad \text{or} \quad x = 1$

11. $(x - 6)^2 = (x + 3)(x - 5)$
$x^2 - 12x + 36 = x^2 - 2x - 15$
$-12x + 36 = -2x - 15$
$36 = 10x - 15$
$51 = 10x$
$\dfrac{51}{10} = x$

13. $u^2 + 1 = \dfrac{13u}{6}$
$6(u^2 + 1) = \dfrac{6}{1} \cdot \dfrac{13u}{6}$
$6u^2 + 6 = 13u$
$6u^2 - 13u + 6 = 0$
$(3u - 2)(2u - 3) = 0$
$3u - 2 = 0 \quad \text{or} \quad 2u - 3 = 0$
$u = \dfrac{2}{3} \quad \text{or} \quad u = \dfrac{3}{2}$

15. $3x(x - 2) = (x - 3)^2$
$3x^2 - 6x = x^2 - 6x + 9$
$2x^2 - 6x = -6x + 9$
$2x^2 = 9$
$x^2 = \dfrac{9}{2}$
$x = \pm\sqrt{\dfrac{9}{2}} = \pm\dfrac{\sqrt{9}}{\sqrt{2}} = \pm\dfrac{3}{\sqrt{2}}$
$= \pm\dfrac{3\sqrt{2}}{\sqrt{2}\sqrt{2}} = \pm\dfrac{3\sqrt{2}}{2}$

17. $\dfrac{x + 3}{x + 6} = \dfrac{x + 2}{x + 4}$
$\dfrac{(x + 6)(x + 4)}{1} \cdot \dfrac{x + 3}{x + 6} = \dfrac{(x + 6)(x + 4)}{1} \cdot \dfrac{x + 2}{x + 4}$
$(x + 4)(x + 3) = (x + 6)(x + 2)$
$x^2 + 7x + 12 = x^2 + 8x + 12$
$7x + 12 = 8x + 12$
$7x = 8x$
$0 = x$

19. $x^2 - 7x + 3 = 0$
$x^2 - 7x = -3$
$+\dfrac{49}{4} +\dfrac{49}{4}$
$x^2 - 7x + \dfrac{49}{4} = -3 + \dfrac{49}{4}$
$\left(x - \dfrac{7}{2}\right)^2 = \dfrac{37}{4}$
$x - \dfrac{7}{2} = \pm\sqrt{\dfrac{37}{4}} = \pm\dfrac{\sqrt{37}}{\sqrt{4}} = \pm\dfrac{\sqrt{37}}{2}$
$x = \dfrac{7}{2} \pm \dfrac{\sqrt{37}}{2} = \dfrac{7 \pm \sqrt{37}}{2}$

21. $y = x^2 - 6x$
Vertex: $x = \dfrac{-b}{2a} = \dfrac{-(-6)}{2(1)} = 3$;
$(3, -9)$
$0 = x^2 - 6x$
$0 = x(x - 6)$
$0 = x$ or $0 = x - 6$
$\phantom{0 = x \text{ or } 0 = }6 = x$
The x-intercepts are 0 and 6.
The y-intercept is 0.

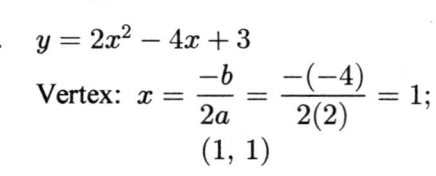

23. $y = -x^2 - 2x + 8$
Vertex: $x = \dfrac{-b}{2a} = \dfrac{-(-2)}{2(-1)} = -1$;
$(-1, 9)$
$0 = -x^2 - 2x + 8$
$0 = x^2 + 2x - 8$
$0 = (x + 4)(x - 2)$
$0 = x + 4$ or $0 = x - 2$
$-4 = x$ or $2 = x$
The x-intercepts are -4 and 2.
The y-intercept is 8.

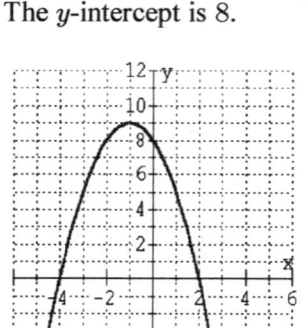

25. $y = 2x^2 - 4x + 3$
Vertex: $x = \dfrac{-b}{2a} = \dfrac{-(-4)}{2(2)} = 1$;
$(1, 1)$

There are no x-intercepts.
The y-intercept is 3.

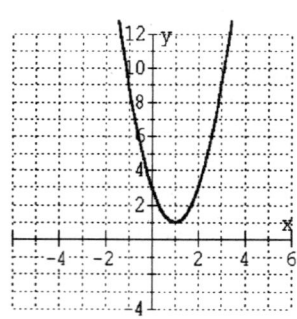

Chapter 11 Practice Test

1. $(x+5)(x-2) = 18$
   $x^2 + 3x - 10 = 18$
   $x^2 + 3x - 28 = 0$
   $(x+7)(x-4) = 0$
   $x+7 = 0$ or $x-4 = 0$
   $x = -7$ or $x = 4$

3. $x^2 - x + 14 = 2x(x-3)$
   $x^2 - x + 14 = 2x^2 - 6x$
   $0 = x^2 - 5x - 14$
   $0 = (x-7)(x+2)$
   $0 = x - 7$ or $0 = x + 2$
   $7 = x$ or $-2 = x$

5. $(x+5)^2 = 10x$
   $x^2 + 10x + 25 = 10x$
   $x^2 + 25 = 0$
   $x^2 = -25$
   $x = \pm\sqrt{-25}$

   Since square roots of negative number are not real, the equation has no real solution.

7. $5x^2 + 15 = 30x$
   $5x^2 - 30x + 15 = 0$
   $5(x^2 - 6x + 3) = 0$
   $x^2 - 6x + 3 = 0$
   $a = 1, b = -6, c = 3$
   $x = \dfrac{-b \pm \sqrt{b^2 - 4ac}}{2a}$
   $x = \dfrac{-(-6) \pm \sqrt{(-6)^2 - 4(1)(3)}}{2(1)}$
   $x = \dfrac{6 \pm \sqrt{36 - 12}}{2} = \dfrac{6 \pm \sqrt{24}}{2}$
   $x = \dfrac{6 \pm 2\sqrt{6}}{2} = \dfrac{\cancel{2}(3 \pm \sqrt{6})}{\cancel{2}}$
   $x = 3 \pm \sqrt{6}$

9. $3x^2 - 12x = 7$
   $x^2 - 4x = \dfrac{7}{3}$
   $ +4 +4$
   $\overline{x^2 - 4x + 4 = \dfrac{7}{3} + 4}$
   $(x-2)^2 = \dfrac{19}{3}$
   $x - 2 = \pm\sqrt{\dfrac{19}{3}}$
   $x - 2 = \pm\dfrac{\sqrt{19}}{\sqrt{3}} = \pm\dfrac{\sqrt{19}\sqrt{3}}{\sqrt{3}\sqrt{3}}$
   $x - 2 = \pm\dfrac{\sqrt{57}}{3}$
   $x = 2 \pm \dfrac{\sqrt{57}}{3} = \dfrac{6 \pm \sqrt{57}}{3}$

   $3x^2 - 12x = 7$
   $3x^2 - 12x - 7 = 0$
   $a = 3, b = -12, c = -7$
   $x = \dfrac{-b \pm \sqrt{b^2 - 4ac}}{2a}$
   $x = \dfrac{-(-12) \pm \sqrt{(-12)^2 - 4(3)(-7)}}{2(3)}$
   $x = \dfrac{12 \pm \sqrt{144 + 84}}{6} = \dfrac{12 \pm \sqrt{228}}{6}$
   $x = \dfrac{12 \pm 2\sqrt{57}}{6} = \dfrac{\cancel{2}(6 \pm \sqrt{57})}{\underset{3}{\cancel{6}}}$
   $x = \dfrac{6 \pm \sqrt{57}}{3}$

11. $y = 6x - x^2$

Vertex: $x = \dfrac{-b}{2a} = \dfrac{-(6)}{2(-1)} = 3$; (3, 9)

$0 = 6x - x^2$
$0 = x(6 - x)$
$0 = x$ or $0 = 6 - x$
$\phantom{0 = x \text{ or } 0 = }x = 6$

The x-intercepts are 0 and 6.
The y-intercept is 0.

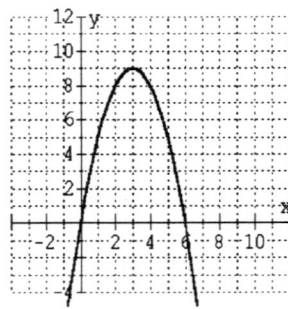

Chapters 10-11
Cumulative Review

1. $\sqrt{36x^{16}y^{12}} = \sqrt{36}\sqrt{x^{16}}\sqrt{y^{12}} = 6x^8y^6$

3. $\dfrac{7}{\sqrt{6}} = \dfrac{7\sqrt{6}}{\sqrt{6}\sqrt{6}} = \dfrac{7\sqrt{6}}{6}$

5. $\dfrac{20}{3-\sqrt{5}} = \dfrac{20(3+\sqrt{5})}{(3-\sqrt{5})(3+\sqrt{5})} = \dfrac{20(3+\sqrt{5})}{9-5} = \dfrac{\overset{5}{\cancel{20}}(3+\sqrt{5})}{\cancel{4}} = 5(3+\sqrt{5})$

7. $\sqrt{120} = \sqrt{4}\sqrt{30} = 2\sqrt{30}$

9. $\sqrt{45t} - \sqrt{20t} = \sqrt{9\cdot 5t} - \sqrt{4\cdot 5t} = \sqrt{9}\sqrt{5t} - \sqrt{4}\sqrt{5t} = 3\sqrt{5t} - 2\sqrt{5t} = \sqrt{5t}$

11. $\sqrt{\dfrac{3}{7}} + \sqrt{21} = \dfrac{\sqrt{3}}{\sqrt{7}} + \sqrt{21} = \dfrac{\sqrt{3}\sqrt{7}}{\sqrt{7}\sqrt{7}} + \sqrt{21} = \dfrac{\sqrt{21}}{7} + \sqrt{21} = \left(\dfrac{1}{7}+1\right)\sqrt{21} = \dfrac{8}{7}\sqrt{21}$

13. $\sqrt{27x^6y^5} - xy\sqrt{12x^4y^3} = \sqrt{9x^6y^4 \cdot 3y} - xy\sqrt{4x^4y^2 \cdot 3y}$
$= \sqrt{9x^6y^4}\sqrt{3y} - xy\sqrt{4x^4y^2}\sqrt{3y}$
$= 3x^3y^2\sqrt{3y} - xy(2x^2y)\sqrt{3y}$
$= 3x^3y^2\sqrt{3y} - 2x^3y^2\sqrt{3y} = x^3y^2\sqrt{3y}$

15. $(3\sqrt{2} - 4\sqrt{5})(4\sqrt{2} - 2\sqrt{5}) = (3\sqrt{2})(4\sqrt{2}) - (3\sqrt{2})(2\sqrt{5}) - (4\sqrt{5})(4\sqrt{2}) + (4\sqrt{5})(2\sqrt{5})$
$= 3\cdot 4\sqrt{2}\sqrt{2} - 3\cdot 2\sqrt{2}\sqrt{5} - 4\cdot 4\sqrt{5}\sqrt{2} + 4\cdot 2\sqrt{5}\sqrt{5}$
$= 12\cdot 2 - 6\sqrt{10} - 16\sqrt{10} + 8\cdot 5$
$= 24 - 6\sqrt{10} - 16\sqrt{10} + 40 = 64 - 22\sqrt{10}$

17. $\dfrac{15}{\sqrt{7}-\sqrt{2}} - \dfrac{10}{\sqrt{2}} = \dfrac{15(\sqrt{7}+\sqrt{2})}{(\sqrt{7}-\sqrt{2})(\sqrt{7}+\sqrt{2})} - \dfrac{10\sqrt{2}}{\sqrt{2}\sqrt{2}} = \dfrac{15(\sqrt{7}+\sqrt{2})}{7-2} - \dfrac{\overset{5}{\cancel{10}}\sqrt{2}}{\cancel{2}}$
$= \dfrac{\overset{3}{\cancel{15}}(\sqrt{7}+\sqrt{2})}{\cancel{5}} - 5\sqrt{2} = 3(\sqrt{7}+\sqrt{2}) - 5\sqrt{2}$
$= 3\sqrt{7} + 3\sqrt{2} - 5\sqrt{2} = 3\sqrt{7} - 2\sqrt{2}$

19. $x^2 - 3x = 10$
$x^2 - 3x - 10 = 0$
$(x-5)(x+2) = 0$
$x - 5 = 0$ or $x + 2 = 0$
$x = 5$ or $x = -2$

21. $$\frac{x}{x-2} = \frac{x+1}{x-3}$$
$$\frac{(x-2)(x-3)}{1}\left(\frac{x}{x-2}\right) = \frac{(x-2)(x-3)}{1}\left(\frac{x+1}{x-3}\right)$$
$$(x-3)x = (x-2)(x+1)$$
$$x^2 - 3x = x^2 + x - 2x - 2$$
$$x^2 - 3x = x^2 - x - 2$$
$$-3x = -x - 2$$
$$-2x = -2$$
$$x = 1$$

23. $$3x^2 - 6x - 2 = x^2 - x - 4$$
$$2x^2 - 6x - 2 = -x - 4$$
$$2x^2 - 5x - 2 = -4$$
$$2x^2 - 5x + 2 = 0$$
$$(2x - 1)(x - 2) = 0$$
$$2x - 1 = 0 \quad \text{or} \quad x - 2 = 0$$
$$2x = 1 \quad \text{or} \quad x = 2$$
$$x = \frac{1}{2} \quad \text{or} \quad x = 2$$

25. $$(x+3)(x-4) = 8$$
$$x^2 - 4x + 3x - 12 = 8$$
$$x^2 - x - 12 = 8$$
$$x^2 - x - 20 = 0$$
$$(x - 5)(x + 4) = 0$$
$$x - 5 = 0 \quad \text{or} \quad x + 4 = 0$$
$$x = 5 \quad \text{or} \quad x = -4$$

27. $$(2x-3)(x-4) = (x-2)(2x-1)$$
$$2x^2 - 8x - 3x + 12 = 2x^2 - x - 4x + 2$$
$$2x^2 - 11x + 12 = 2x^2 - 5x + 2$$
$$-11x + 12 = -5x + 2$$
$$12 = 6x + 2$$
$$10 = 6x$$
$$\frac{5}{3} = x$$

29. $$(s-2)^2 = 10$$
$$s - 2 = \pm\sqrt{10}$$
$$s = 2 \pm \sqrt{10}$$

31. $$\sqrt{x} - 3 = 4$$
$$\sqrt{x} = 7$$
$$\left(\sqrt{x}\right)^2 = 7^2$$
$$x = 49$$

33. $$3\sqrt{t+4} - 1 = 7$$
$$3\sqrt{t+4} = 8$$
$$\sqrt{t+4} = \frac{8}{3}$$
$$\left(\sqrt{t+4}\right)^2 = \left(\frac{8}{3}\right)^2$$
$$t + 4 = \frac{64}{9}$$
$$t = \frac{28}{9}$$

35. $$2x^2 - 10x + 6 = 0$$
$$2x^2 - 10x = -6$$
$$x^2 - 5x = -3$$
$$ +\frac{25}{4} +\frac{25}{4}$$
$$x^2 - 5x + \frac{25}{4} = -3 + \frac{25}{4}$$
$$\left(x - \frac{5}{2}\right)^2 = \frac{13}{4}$$
$$x - \frac{5}{2} = \pm\sqrt{\frac{13}{4}} = \pm\frac{\sqrt{13}}{\sqrt{4}}$$
$$x - \frac{5}{2} = \pm\frac{\sqrt{13}}{2}$$
$$x = \frac{5}{2} \pm \frac{\sqrt{13}}{2} = \frac{5 \pm \sqrt{13}}{2}$$

37. $y = (2x-3)^2$
$y = 4x^2 - 12x + 9$
Vertex: $x = \dfrac{-b}{2a} = \dfrac{-(-12)}{2(4)} = \dfrac{3}{2}$;
$\left(\dfrac{3}{2}, 0\right)$
$0 = (2x-3)^2$
$0 = 2x - 3$
$3 = 2x$
$\dfrac{3}{2} = x$
The x-intercept is $\dfrac{3}{2}$.
The y-intercept is 9.

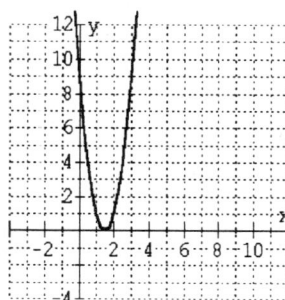

39. $y = x^2 - x - 1$
Vertex: $x = \dfrac{-b}{2a} = \dfrac{-(-1)}{2(1)} = \dfrac{1}{2}$; $\left(\dfrac{1}{2}, -\dfrac{5}{4}\right)$
$0 = x^2 - x - 1$
$x = \dfrac{-b \pm \sqrt{b^2 - 4ac}}{2a}$
$= \dfrac{-(-1) \pm \sqrt{1+4}}{2}$
$= \dfrac{1 \pm \sqrt{5}}{2}$
The x-intercepts are $1 + \sqrt{5} \approx 1.6$ and $1 - \sqrt{5} \approx -0.6$.
The y-intercept is -1.

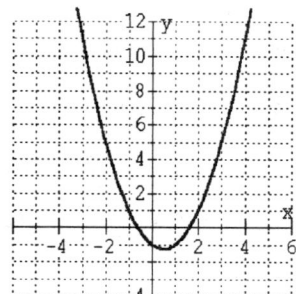

41. Let $x =$ length of the third side of the triangle.
By the Pythagorean Theorem,
$8^2 + x^2 = 17^2$
$64 + x^2 = 289$
$x^2 = 225$
$x = \pm 15$
We reject $x = -15$, since length cannot be negative. So $x = 15$. Since the length of the third side is 15, the perimeter of the triangle is $8 + 15 + 17 = 40$.

43. Let $x =$ Kim's average speed on the way there.
Then $x - 30 =$ Kim's average speed on the way back.
$\dfrac{280}{x} + \dfrac{280}{x-30} = 11$
$x(x-30)\left(\dfrac{280}{x} + \dfrac{280}{x-30}\right) = x(x-30)(11)$
$\dfrac{\cancel{x}(x-30)}{1} \cdot \dfrac{280}{\cancel{x}} + \dfrac{x\cancel{(x-30)}}{1} \cdot \dfrac{280}{\cancel{x-30}} = 11x(x-30)$
$280(x-30) + 280x = 11x^2 - 330x$
$280x - 8400 + 280x = 11x^2 - 330x$
$560x - 8400 = 11x^2 - 330x$
$-11x^2 + 890x - 8400 = 0$
$11x^2 - 890x + 8400 = 0$
$(x-70)(11x-120) = 0$
$x - 70 = 0$ or $11x - 120 = 0$
$x = 70$ $\qquad 11x = 120$
$\qquad\qquad\qquad x = \dfrac{120}{11} = 10.9$
Reject $x = 10.9$, since this would make Kim's speed negative on the way back. So $x = 70$. Therefore, Kim's average speed on the way there is 70 mph.

Chapters 10 - 11
Cumulative Practice Test

1. $\sqrt{72} = \sqrt{36 \cdot 2} = \sqrt{36}\sqrt{2} = 6\sqrt{2}$

3. $3\sqrt{20} - 4\sqrt{45} = 3\sqrt{4 \cdot 5} - 4\sqrt{9 \cdot 5} = 3\sqrt{4}\sqrt{5} - 4\sqrt{9}\sqrt{5}$
$= 3(2\sqrt{5}) - 4(3\sqrt{5}) = 6\sqrt{5} - 12\sqrt{5} = -6\sqrt{5}$

5. $\dfrac{8}{\sqrt{5}} = \dfrac{8\sqrt{5}}{\sqrt{5}\sqrt{5}} = \dfrac{8\sqrt{5}}{5}$

7. $\dfrac{24}{4-\sqrt{7}} = \dfrac{24(4+\sqrt{7})}{(4-\sqrt{7})(4+\sqrt{7})} = \dfrac{24(4+\sqrt{7})}{16-7} = \dfrac{\overset{8}{\cancel{24}}(4+\sqrt{7})}{\underset{3}{\cancel{9}}} = \dfrac{8(4+\sqrt{7})}{3}$

9. $(5\sqrt{t}+\sqrt{3})(2\sqrt{t}-4\sqrt{3}) = (5\sqrt{t})(2\sqrt{t}) - (5\sqrt{t})(4\sqrt{3}) + \sqrt{3}(2\sqrt{t}) - \sqrt{3}(4\sqrt{3})$
$= 10t - 20\sqrt{3t} + 2\sqrt{3t} - 12 = 10t - 18\sqrt{3t} - 12$

11. $(x+3)(x-5) = 9$
$x^2 - 5x + 3x - 15 = 9$
$x^2 - 2x - 15 = 9$
$x^2 - 2x - 24 = 0$
$(x-6)(x+4) = 0$
$x - 6 = 0 \quad \text{or} \quad x + 4 = 0$
$x = 6 \quad \text{or} \quad x = -4$

13. $\dfrac{x}{2x-5} = \dfrac{x+4}{2x}$
$\dfrac{(2x-5)(2x)}{1}\left(\dfrac{x}{2x-5}\right) = \dfrac{(2x-5)(2x)}{1}\left(\dfrac{x+4}{2x}\right)$
$(2x)(x) = (2x-5)(x+4)$
$2x^2 = 2x^2 + 8x - 5x - 20$
$2x^2 = 2x^2 + 3x - 20$
$0 = 3x - 20$
$20 = 3x$
$\dfrac{20}{3} = x$

15. $2x^2 + 1 = 5x$
$2x^2 - 5x + 1 = 0$
$a = 2, b = -5, c = 1$
$x = \dfrac{-b \pm \sqrt{b^2 - 4ac}}{2a}$
$x = \dfrac{-(-5) \pm \sqrt{(-5)^2 - 4(2)(1)}}{2(2)}$
$x = \dfrac{5 \pm \sqrt{17}}{4}$

17. $(2x+5)^2 = 4x(x-3)$
$4x^2 + 20x + 25 = 4x^2 - 12x$
$20x + 25 = -12x$
$25 = -32x$
$-\dfrac{25}{32} = x$

19. Let $x =$ length of the rectangle (in inches).
    By the Pythagorean Theorem,
    $x^2 + 5^2 = 10^2$
    $x^2 + 25 = 100$
    $\quad\quad x^2 = 75$
    $\quad\quad\quad x = \pm\sqrt{75}$
    $\quad\quad\quad x = \pm 5\sqrt{3}$
    We reject $x = -5\sqrt{3}$, since length cannot be negative. So $x = 5\sqrt{3}$. Then the area of the rectangle is $(length) \times (width) = (5\sqrt{3})(5) = 25\sqrt{3}$ sq. in.

21. $y = x^2 + 4x - 5$
    Vertex: $x = \dfrac{-b}{2a} = \dfrac{-(4)}{2(1)} = -2;\ (-2, -9)$
    $0 = x^2 + 4x - 5$
    $0 = (x+5)(x-1)$
    $\quad 0 = x + 5 \quad$ or $\quad 0 = x - 1$
    $-5 = x \quad\quad$ or $\quad 1 = x$
    The x-intercepts are -5 and 1.
    The y-intercept is -5.

    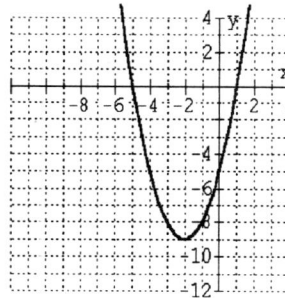

Supplementary Chapter on Geometry

Exercises G.1

1. $\angle AEB$ and $\angle BED$
   $\angle EAD$ and $\angle DAB$

3. (a) $60°$ (b) $30°$ (c) $45°$ (d) $72°$ (e) $1°$

5. Let $x =$ the angle.
   Then $180 - x =$ its supplement.
   $x = 3(180 - x) - 28$
   $x = 540 - 3x - 28$
   $x = 512 - 3x$
   $4x = 512$
   $x = 128$
   The angle is $128°$.

7. Let $x =$ the angle.
   Then $90 - x =$ its complement.
   $x = 2(90 - x) + 12$
   $x = 180 - 2x + 12$
   $x = 192 - 2x$
   $3x = 192$
   $x = 64$
   The angle is $64°$.

9. $x + (3x + 10) = 90$
   $4x + 10 = 90$
   $4x = 80$
   $x = 20$

11. $5x + 90 + 10x + 30 = 180$
    $15x + 120 = 180$
    $15x = 60$
    $x = 4$

13. $\angle 1 = 145°$ (supplement of $\angle 4$)
    $\angle 2 = 35°$ ($\angle 2$ and $\angle 4$ are vertical angles)
    $\angle 3 = 145°$ (supplement of $\angle 4$)

15. $\angle 1 = 65°$ ($\angle 1 + 90 + 25 = 180$)
    $\angle 2 = 115°$ (supplement of $\angle 1$)
    $\angle 3 = 65°$ ($\angle 1$ and $\angle 3$ are vertical angles)

17. $\angle 1 = \angle 3 = \angle 6 = \angle 8 = 55°$
    $\angle 2 = \angle 5 = \angle 7 = 125°$

19. $2x + (x + 12) = 180$
    $3x + 12 = 180$
    $3x = 168$
    $x = 56$

21. $\angle 1 = \angle 4 = \angle 5 = \angle 7 = \angle 9 = \angle 11 = \angle 14 = \angle 16 = 140°$
    $\angle 3 = \angle 6 = \angle 8 = \angle 10 = \angle 12 = \angle 13 = \angle 15 = 40°$

23. $\angle 1 = 50°, \angle 2 = 130°,$
    $\angle 3 = 50°, \angle 4 = 130°$

Exercises G.2

1. $\angle C = 180 - (34 + 23)$
   $= 180 - 57 = 123°$

3. Since $AC = BC$, $\angle A = \angle B$.
   Let $x =$ the number of degrees in $\angle A$ and in $\angle B$.
   $x + x + 40 = 180$
   $2x + 40 = 180$
   $2x = 140$
   $x = 70$
   So $\angle A = 70°$ and $\angle B = 70°$.

5. $x + 4x + (2x - 16) = 180$
   $7x - 16 = 180$
   $7x = 196$
   $x = 28$

7. $x + (2x + 20) + (2x + 20) = 180$
   $5x + 40 = 180$
   $5x = 140$
   $x = 28$

9. $x + 23 + 90 = 180$
   $x + 113 = 180$
   $x = 67$

11. $x + (2x + 12) + 90 = 180$
    $3x + 102 = 180$
    $3x = 78$
    $x = 26$

13. $|\overline{AB}|^2 = 3^2 + 4^2$
    $|\overline{AB}|^2 = 9 + 16$
    $|\overline{AB}|^2 = 25$
    $|\overline{AB}| = 5$

15. $|\overline{AC}|^2 + 4^2 = 9^2$
    $|\overline{AC}|^2 + 16 = 81$
    $|\overline{AC}|^2 = 65$
    $|\overline{AC}| = \sqrt{65}$

17. Let $x = |\overline{AC}| = |\overline{BC}|$.
    $x^2 + x^2 = 8^2$
    $2x^2 = 64$
    $x^2 = 32$
    $x = \sqrt{32} = 4\sqrt{2}$
    So $|\overline{AC}| = 4\sqrt{2}$ and $|\overline{BC}| = 4\sqrt{2}$.

19. $|\overline{AC}|^2 = 4^2 + 8^2$
    $|\overline{AC}|^2 = 16 + 64$
    $|\overline{AC}|^2 = 80$
    $|\overline{AC}| = \sqrt{80} = 4\sqrt{5}$

21. Let $x =$ distance between top of the ladder and the ground (in feet).

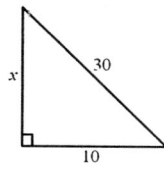

   $x^2 + 10^2 = 30^2$
   $x^2 + 100 = 900$
   $x^2 = 800$
   $x = \sqrt{800} = 20\sqrt{2}$

   The top of the ladder is $20\sqrt{2}$ (approximately 28.3) feet above the ground.

23. Let $d =$ distance between planes

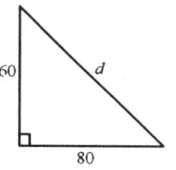

   $d^2 = 60^2 + 80^2$
   $d^2 = 3600 + 6400$
   $d^2 = 10,000$
   $d = \sqrt{10,000} = 100$ miles

25. In 3 hours, one boat traveled $3(20) = 60$ nautical miles, the other boat traveled $3(32) = 96$ nautical miles.

   $d^2 = 96^2 + 60^2$
   $d^2 = 9216 + 3600$
   $d^2 = 12,816$
   $d = \sqrt{12,816} \approx 113.2$ nautical miles.

27. Let $d =$ length of envelope's diagonal

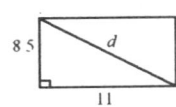

$d^2 = 8.5^2 + 11^2$
$d^2 = 72.25 + 121$
$d^2 = 193.25$
$d = \sqrt{193.25} \approx 13.9$
$15 > 13.9$, so the ruler will not fit.

29. By alternate interior angles, the unknown angle in the triangle $= x$.
$x + 63 + 90 = 180$
$x + 153 = 180$
$x = 27$

31. Referring to the sketch, $a = 50°$, since it is the supplement of $130°$.

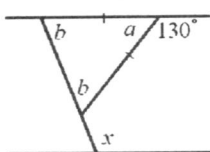

$50 + b + b = 180$
$50 + 2b = 180$
$2b = 130$
$b = 65$
Since the lines are parallel, $x + b = 180$
$x + 65 = 180$
$x = 115$

33. Referring to the sketch, $a + 125 = 180$, (since the lines are parallel). So $a = 55$.

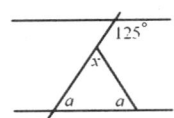

$55 + 55 + x = 180$
$110 + x = 180$
$x = 70$

35. Referring to the sketch, $a = x$ and $b + x = x + 70$ by alternate interior angles. From the second equation, we get $b = 70$. Then $a + 70 + 70 = 180$, so $a = 40$. Therefore, $x = 40$.

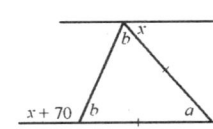

Exercises G.3

1. SSS
3. ASA
5. ASA
7. $\angle DBE \cong \angle ABC$ (vertical angles) SAS
9. $\overline{CD}$ is in both triangles. SAS
11. $\overline{DB}$ is in both triangles. SAS
13. Looking at parallel lines cut by transversal $\overline{AC}$:
   $\angle DCA \cong \angle CAB$ (alternate interior angles)
   $\angle DAC \cong \angle ACB$ (alternate interior angles)
   $\overline{AC}$ is in both triangles. ASA
15. $\overline{CE} \cong \overline{CA}$ (given)
    $\angle E \cong \angle A$ (given)
    $\angle ECD \cong \angle ACB$ (vertical angles)
    $\triangle ACB \cong \triangle ECD$ by ASA
    Hence corresponding angles are congruent, $\angle D \cong \angle B$

17. $\overline{AC} \cong \overline{CB}$ (given)
    $\angle CAD \cong \angle CBE$ (given)
    $\angle C$ is in both triangles.
    $\triangle ADC \cong \triangle BEC$ by ASA

19. In $\triangle ACD$ and $\triangle BCE$, we know $\angle CAD \cong \angle CBE$ (given). Also $\angle C$ is in both triangles. Therefore, $\angle CEB \cong \angle CDA$ since if two sets of angles in a pair of triangles are congruent then the third set of angles must also be congruent. Thus by ASA, $\triangle ECB \cong \triangle DCA$. Hence since corresponding sides of congruent triangles are congruent, $\overline{EB} \cong \overline{AD}$.

21. By SAS, we need $\overline{EB} \cong \overline{CB}$.
    By ASA, we need $\angle EAB \cong \angle CDB$.

23. By ASA, we need $\angle DAE \cong \angle CBE$.
    By SAS, we have everything necessary.
    $\overline{AD} \cong \overline{BC}$
    $\angle ADC \cong \angle BCD$ (both are right angles)
    Both triangles contain $\overline{DC}$.

Exercises G.4

1. $\angle A \cong \angle D$ (given)
    $\angle B \cong \angle E$ (given)
    Hence, $\triangle ACB \cong \triangle DFE$.
    Corresponding angles:
    $\angle A$ & $\angle D$; $\angle B$ & $\angle E$; $\angle C$ & $\angle F$

3. $\overline{BA}$ is a transversal that cuts parallel line segments, $\overline{CA}$ and $\overline{ED}$. Thus $\angle BDE \cong \angle BAC$ by corresponding angles. Both triangles contain $\angle B$. Hence $\triangle CAB \cong \triangle EDB$. Corresponding angles: $\angle EDB$ & $\angle CBA$; $\angle BDE$ & $\angle BAC$; $\angle BED$ & $\angle BCA$

5. $\angle CAB \cong \angle FDE$ (both are right angles)
    $\angle CBA \cong \angle FED$ (given)
    Thus, $\triangle ABC \cong \triangle DEF$.
    Corresponding angles: $\angle CAB$ & $\angle FDE$; $\angle ABC$ & $\angle DEF$; $\angle ACB$ & $\angle DFE$

7. $\dfrac{|\overline{DE}|}{|\overline{DF}|} = \dfrac{|\overline{AB}|}{|\overline{AC}|}$
    $\dfrac{|\overline{DE}|}{15} = \dfrac{8}{5}$, so $|\overline{DE}| = 24$
    $\dfrac{|\overline{EF}|}{|\overline{DF}|} = \dfrac{|\overline{BC}|}{|\overline{AC}|}$
    $\dfrac{|\overline{EF}|}{15} = \dfrac{4}{5}$, so $|\overline{EF}| = 12$

9. $\dfrac{|\overline{AC}|}{|\overline{AB}|} = \dfrac{|\overline{FE}|}{|\overline{ED}|}$
    $\dfrac{8}{6} = \dfrac{10}{|\overline{ED}|}$, so $|\overline{ED}| = \dfrac{15}{2}$
    $\dfrac{|\overline{AC}|}{|\overline{CB}|} = \dfrac{|\overline{FE}|}{|\overline{FD}|}$
    $\dfrac{8}{10} = \dfrac{10}{|\overline{FD}|}$, so $|\overline{FD}| = \dfrac{25}{2}$

11. $\dfrac{|\overline{AC}|}{24} = \dfrac{10}{12}$, so $|\overline{AC}| = 20$

13. $\dfrac{|\overline{EA}|}{|\overline{EC}|} = \dfrac{|\overline{DB}|}{|\overline{DC}|}$

    $\dfrac{|\overline{EA}|}{18} = \dfrac{6}{10}$, so $|\overline{EA}| = 10.8$

15. $\dfrac{|\overline{ED}|}{|\overline{DC}|} = \dfrac{|\overline{BA}|}{|\overline{AC}|}$

    $\dfrac{|\overline{ED}|}{8} = \dfrac{6}{4}$, so $|\overline{ED}| = 12$

17. $|\overline{AB}|^2 = 5^2 + 6^2$

    $|\overline{AB}|^2 = 25 + 36 = 61$

    So, $|\overline{AB}| = \sqrt{61}$

    $\dfrac{|\overline{BC}|}{|\overline{AC}|} = \dfrac{|\overline{EC}|}{|\overline{DC}|}$

    $\dfrac{6}{5} = \dfrac{10}{|\overline{DC}|}$

    So, $|\overline{DC}| = \dfrac{25}{3}$

    $\dfrac{|\overline{BC}|}{|\overline{AB}|} = \dfrac{|\overline{EC}|}{|\overline{DE}|}$

    $\dfrac{6}{\sqrt{61}} = \dfrac{10}{|\overline{DE}|}$

    So, $|\overline{DE}| = \dfrac{\cancel{10}^5 \sqrt{61}}{\cancel{6}_3}$

19. Let $x =$ height of the flagpole (in feet).

    $\dfrac{x}{18} = \dfrac{6}{4}$

    $\dfrac{x}{18} = \dfrac{3}{2}$

    $\cancel{18}^9 \cdot \dfrac{x}{\cancel{18}_1} = \cancel{18}^9 \left(\dfrac{3}{\cancel{2}_1}\right)$

    $x = 27$

    The flagpole is 27 ft high.

21. $\dfrac{4}{7} = \dfrac{5.8}{x}$

    $x = \dfrac{7(5.8)}{4} = 10.15 \text{ cm}$

23. $\dfrac{1}{25} = \dfrac{1.4}{x}$

    $x = 35$ in.

25. Let $w =$ widest part.

    $\dfrac{28}{1} = \dfrac{8.3}{w}$

    $w = 0.3$ cm

    Let $n =$ narrowest part.

    $\dfrac{28}{1} = \dfrac{3.2}{n}$

    $n = 0.1$ cm

27. $\overline{AC} \cong \overline{AB}$, hence,

    $\overline{AC} = 6$

    $\dfrac{|\overline{BC}|}{6} = \dfrac{\sqrt{2}}{1}$

    $|\overline{BC}| = 6\sqrt{2}$

29. $\dfrac{|\overline{AB}|}{12} = \dfrac{\sqrt{3}}{1}$

    So, $|\overline{AB}| = 12\sqrt{3}$

    $\dfrac{|\overline{BC}|}{12} = \dfrac{2}{1}$

    So, $|\overline{BC}| = 24$

31. $\overline{AB} \cong \overline{AC}$

    $\dfrac{|\overline{AB}|}{20} = \dfrac{1}{\sqrt{2}}$

    $|\overline{AB}| = \dfrac{20}{\sqrt{2}} = \dfrac{20\sqrt{2}}{2} = 10\sqrt{2}$

33. $\dfrac{|\overline{AB}|}{12} = \dfrac{1}{\sqrt{3}}$

So, $|\overline{AB}| = \dfrac{12}{\sqrt{3}} = \dfrac{12\sqrt{3}}{3} = 4\sqrt{3}$

$\dfrac{|\overline{BC}|}{12} = \dfrac{2}{\sqrt{3}}$

So, $|\overline{BC}| = \dfrac{24}{\sqrt{3}} = \dfrac{24\sqrt{3}}{3} = 8\sqrt{3}$

Exercises G.5

1. $\angle C = 360 - (110 + 130 + 40) = 360 - 280 = 80°$

3. $\angle C = 60°$ (opposite angles in a parallelogram are equal)
   $\angle B = 120°$ (adjacent angles in a parallelogram are supplementary)
   $\angle D = 120°$ (opposite angles in a parallelogram are equal)

5. $2x + 3 = x + 8$
   $x + 3 = 8$
   $x = 5$

7. $5x - 18 + 6x = 180$
   $11x - 18 = 180$
   $11x = 198$
   $x = 18$

9. Let $x =$ length of the diagonal (in cm).
   $x^2 = 5^2 + 8^2$
   $x^2 = 25 + 64$
   $x^2 = 89$
   $x = \sqrt{89}$

   The length of the diagonal is $\sqrt{89}$ (approximately 9.4) cm.

11. Let $x =$ length of the diagonal (in inches).
    $x^2 = 4^2 + 4^2$
    $x^2 = 16 + 16$
    $x^2 = 32$
    $x = \sqrt{32} = 4\sqrt{2}$

    The length of the diagonal is $4\sqrt{2}$ (approximately 5.7) in.

13. Let $x =$ length of the side (in cm).
    $x^2 + x^2 = 5^2$
    $2x^2 = 25$
    $x^2 = \dfrac{25}{2}$
    $x = \sqrt{\dfrac{25}{2}} = \dfrac{\sqrt{25}}{\sqrt{2}}$
    $= \dfrac{5}{\sqrt{2}} = \dfrac{5\sqrt{2}}{2}$

    The length of the side is $\dfrac{5\sqrt{2}}{2}$ (approximately 3.5) cm.

15. $\overline{AD} \cong \overline{CB}$ (given)
    $\angle DAB \cong \angle CBA$ since, base angles of isosceles trapezoid are congruent.
    $\overline{AB}$ is in both triangles.
    $\triangle ADB \cong \triangle BCA$ by SAS.

17. ∠DCB ≅ ∠CDA (given)
    $\overline{AD} \cong \overline{CB}$ (given)
    $\overline{DC}$ is in both triangles.
    △ADC ≅ △BCD by SAS.

19. $\overline{AD} \cong \overline{BC}$ (opposite sides of parallelogram)
    $\overline{FC} \parallel \overline{AE}$ and they are cut by transversal $\overline{DA}$.
    ∠FDA ≅ ∠DAB (alternate interior angles)
    ∠DAB ≅ ∠DCB (opposite angles in parallelogram)
    Also, $\overline{FC} \parallel \overline{AE}$ and they are cut by transversal $\overline{CB}$.
    ∠DCB ≅ ∠CBE (alternate interior angles)
    Putting it together,
    ∠FDA ≅ ∠DAB ≅ ∠DCB ≅ ∠CBE
    hence, ∠FDA ≅ ∠CBE.
    Since two corresponding pairs of angles are congruent in two right triangles, the third pair of angles must be congruent. Thus ∠FAD ≅ ∠ECB.
    △AFD ≅ △CEB by ASA.

21. Let x = length of rhombus' side.

    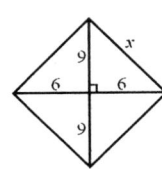

    $x^2 = 6^2 + 9^2$
    $x^2 = 36 + 81$
    $x^2 = 117$
    $x = \sqrt{117} = 3\sqrt{13}$
    Each side is $3\sqrt{13}$ inches.

23. Let y = base of right triangle.

    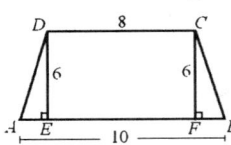

    $y = 8 - 5 = 3$
    $x^2 = 3^2 + 4^2$
    $x^2 = 9 + 16$
    $x^2 = 25$
    $x = 5$

25. Each side of the square is 8 units.
    $\overline{CE} = 8 - 3 = 5$
    $x^2 = 5^2 + 8^2$
    $x^2 = 25 + 64$
    $x^2 = 89$
    $x = \sqrt{89} \approx 9.4$

27. $y = 12 - 5 = 7$

    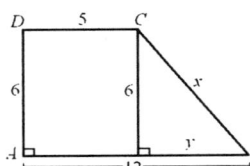

    $x^2 = 6^2 + 7^2$
    $x^2 = 36 + 49 = 85$
    $x = \sqrt{85} \approx 9.2$

29. Consider the diagram,

    Therefore, $\overline{AE} \cong \overline{FB}$.
    $|\overline{AE}| + |\overline{FB}| = 10 - 8 = 2$
    $|\overline{FB}| + |\overline{FB}| = 2$
    $2|\overline{FB}| = 2$
    $|\overline{FB}| = 1$

    ∠DAE ≅ ∠CBF (base angles of isosceles trapezoid)
    Hence, ∠EDA ≅ ∠FCB (complements to congruent angles)
    Thus by ASA, △AED ≅ △BFC.

    Then in the right triangle △CFB,
    $x^2 = 1^2 + 6^2 = 1 + 36 = 37$
    $x = \sqrt{37};\ x \approx 6.1$

Exercises G.6

1. $p = 2(5) + 2(8) = 10 + 16 = 26$ ft.
   $A = (5)(8) = 40$ sq. ft.

3. $p = 2(6) + 2(12) = 12 + 24 = 36$ in.
   $A = (6)(12) = 72$ sq. in.

5. $p = 4(3) = 12$ in.
   $A = 3^2 = 9$ sq. in.

7. $A = (8)(6) = 48$

9. $A = (10)(6) = 60$

11. $A = \frac{1}{2}(8)(6) = 24$

13. $A = \frac{1}{2}(12)(4) = 24$

15. $A = \frac{1}{2}(6)(4) = 12$

17. $A = \frac{1}{2}(5 + 12)(6) = 51$

19. $|\overline{AD}|^2 = 5^2 + 12^2 = 25 + 144 = 169$
    $|\overline{AD}| = 13$
    Then $p = 2(13) + 2(15) = 56$
    $A = (15)(12) = 180$

21. Let $x =$ length of the hypotenuse.
    $x^2 = 5^2 + 8^2 = 25 + 64 = 89$
    $x = \sqrt{89}$
    Then $p = 5 + 8 + \sqrt{89} = 13 + \sqrt{89} \approx 22.4$
    $A = \frac{1}{2}(5)(8) = 20$

23. Let $x =$ length of the missing leg.
    $x^2 + 7^2 = 9^2$
    $x^2 + 49 = 81$
    $x^2 = 32$
    $x = \sqrt{32} = 4\sqrt{2}$
    Then $p = 7 + 9 + 4\sqrt{2} = 16 + 4\sqrt{2}$
    (approximately 21.7)
    $A = \frac{1}{2}(4\sqrt{2})(7) = 14\sqrt{2}$
    (approximately 19.8)

25. Let $x =$ height of the rectangle (in cm).
    $x^2 + 8^2 = 12^2$
    $x^2 + 64 = 144$
    $x^2 = 80$
    $x = \sqrt{80} = 4\sqrt{5}$
    Then $p = 2(4\sqrt{5}) + 2(8) = 8\sqrt{5} + 16$
    (approximately 33.9) cm.
    $A = (8)(4\sqrt{5}) = 32\sqrt{5}$
    (approximately 71.6) sq. cm.

27. Let $x =$ side of the square (in mm).
    $x^2 + x^2 = 10^2$
    $2x^2 = 100$
    $x^2 = 50$
    $x = \sqrt{50} = 5\sqrt{2}$
    Then $p = 4(5\sqrt{2})$
    $= 20\sqrt{2} (\approx 28.3)$ mm.
    $A = (5\sqrt{2})^2 = 50$ sq. mm

29. Drop a perpendicular from D to AB.
    Call its foot E.

    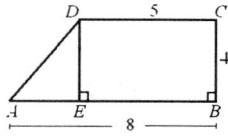

    Then, $|\overline{EB}| = 5$, so $|\overline{AE}| = 3$.
    Using the Pythagorean Theorem,
    $|\overline{AD}|^2 = 3^2 + 4^2 = 9 + 16 = 25$, so $|\overline{AD}| = 5$.
    Then $p = 8 + 4 + 5 + 5 = 22$
    and $A = \frac{1}{2}(5 + 8)(4) = 26$.

31. $p = 10 + 10 + 10 + 26 = 56$
    Drop a perpendicular from both D and C to AB, calling their feet E and F, respectively.

    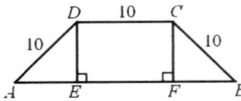

    Since $|\overline{EF}| = 10$ and $|\overline{AE}| = |\overline{FB}|$, it follows that $|\overline{AE}| = 8$. Then
    $|\overline{DE}|^2 + 8^2 = 10^2$
    $|\overline{DE}|^2 + 64 = 100$
    $\quad\quad |\overline{DE}|^2 = 36$
    $\quad\quad |\overline{DE}| = 6$
    $A = \frac{1}{2}(10 + 26)(6) = 108$

33. Area of $ABCDE$
    $= $ Area of rectangle $ABDE - $ Area of $\triangle BCD$
    $|\overline{BD}|^2 = 6^2 + 6^2$
    $|\overline{BD}|^2 = 36 + 36$
    $|\overline{BD}|^2 = 72$
    $|\overline{BD}| = \sqrt{72} = 6\sqrt{2}$
    Then area of rectangle $ABDE$
    $= 9(6\sqrt{2}) = 54\sqrt{2}$ and
    Area of $\triangle BCD = \frac{1}{2}(6)(6) = 18$.
    So Area of $ABCDE = 54\sqrt{2} - 18$ (≈ 58.4)

35. From the sketch, $|\overline{DG}| = 4$ and $|\overline{FC}| = 2$.
    Area of $AEFG = $ (Area of rectangle $ABCD$) $-$ (Area of $\triangle ADG$)
    $\quad\quad\quad\quad\quad\quad\quad\quad\quad\quad\quad\quad\quad\quad -$ (Area of $\triangle GCF$) $-$ (Area of $\triangle FBE$)

    Area of rectangle $ABCD = (10)(8) = 80$
    Area of $\triangle ADG = \frac{1}{2}(8)(4) = 16$
    Area of $\triangle GCF = \frac{1}{2}(6)(2) = 6$
    Area of $\triangle FBE = \frac{1}{2}(6)(7) = 21$
    So Area of $AEFG = 80 - 16 - 6 - 21 = 37$.

37. From the sketch, $|\overline{DF}| = 6$ and $|\overline{DG}| = 9$.
    Area of the figure $= $ (Area of rectangle $ABCD$) $+$ (Area of $\triangle FDG$) $+$ (Area of $\triangle ECG$)
    Area of rectangle $ABCD = (12)(8) = 96$
    Area of $\triangle FDG = \frac{1}{2}(6)(9) = 27$
    Area of $\triangle ECG = \frac{1}{2}(3)(6) = 9$
    So Area of the figure $= 96 + 27 + 9 = 132$.

39. From the sketch, $|\overline{DE}| = 9$.
    Area of shaded region $= $ (Area of rectangle $ACDF$) $-$ (Area of trapezoid $BCDE$)
    $\quad\quad\quad\quad\quad\quad\quad\quad\quad\quad\quad\quad\quad\quad\quad\quad -$ (Area of $\triangle EFG$)
    Area of rectangle $ACDF = (12)(6) = 72$
    Area of trapezoid $BCDE = \frac{1}{2}(4 + 9)(6) = 39$
    Area of $\triangle EFG = \frac{1}{2}(2)(3) = 3$
    So Area of shaded region $= 72 - 39 - 3 = 30$.

41. $\dfrac{28}{42} = \dfrac{2}{3}$ = scaling factor

$\dfrac{2}{3}$ is the scaling factor for perimeter.

For perimeter:
$\dfrac{2}{3} = \dfrac{98}{p}$
$2p = 294$
$p = 147$ in.

$\left(\dfrac{2}{3}\right)^2 = \dfrac{4}{9}$ is scaling factor for area

$\dfrac{4}{9} = \dfrac{420}{A}$
$4A = 3780$
$A = 945$ sq. in.

43. 1st polygon: 15 in.
2nd polygon: 2 ft = 24 in.
$\dfrac{15}{24} = \dfrac{5}{8}$ = scaling factor

$\dfrac{5}{8}$ is the scaling factor for perimeter.

For perimeter:
$\dfrac{5}{8} = \dfrac{80}{p}$
$5p = 640$
$p = 128$ in.

$\left(\dfrac{5}{8}\right)^2 = \dfrac{25}{64}$ is scaling factor for area

$\dfrac{25}{64} = \dfrac{2000}{A}$
$25A = 128,000$
$A = 5120$ sq. in.

45. $\dfrac{1}{20}$ = scaling factor

$\left(\dfrac{1}{20}\right)^2 = \dfrac{1}{400}$ is scaling factor for area

$\dfrac{1}{400} = \dfrac{378}{A}$

$A = 151,200$ sq. in. $= \dfrac{151,200}{144}$ sq. ft $= 1050$ sq. ft

47. We recognize that since $\triangle ABC$ is a right triangle, $\angle 1 + \angle 2 = 90°$.

   (a) The entire figure is a square since the angles are all right angles and the sides are all of equal length (each side is of length $a + b$).
   Looking at the inside figure, we can see that $\angle 1 + \angle 2 + \angle 3 = 180°$. Since we know that $\angle 1 + \angle 2 = 90°$, it follows that $\angle 3 = 90°$. Similarly, all of the angles of the inside figure are right angles and all its sides are of equal length. Therefore, the inside figure is a square.

   (b) Since the side of the outer square is of length $a + b$, its area is $(a + b)^2$.

   (c) Since the side of the inner square is of length c, its area is c^2.

   (d) Since the area of a right triangle is $\dfrac{1}{2}$ the product of its legs, the area of each triangle is $\dfrac{1}{2}ab$.

   (e) The figure clearly shows that the area of the outer square is equal to the area of the inner square plus the area of the four inner triangles.

   (f) Using the area relationship described in part (e), we have

$$(a+b)^2 = c^2 + 4 \cdot \frac{1}{2}(ab) \leftarrow$$ *This says that the area of the outer square equals the area of the inner square plus the area of the four inner triangles.*
$$a^2 + 2ab + b^2 = c^2 + 2ab$$
$$a^2 + b^2 = c^2$$

Thus we have proven the Pythagorean Theorem which says that in a right $\triangle ABC$, we have $a^2 + b^2 = c^2$.

Exercises G.7

1. $\widehat{AB} = 35°$

3. $x + (2x - 30) = 360$
   $3x - 30 = 360$
   $3x = 390$
   $x = 130$
   So $\widehat{AB} = 130°$

5. $C = \pi d$
   $C = \pi(6) = 6\pi$ inches

7. $C = 2\pi r$
   $C = 2\pi(4) = 8\pi$ feet

9. Length of arc $= \frac{30}{360}(18\pi) = \frac{3\pi}{2}$ inches

11. Length of $\widehat{AB} = \frac{65}{360}(12\pi) = \frac{13\pi}{6}$
    The perimeter of sector $= 6 + 6 + \frac{13\pi}{6}$
    $= 12 + \frac{13\pi}{6}$
    The area of sector $= \frac{65}{360}\pi(6^2) = \frac{13\pi}{2}$

13. Length of $\widehat{AB} = \frac{10}{360}(10\pi) = \frac{5\pi}{18}$
    The perimeter of sector $= 5 + 5 + \frac{5\pi}{18}$
    $= 10 + \frac{5\pi}{18}$
    The area of sector $= \frac{10}{360}\pi(5^2) = \frac{25\pi}{36}$

15. $A = \pi r^2 = \pi(3)^2 = 9\pi$ sq. in.

17. $r = \frac{1}{2}d = \frac{1}{2}(12) = 6$
    $A = \pi r^2 = \pi(6)^2 = 36\pi$ sq. ft

19. Length of $\widehat{AB} = \frac{1}{2}(20\pi) = 10\pi$. Length of $\overline{AB} = 20$.

    Perimeter of semicircle $= (10\pi + 20)$ inches. Area of semicircle $= \frac{1}{2}(100\pi) = 50\pi$ sq. in.

21. $\frac{80}{360}\pi(5^2) = \frac{50\pi}{9}$ sq. in.

23. Radius of the semicircle $= 3$, so the area of the semicircle is $\frac{1}{2}(9\pi) = \frac{9}{2}\pi$. Since the area of the rectangle $= (6)(8) = 48$, the area of the figure $= 48 + \frac{9}{2}\pi$. Length of the semicircular arc $= \frac{1}{2}(6\pi) = 3\pi$. So the perimeter of the figure $= 8 + 6 + 8 + 3\pi = 22 + 3\pi$.

25. Area of shaded region = (Area of square $ABCD$) − (Area of circle).
Area of square $ABCD = 8^2 = 64$. Area of circle = $\pi(4)^2 = 16\pi$.
So Area of shaded region = $(64 - 16\pi)$ sq. cm.

27. Radius of large semicircle = $6 + 8 = 14$, so area of large semicircle = $\dfrac{1}{2} \cdot \pi(14)^2 = 98\pi$.

 Radius of small semicircle = 6, so area of small semicircle = $\dfrac{1}{2} \cdot \pi(6)^2 = 18\pi$.

 So area of figure = $98\pi - 18\pi = 80\pi$.

29. $|\overline{AB}|^2 = 6^2 + 8^2 = 36 + 64 = 100$. So $|\overline{AB}| = 10$
Then length of $\widehat{AB} = \dfrac{1}{2}(10\pi) = 5\pi$. Perimeter = $6 + 8 + 5\pi = 14 + 5\pi$

31. Since $|\overline{AB}| = |\overline{EB}| - |\overline{EA}| = 30 - 20 = 10$, $\widehat{AB}$ is a semicircle with diameter = 10.
Length of $\widehat{AB} = \dfrac{1}{2}(10\pi) = 5\pi$. Perimeter = $|\overline{AE}| + |\overline{ED}| + |\overline{DC}| + |\overline{CB}|$ + length of $\widehat{AB}$
$= 20 + 10 + 30 + 10 + 5\pi = (70 + 5\pi)$ cm.

 Area = (Area of rectangle $BCDE$) + (area of semicircle) = $(30)(10) + \dfrac{1}{2} \cdot \pi(5)^2$
$= \left(300 + \dfrac{25\pi}{2}\right)$ sq. cm.

33. A round 12-inch pie has a radius of 6 inches and therefore has an area of $\pi(6)^2 = 36\pi$ sq. in. Each round 8-inch pie has a radius of 4 inches and therefore has an area of $\pi(4)^2 = 16\pi$ sq. in. This means that the total area of the two 8-inch pies is $2(16\pi) = 32\pi$ sq. in. As a result, one 12-inch pie contains more pizza than two 8-inch pies. Since the 12-inch pie costs $9, each dollar buys $\dfrac{36\pi}{9} = 4\pi$ sq. in. of pizza. Since the two 8-inch pies costs 2($4) or $8, each dollar buys $\dfrac{32\pi}{8} = 4\pi$ sq. in. of pizza. Therefore, both options have the same value.

Exercises G.8

1. $S = 2LW + 2LH + 2HW$
$= 2(12)(7) + 2(12)(6) + 2(6)(7)$
$= 396$ sq. cm
$V = LWH$
$= (12)(7)(6) = 504$ cu. cm

3. $S = 2\pi rh + 2\pi r^2$
$= 2\pi(2)(8) + 2\pi(2^2)$
$= 40\pi$ sq. ft
$V = \pi r^2 h$
$= \pi(2^2)(8) = 32\pi$ cu. ft

5. $s^2 = 3^2 + 4^2 = 25$
$s = 5$
$S = \pi rs + \pi r^2$
$= \pi(3)(5) + \pi(3^2) = 24\pi$ sq. m
$V = \dfrac{1}{3}\pi r^2 h$
$= \dfrac{1}{3}\pi(3^2)(4) = 12\pi$ cu. m

7. $S = 4\pi r^2$
$= 4\pi(8.3)^2 = 275.6\pi$ sq. ft
$V = \dfrac{4}{3}\pi r^3$
$= \dfrac{4}{3}\pi(8.3)^3 = 762.4\pi$ cu. ft

9. $S = 4\pi r^2$
   $= 4\pi(3)^2 = 113.1$ sq. ft
   $V = \dfrac{4}{3}\pi r^3$
   $= \dfrac{4}{3}\pi(3)^3 = 113.1$ cu. ft

11. $S = 2\pi rh + 2\pi r^2$
    $= 2\pi(4)(5) + 2\pi(4^2) = 226.2$ sq. cm
    $V = \pi r^2 h$
    $= \pi(4^2)(5) = 251.3$ cu. cm

13. $S = 2LW + 2LH + 2HW$
    $= 2(15)(2.4) + 2(15)(1.8) + 2(1.8)(2.4)$
    $= 134.6$ sq. m
    $V = LWH$
    $= (15)(2.4)(1.8) = 64.8$ cu. m

15. $S = \pi rs + \pi r^2$
    $= \pi(8)(12) + \pi(8^2) = 502.7$ sq. ft
    $h^2 = 12^2 - 8^2 = 80;$ So, $h = \sqrt{80}$
    $V = \dfrac{1}{3}\pi r^2 h$
    $= \dfrac{1}{3}\pi(8^2)(\sqrt{80}) = 599.5$ cu. ft

17. $r = \dfrac{1}{2}d = \dfrac{1}{2}(25) = 12.5$
    $S = 2\pi rh$ (open)
    $= 2\pi(12.5)(60) = 4712.4$ sq. mm
    $V = \pi r^2 h$
    $= \pi(12.5^2)(60) = 29,452.4$ cu. mm

19. $LS = 2\pi rh$
    $96\pi = 2\pi(3)(h)$
    $16 = h$
    $V = \pi r^2 h$
    $= \pi(3^2)(16) = 144\pi$ cu. m

21. $S = 4\pi r^2$
    $100\pi = 4\pi r^2$
    $25 = r^2$
    $5 = r$
    $V = \dfrac{4}{3}\pi r^3$
    $= \dfrac{4}{3}\pi(5)^3 = \dfrac{500\pi}{3}$ cu. cm

23. $V = \dfrac{1}{3}\pi r^2 h$
    $96\pi = \dfrac{1}{3}\pi r^2(8)$
    $36 = r^2$
    $6 = r$
    $s^2 = 6^2 + 8^2 = 100$
    $s = 10$
    $LS = \pi rs = \pi(6)(10) = 60\pi$ sq. in.

25. $V = LWH$
    $= (20)(3.75)(0.2)$
    $= 15$ cu. yd

27. $S = 2LW + 2LH + 2HW$
    $= 2(3.8)(16) + 2(3.8)(4.5) + 2(4.5)(16)$
    $= 299.8$ sq. ft
    Cost $= 1.26(299.8) = \$377.75$

29. Volume of the block:
    $V = LWH = (60)(12)(3) = 21,600$ cu. in.

    Volume of cylinder:
    $r = \dfrac{1}{2}d = \dfrac{1}{2}(2) = 1$
    $V = \pi r^2 h = \pi(1)^2(8) = 8\pi$ cu. in.

    Number of cylinders $= \dfrac{21,600}{8\pi} = 859$

31. $r = \dfrac{1}{2}d = \dfrac{1}{2}(3.5) = 1.75$
    $\dfrac{1}{2}S = \dfrac{1}{2}(4\pi r^2) = 2\pi(1.75)^2 = 19.24$
    Cost $= 1.85(19.24) \approx \$35.60$

33. $s^2 = 15^2 + 7.5^2 = 281.25$
    $s = \sqrt{281.25} \approx 16.77$
    $LS = \pi rs = \pi(15)(16.77) = 790.3$ sq. cm

35. Volume of hemisphere:
$$r = \frac{1}{2}d = \frac{1}{2}(18) = 9$$
$$V_1 = \frac{1}{2}\left(\frac{4}{3}\pi r^3\right) = \frac{2}{3}\pi(9)^3 = 1526.8$$
Volume of cylinder:
$$V_2 = \pi r^2 h$$
$$= \pi(9)^2(38) = 9669.8$$
Total volume $= V_1 + V_2$
$$= 1526.8 + 9669.8 = 11,196.6 \text{ cu. ft}$$

37. Volume of cone:
$$V_1 = \frac{1}{3}\pi r^2 h$$
$$= \frac{1}{3}\pi(6)^2(15) = 180\pi$$
Volume of cylinder:
$$V_2 = \pi r^2 h$$
$$= \pi(6)^2(15) = 540\pi$$
Volume $= V_2 - V_1$
$$= 540\pi - 180\pi = 360\pi \text{ cu. cm}$$

39. (a) Each of the six faces of a cube is a square of side x. Since the area of each square face is x^2, the surface area of the cube is $6x^2$.
    (b) If the edge of a cube is doubled in length, the surface area is multiplied by 4. The volume is multiplied by 8.
    (c) If the edge of a cube is tripled in length, the surface area is multiplied by 9. The volume is multiplied by 27.
    (d) If the edge of a cube has its length multiplied by k, the surface area is multiplied by k^2. The volume is multiplied by k^3.

Chapter G Review Exercises

1. $\angle 1 = 180° - 140° = 40°$ (supplements)
   $\angle 2 = 90° - 40° = 50°$ (complements)
   $\angle 3 = \angle 2 = 50°$ (alternate interior angles)
   $\angle 4 = \angle 1 = 40°$ (corresponding angles)

3. Let $y =$ base angle measure
   Recall base angles of isosceles triangle have equal measures.

   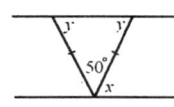

   $$2y + 50 = 180$$
   $$2y = 130$$
   $$y = 65$$
   $x = y = 65$ (alternate interior angles)
   $\angle z = 70°$ (corresponding angles)
   $\angle y \cong \angle z$ (alternate interior angles)
   $\angle y = 70°$
   $$2y + x = 180$$
   $$2(70) + x = 180$$
   $$140 + x = 180$$
   $$x = 40$$

5. Let $y =$ base angle measure.
   Recall base angles of isosceles triangle have equal measures.

   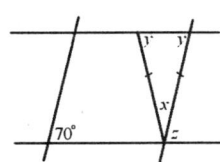

7. Let y = base angle measure
Recall base angles of isosceles triangle have equal measures.

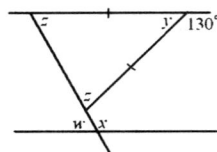

$y = 180° - 130° = 50°$ (supplements)
$2z + y = 180$
$2z + 50 = 180$
$2z = 130$
$z = 65$
$w = z = 65$ (alternate interior angles)
$x = 180° - 65° = 115°$ (supplements)

9. $|\overline{AC}|^2 + 5^2 = 13^2$
$|\overline{AC}|^2 = 13^2 - 5^2 = 144$
$|\overline{AC}| = 12$

11. $|\overline{AB}|^2 = 3^2 + 4^2 = 25$
$|\overline{AB}| = 5$
$\angle A = 90° - 53° = 37°$
$\angle EBD = 180° - 90° - 53° = 37°$
$\angle E = 90° - 37° = 53°$

Since the angles in $\triangle ABC$ are congruent to the angles in $\triangle BED$, $\triangle ABC$ is similar to $\triangle BED$.

$\dfrac{6}{|\overline{DB}|} = \dfrac{3}{4}$ $\qquad$ $\dfrac{6}{|\overline{EB}|} = \dfrac{3}{5}$

$|\overline{DB}| = 8$ $\qquad$ $|\overline{EB}| = 10$

13. In the figure, $\triangle ACE$ and $\triangle ABD$ both contain $\angle EAC$. Furthermore, $\angle AEC \cong \angle ADB$ since both are complements of $\angle EAC$. Thus, $\triangle ACE$ and $\triangle ABD$ are similar triangles.

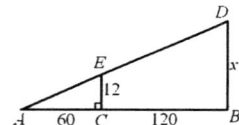

$\dfrac{x}{60+120} = \dfrac{12}{60}$
$\dfrac{x}{180} = \dfrac{12}{60}$
$x = 36$

15. Since this is a 45° - 45° - 90° triangle, $a = 8$.
$b^2 = 8^2 + 8^2$
$b^2 = 64 + 64 = 128$
$b = \sqrt{128} = 8\sqrt{2}$

17. $y = 9 - 5 = 4$

$x^2 = 3^2 + 4^2 = 25$
$x = 5$

19. Let x = side length

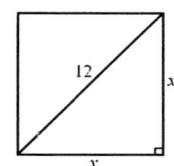

$$x^2 + x^2 = 12^2$$
$$2x^2 = 144$$
$$x^2 = 72$$
$$x = \sqrt{72} = 6\sqrt{2}$$
$$P = 4s = 4(6\sqrt{2}) = 24\sqrt{2} \text{ in.}$$
$$A = s^2 = \left(6\sqrt{2}\right)^2 = 72 \text{ sq. in.}$$

23. $P = 6 + 10 + 6 + 10 = 32$
$A = bh = 10(5) = 50$

25. $y = 16 - 10 = 6$

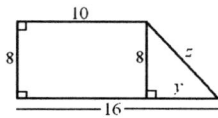

$$z^2 = 6^2 + 8^2 = 100$$
$$z = 10$$
$$P = 8 + 16 + 10 + 10 = 44$$
$$A = \frac{1}{2}h(b_1 + b_2)$$
$$A = \frac{1}{2}(8)(10 + 16) = 104$$

29. Area of rectangle $= wl = 10(20) = 200$
$|\overline{EF}| = 20 - 5 - 7 = 8$
Area of $\triangle GEF = \frac{1}{2}bh = \frac{1}{2}(8)(10) = 40$
Shaded area $= 200 - 40 = 160$

21. Let x = other leg length

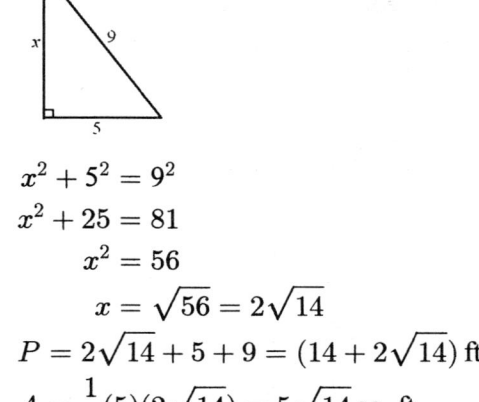

$$x^2 + 5^2 = 9^2$$
$$x^2 + 25 = 81$$
$$x^2 = 56$$
$$x = \sqrt{56} = 2\sqrt{14}$$
$$P = 2\sqrt{14} + 5 + 9 = (14 + 2\sqrt{14}) \text{ ft}$$
$$A = \frac{1}{2}(5)(2\sqrt{14}) = 5\sqrt{14} \text{ sq. ft}$$

27. Since this is a 30° - 60° right triangle

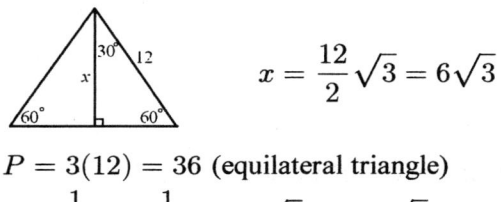

$$x = \frac{12}{2}\sqrt{3} = 6\sqrt{3}$$

$P = 3(12) = 36$ (equilateral triangle)
$A = \frac{1}{2}bh = \frac{1}{2}(12)(6\sqrt{3}) = 36\sqrt{3}$

31. $|\overline{DF}| = 6$
$|\overline{DG}| = 12 - 3 = 9$
Area of $\triangle FDG = \frac{1}{2}(9)(6) = 27$
Area of $\triangle GCE = \frac{1}{2}(3)(6) = 9$
Area of rectangle $ABCD = (8)(12) = 96$
Total area $= 27 + 9 + 96 = 132$

33. (a) $A = \frac{1}{2}(6)(5) = 15$

(b) $|\overline{DE}| = 12 - 5 = 7$
$A = \frac{1}{2}(14)(7) = 49$

(c) $|\overline{FG}| = 10 + 6 - 14 = 2$
$A = \frac{1}{2}(12)(2 + 10) = 72$

(d) Area of rectangle $ACEG = 12(10 + 6)$
$= 192$
Shaded area = area of rectangle $ACEG$
– area of $\triangle BCD$ – area of $\triangle DEF$
– area of trapezoid $ABFG$
$= 192 - 15 - 49 - 72 = 56$

35. (a) $A = \frac{1}{2}(8)(6) = 24$

(b) $|\overline{EF}| = (8 + 12) - 5 = 15$
$|\overline{FG}| = 15 - 6 = 9$
$A = \frac{1}{2}(15)(9) = 67.5$

(c) $A = \frac{1}{2}(15)(5 + 12) = 127.5$

(d) $A = 15(8 + 12) = 300$

(e) Shaded area = area of rectangle $ACDF$
– area of $\triangle ABG$ – area of $\triangle EFG$
– area of trapezoid $BCDE$
$= 300 - 24 - 67.5 - 127.5 = 81$

37. (a) $A = \frac{1}{2}(8)(12) = 48$

(b) $|\overline{ED}| = 16 + 8 - 6 = 18$
$A = \frac{1}{2}(18)(12) = 108$

(c) $|\overline{FG}| = 12 - 8 = 4$
$A = \frac{1}{2}(6)(4) = 12$

(d) Area of rectangle $ACDF = (16 + 8)(12)$
$= 288$
Shaded area = area of rectangle $ACDF$
– area of $\triangle BCD$ – area of $\triangle BDE$ – area of $\triangle EFG$
$= 288 - 48 - 108 - 12 = 120$

39. $C = 2\pi r = 2\pi(16) = 32\pi$ in.
$A = \pi r^2 = \pi(16)^2 = 256\pi$ sq. in.

41. $A = \frac{120}{360}(\pi r^2)$
$= \frac{1}{3}\pi(12)^2 = 48\pi$

43. Area of sector $= \frac{90}{360}(\pi r^2) = \frac{1}{4}\pi(8)^2 = 16\pi$
Area of triangle $= \frac{1}{2}(8)(8) = 32$
Shaded area = area of sector – area of triangle $= 16\pi - 32$

45. $|\overline{DF}| = |\overline{EC}|$
$|\overline{DF}| + 12 + |\overline{EC}| = 36$
$|\overline{DF}| + 12 + |\overline{DF}| = 36$
$2|\overline{DF}| = 24$
$|\overline{DF}| = 12$
$C = \pi d = \pi(12) = 12\pi$
$P = 12\pi + 12 + 10 + 36 + 10 = 68 + 12\pi$

47. $r = \frac{1}{2}d = \frac{1}{2}(12) = 6$
Area of 2 semicircles = Area of circle
$= \pi r^2 = \pi(6)^2 = 36\pi$
Area of square $= 12^2 = 144$
Shaded area = area of square – area of circle
$= 144 - 36\pi$

49. Area of semicircle $AB = \frac{1}{2}\pi(6)^2 = 18\pi$
Area of semicircle $CD = \frac{1}{2}\pi(4)^2 = 8\pi$
Shaded area $= 18\pi - 8\pi = 10\pi$

51. $S = 2LW + 2LH + 2HW$
$= 2(3)(5) + 2(3)(8) + 2(8)(5)$
$= 158$ sq. cm
$A = LWH$
$= (3)(8)(5) = 120$ cu. cm

283

53. $S = 2\pi rh + 2\pi r^2$
    $= 2\pi(3)(10) + 2\pi(3)^2$
    $= 60\pi + 18\pi = 78\pi$ sq. ft
    $A = \pi r^2 h$
    $= \pi(3^2)(10) = 90\pi$ cu. ft

Chapter G Practice Test

1. $y = 180° - 70° - 70° = 40°$

    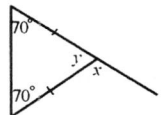

    $x = 180° - 40° = 140°$

3. $\angle 1 = 90° - 37° = 53°$
    $\angle 2 = 180° - 90° - 37° = 53°$
    $\angle 3 = 90° - 53° = 37°$
    Since these are similar triangles,
    $\dfrac{x}{9} = \dfrac{8}{6}$
    $x = 12$

5. $h = \dfrac{6}{2}\sqrt{3} = 3\sqrt{3}$ 　　　$A = \dfrac{1}{2}bh$

    $= \dfrac{1}{2}(6)(3\sqrt{3}) = 9\sqrt{3}$ sq. in.

7. $|\overline{DC}| = \dfrac{1}{2}(12) = 6$

    radius of semicircle $= \dfrac{1}{2}(6) = 3$

    Area of 2 semicircles $= \pi(3^2) = 9\pi$

    Area of rectangle $= 12(18) = 216$

    Area of semicircle $EA = \dfrac{1}{2}\pi(6^2) = 18\pi$

    Shaded area $=$ area of semicircle $EA +$ area of rectangle $ABDE -$ area of 2 semicircles
    $= 18\pi + 216 - 9\pi = 216 + 9\pi$

9. $|\overline{AB}|^2 = 6^2 + 8^2 = 100$
    $|\overline{AB}| = 10$

    radius of semicircle $= \dfrac{1}{2}(10) = 5$

    Area of semicircle $= \dfrac{1}{2}\pi(5^2) = \dfrac{25}{2}\pi$

    Area of triangle $= \dfrac{1}{2}(6)(8) = 24$

    Shaded area $=$ area of semicircle $-$ area of triangle
    $= \dfrac{25}{2}\pi - 24$

Appendix A
A Review of Arithmetic

Exercises A.1

1. $\dfrac{6}{9} = \dfrac{2}{3} = \dfrac{60}{90}$

3. $\dfrac{14}{35} = \dfrac{2}{5} = \dfrac{140}{350}$

5. $\dfrac{33}{44} = \dfrac{3}{4} = \dfrac{66}{88}$

7. $5 = \dfrac{5}{1} = \dfrac{10}{2} = \dfrac{20}{4}$

9. $1 = \dfrac{7}{7} = \dfrac{12}{12}$

11. $\dfrac{8}{10} = \dfrac{2 \cdot 4}{2 \cdot 5} = \dfrac{4}{5}$

13. $\dfrac{6}{42} = \dfrac{6 \cdot 1}{6 \cdot 7} = \dfrac{1}{7}$

15. $\dfrac{18}{36} = \dfrac{18 \cdot 1}{18 \cdot 2} = \dfrac{1}{2}$

17. $\dfrac{20}{49}$ cannot be reduced.

19. $\dfrac{28}{72} = \dfrac{4 \cdot 7}{4 \cdot 18} = \dfrac{7}{18}$

21. $\dfrac{90}{15} = \dfrac{15 \cdot 6}{15 \cdot 1} = \dfrac{6}{1} = 6$

23. $\dfrac{220}{80} = \dfrac{20 \cdot 11}{20 \cdot 4} = \dfrac{11}{4}$

25. $\dfrac{81}{100}$ cannot be reduced.

27. $\dfrac{5}{6} = \dfrac{5 \cdot 4}{6 \cdot 4} = \dfrac{20}{24}$ So $? = 20$.

29. $\dfrac{15}{7} = \dfrac{15 \cdot 3}{7 \cdot 3} = \dfrac{45}{21}$ So $? = 21$.

31. $\dfrac{16}{25} = \dfrac{16 \cdot 4}{25 \cdot 4} = \dfrac{64}{100}$ So $? = 64$.

33. $\dfrac{1}{18} = \dfrac{1 \cdot 6}{18 \cdot 6} = \dfrac{6}{108}$ So $? = 6$.

35. $\dfrac{1250}{1500} = \dfrac{250 \cdot 5}{250 \cdot 6} = \dfrac{5}{6}$ was on the computer

    $\dfrac{250}{1500} = \dfrac{250 \cdot 1}{250 \cdot 6} = \dfrac{1}{6}$ was on the software

37. $\dfrac{54}{60} = \dfrac{6 \cdot 9}{6 \cdot 10} = \dfrac{9}{10}$ hit the bull's-eye

    $\dfrac{6}{60} = \dfrac{6 \cdot 1}{6 \cdot 10} = \dfrac{1}{10}$ missed the bull's-eye

39. $\dfrac{24}{28} = \dfrac{4 \cdot 6}{4 \cdot 7} = \dfrac{6}{7}$ was correct

    $\dfrac{4}{28} = \dfrac{4 \cdot 1}{4 \cdot 7} = \dfrac{1}{7}$ was incorrect

41. $\dfrac{45}{45 + 30 + 25} = \dfrac{45}{100} = \dfrac{5 \cdot 9}{5 \cdot 20}$

    $= \dfrac{9}{20}$ are cheese

    $\dfrac{30}{100} = \dfrac{10 \cdot 3}{10 \cdot 10} = \dfrac{3}{10}$ are blueberry

    $\dfrac{25}{100} = \dfrac{25 \cdot 1}{25 \cdot 4} = \dfrac{1}{4}$ are cherry

43. $\dfrac{48}{48 + 56 + 16} = \dfrac{48}{120} = \dfrac{24 \cdot 2}{24 \cdot 5}$

    $= \dfrac{2}{5}$ are CDs

    $\dfrac{56}{120} = \dfrac{8 \cdot 7}{8 \cdot 15} = \dfrac{7}{15}$ are cassettes

    $\dfrac{16}{120} = \dfrac{8 \cdot 2}{8 \cdot 15} = \dfrac{2}{15}$ are LPs

Exercises A.2

1. $\dfrac{2}{3} \cdot \dfrac{5}{7} = \dfrac{2 \cdot 5}{3 \cdot 7} = \dfrac{10}{21}$

3. $\dfrac{1}{4} \cdot \dfrac{1}{9} = \dfrac{1 \cdot 1}{4 \cdot 9} = \dfrac{1}{36}$

5. $\dfrac{\overset{2}{\cancel{8}}}{\underset{3}{\cancel{9}}} \cdot \dfrac{\cancel{3}}{\cancel{4}} = \dfrac{2}{3}$

7. $\dfrac{\cancel{5}}{\underset{3}{\cancel{6}}} \cdot \dfrac{\overset{2}{\cancel{4}}}{\underset{3}{\cancel{15}}} = \dfrac{2}{9}$

9. $\dfrac{5}{6} \div \dfrac{4}{15} = \dfrac{5}{\underset{2}{\cancel{6}}} \cdot \dfrac{\overset{5}{\cancel{15}}}{4} = \dfrac{25}{8}$

11. $\dfrac{\overset{3}{\cancel{15}}}{\underset{3}{\cancel{18}}} \cdot \dfrac{\overset{4}{\cancel{24}}}{\underset{5}{\cancel{25}}} = \dfrac{4}{5}$

13. $\dfrac{15}{18} \div \dfrac{24}{25} = \dfrac{\overset{5}{\cancel{15}}}{18} \cdot \dfrac{25}{\underset{8}{\cancel{24}}} = \dfrac{125}{144}$

15. $\dfrac{\cancel{3}}{\cancel{5}} \cdot \dfrac{\cancel{5}}{\cancel{3}} = \dfrac{1}{1} = 1$

17. $\dfrac{3}{5} \div \dfrac{5}{3} = \dfrac{3}{5} \cdot \dfrac{3}{5} = \dfrac{9}{25}$

19. $18 \cdot \dfrac{3}{2} = \dfrac{\overset{9}{\cancel{18}}}{1} \cdot \dfrac{3}{\cancel{2}} = \dfrac{27}{1} = 27$

21. $18 \div \dfrac{3}{2} = \dfrac{\overset{6}{\cancel{18}}}{1} \cdot \dfrac{2}{\cancel{3}} = \dfrac{12}{1} = 12$

23. $\dfrac{3}{2} \div 18 = \dfrac{3}{2} \div \dfrac{18}{1} = \dfrac{\cancel{3}}{2} \cdot \dfrac{1}{\underset{6}{\cancel{18}}} = \dfrac{1}{12}$

25. $\dfrac{\overset{7}{\cancel{28}}}{\underset{5}{\cancel{45}}} \cdot \dfrac{\overset{7}{\cancel{63}}}{\underset{10}{\cancel{40}}} = \dfrac{49}{50}$

27. $\dfrac{\cancel{3}}{\cancel{5}} \cdot \dfrac{\overset{4}{\cancel{12}}}{7} \cdot \dfrac{\overset{2}{\cancel{10}}}{\underset{\cancel{3}}{\cancel{9}}} = \dfrac{8}{7}$

29. $\dfrac{12}{25} \cdot \dfrac{5}{2} \div \dfrac{6}{5} = \left(\dfrac{\overset{6}{\cancel{12}}}{\underset{5}{\cancel{25}}} \cdot \dfrac{\cancel{5}}{\cancel{2}}\right) \div \dfrac{6}{5} = \dfrac{6}{5} \div \dfrac{6}{5} = \dfrac{\cancel{6}}{\cancel{5}} \cdot \dfrac{\cancel{5}}{\cancel{6}} = \dfrac{1}{1} = 1$

31. The reciprocal of $\dfrac{5}{9}$ is $\dfrac{9}{5}$.

33. $\dfrac{4}{7}$ is the reciprocal of $\dfrac{7}{4}$.

35. $\dfrac{4}{\cancel{5}} \cdot \dfrac{\overset{4,500}{\cancel{22,500}}}{1} = \dfrac{18,000}{1} = 18,000$

Allison earns $18,000.

37. $12 \div \dfrac{5}{6} = \dfrac{12}{1} \cdot \dfrac{6}{5} = \dfrac{72}{5} = 14\dfrac{2}{5}$ pieces

39. $\dfrac{1}{\cancel{4}} \cdot \dfrac{\overset{75}{\cancel{300}}}{1} = \dfrac{75}{1} = 75 \qquad \dfrac{3}{\cancel{5}} \cdot \dfrac{\overset{60}{\cancel{300}}}{1} = \dfrac{180}{1} = 180$

So 75 hours are for assembly and 180 hours are for testing. Therefore, $300 - (75 + 180) = 300 - 255 = 45$ hours are left for packing.

41. $\dfrac{400}{1} \div \dfrac{2}{5} = \dfrac{\overset{200}{\cancel{400}}}{1} \cdot \dfrac{5}{\cancel{2}} = 1000 \qquad \dfrac{200}{1} \div \dfrac{5}{8} = \dfrac{\overset{40}{\cancel{200}}}{1} \cdot \dfrac{8}{\cancel{5}} = 320$

So 1000 $\frac{2}{5}$-gram tablets are made and 320 $\frac{5}{8}$-gram tablets are made. Therefore, $1000 + 320 = 1320$ tablets are made altogether.

Exercises A.3

1. $\dfrac{2}{7} + \dfrac{3}{7} = \dfrac{2+3}{7} = \dfrac{5}{7}$

3. $\dfrac{5}{8} - \dfrac{3}{8} + \dfrac{1}{8} = \dfrac{5-3+1}{8} = \dfrac{3}{8}$

5. $\dfrac{5}{12} + \dfrac{1}{12} = \dfrac{5+1}{12} = \dfrac{6}{12}$
$= \dfrac{\cancel{6} \cdot 1}{\cancel{6} \cdot 2} = \dfrac{1}{2}$

7. $\dfrac{3}{10} + \dfrac{7}{10} = \dfrac{3+7}{10} = \dfrac{10}{10} = 1$

9. $\dfrac{1}{2} + \dfrac{1}{4} = \dfrac{1 \cdot 2}{2 \cdot 2} + \dfrac{1}{4}$
$= \dfrac{2}{4} + \dfrac{1}{4} = \dfrac{2+1}{4} = \dfrac{3}{4}$

11. $\dfrac{5}{6} - \dfrac{3}{8} = \dfrac{5 \cdot 4}{6 \cdot 4} - \dfrac{3 \cdot 3}{8 \cdot 3}$
$= \dfrac{20}{24} - \dfrac{9}{24} = \dfrac{20-9}{24} = \dfrac{11}{24}$

13. $\dfrac{5}{\underset{2}{\cancel{6}}} \cdot \dfrac{\cancel{3}}{8} = \dfrac{5}{16}$

15. $\dfrac{7}{12} + \dfrac{1}{3} = \dfrac{7}{12} + \dfrac{1 \cdot 4}{3 \cdot 4}$
$= \dfrac{7}{12} + \dfrac{4}{12} = \dfrac{7+4}{12} = \dfrac{11}{12}$

17. $\dfrac{1}{2} + \dfrac{1}{3} + \dfrac{1}{4} = \dfrac{1 \cdot 6}{2 \cdot 6} + \dfrac{1 \cdot 4}{3 \cdot 4} + \dfrac{1 \cdot 3}{4 \cdot 3}$
$= \dfrac{6}{12} + \dfrac{4}{12} + \dfrac{3}{12}$
$= \dfrac{6+4+3}{12} = \dfrac{13}{12}$

19. $\dfrac{4}{15} + \dfrac{7}{20} = \dfrac{4 \cdot 4}{15 \cdot 4} + \dfrac{7 \cdot 3}{20 \cdot 3}$
$= \dfrac{16}{60} + \dfrac{21}{60} = \dfrac{16+21}{60} = \dfrac{37}{60}$

21. $\dfrac{3}{28} + \dfrac{2}{35} = \dfrac{3 \cdot 5}{28 \cdot 5} + \dfrac{2 \cdot 4}{35 \cdot 4}$
$= \dfrac{15}{140} + \dfrac{8}{140}$
$= \dfrac{15+8}{140} = \dfrac{23}{140}$

23. $5 + \dfrac{3}{5} = \dfrac{5 \cdot 5}{1 \cdot 5} + \dfrac{3}{5}$
$= \dfrac{25}{5} + \dfrac{3}{5} = \dfrac{25+3}{5} = \dfrac{28}{5}$

25. $5 \div \dfrac{3}{5} = \dfrac{5}{1} \cdot \dfrac{5}{3} = \dfrac{25}{3}$

27. $3\dfrac{3}{4} = 3 + \dfrac{3}{4}$
$= \dfrac{3}{1} + \dfrac{3}{4}$
$= \dfrac{3 \cdot 4}{1 \cdot 4} + \dfrac{3}{4}$
$= \dfrac{12}{4} + \dfrac{3}{4} = \dfrac{12+3}{4} = \dfrac{15}{4}$

29. $12\dfrac{4}{5} = 12 + \dfrac{4}{5}$
$= \dfrac{12}{1} + \dfrac{4}{5}$
$= \dfrac{12 \cdot 5}{1 \cdot 5} + \dfrac{4}{5}$
$= \dfrac{60}{5} + \dfrac{4}{5} = \dfrac{60+4}{5} = \dfrac{64}{5}$

31. $\dfrac{23}{5} = \dfrac{20+3}{5}$

$= \dfrac{20}{5} + \dfrac{3}{5}$

$= 4 + \dfrac{3}{5} = 4\dfrac{3}{5}$

33. $\dfrac{43}{6} = \dfrac{42+1}{6}$

$= \dfrac{42}{6} + \dfrac{1}{6}$

$= 7 + \dfrac{1}{6} = 7\dfrac{1}{6}$

35. $\dfrac{3}{4} = \dfrac{3 \cdot 15}{4 \cdot 15} = \dfrac{45}{60}$

$\dfrac{4}{5} = \dfrac{4 \cdot 12}{5 \cdot 12} = \dfrac{48}{60}$

$\dfrac{2}{3} = \dfrac{2 \cdot 20}{3 \cdot 20} = \dfrac{40}{60}$

Therefore, $\dfrac{2}{3}$ is the smallest of the three fractions.

37. $\dfrac{5}{12} = \dfrac{5 \cdot 5}{12 \cdot 5} = \dfrac{25}{60}$

$\dfrac{7}{15} = \dfrac{7 \cdot 4}{15 \cdot 4} = \dfrac{28}{60}$

$\dfrac{9}{20} = \dfrac{9 \cdot 3}{20 \cdot 3} = \dfrac{27}{60}$

Therefore, $\dfrac{5}{12}$ is the smallest of the three fractions.

39. $29\dfrac{1}{3} + 17\dfrac{1}{3} - 36\dfrac{1}{2} = \dfrac{88}{3} + \dfrac{52}{3} - \dfrac{73}{2}$

$= \dfrac{88+52}{3} - \dfrac{73}{2}$

$= \dfrac{140}{3} - \dfrac{73}{2}$

$= \dfrac{140 \cdot 2}{3 \cdot 2} - \dfrac{73 \cdot 3}{2 \cdot 3}$

$= \dfrac{280}{6} - \dfrac{219}{6}$

$= \dfrac{280 - 219}{6}$

$= \dfrac{61}{6} = 10\dfrac{1}{6}$

$10\dfrac{1}{6}$ cubic yards of cement are left over.

41. $240 \div 22\dfrac{1}{2} = \dfrac{240}{1} \div \dfrac{45}{2}$

$= \dfrac{\overset{16}{\cancel{240}}}{1} \cdot \dfrac{2}{\underset{3}{\cancel{45}}}$

$= \dfrac{32}{3} \text{ or } 10\dfrac{2}{3}$

On a 240-mile trip, the car uses $10\dfrac{2}{3}$ gallons of gasoline.

43. $28 \cdot 2\dfrac{1}{4} = \dfrac{\overset{7}{\cancel{28}}}{1} \cdot \dfrac{9}{\underset{}{\cancel{4}}}$

$= \dfrac{63}{1} = 63$

She has sold 63 acres.

45. $27\dfrac{1}{2} \div 4\dfrac{1}{2} = \dfrac{55}{2} \div \dfrac{9}{2}$

$= \dfrac{55}{\cancel{2}} \cdot \dfrac{\cancel{2}}{9}$

$= \dfrac{55}{9} \text{ or } 6\dfrac{1}{9}$

Raul's rate of speed is $6\dfrac{1}{9}$ miles per hour.

Exercises A.4

1.   4.7
       3.5
    +21.7

     29.9

3.   15.87
    − 6.35

      9.52

5.   21.620
      4.100
    +57.236

     82.956

7.   6.5
     0.003
    +2.08

     8.583

9.   9.27
    −7.85

     1.42

11.   3.9
    × 5.8

     312
    195

    22.62

13. 6.00
    −0.03

     5.97

15.   13.05
    × 2.63

      3915
     7830
    2610

    34.3215

17.   5.16
    × 0.37

     3612
    1548

    1.9092

19.   7.92
    5)39.60
      35
      --
      46
      45
      --
       10
       10
       --
        0

21. 1.5)630 becomes   420
                      15)6300
                         60
                         --
                         30
                         30
                         --
                          00
                          00
                          --
                           0

23. 0.16)3.28 becomes   20.5
                        16)328.0
                          32
                          --
                           80
                           80
                           --
                            0

25.   20.05
    × 0.004

    0.0802

27.   0.032
    × 0.05

    0.00160

29. $0.04\overline{)0.28}$ becomes

$\quad\quad\quad\quad\quad\quad\quad\quad 7$
$\quad\quad\quad\quad\quad\quad 4\overline{)28}$
$\quad\quad\quad\quad\quad\quad\ \ \underline{28}$
$\quad\quad\quad\quad\quad\quad\ \ \ \ 0$

31. $\quad\quad 0.01 \quad\quad\quad\quad 0.0002$
$\quad\ \ \underline{\times\ 0.02} \quad\quad\ \ \underline{\times\ \ 0.03}$
$\quad\quad 0.0002 \quad\quad\ 0.000006$

33. $0.015\overline{)3.900}$ becomes

$\quad\quad\quad\quad\quad\quad\quad\quad\ \ 260$
$\quad\quad\quad\quad\quad\quad 15\overline{)3900}$
$\quad\quad\quad\quad\quad\quad\ \ \underline{30}$
$\quad\quad\quad\quad\quad\quad\ \ \ \ 90$
$\quad\quad\quad\quad\quad\quad\ \ \ \ \underline{90}$
$\quad\quad\quad\quad\quad\quad\ \ \ \ \ \ 00$
$\quad\quad\quad\quad\quad\quad\ \ \ \ \ \ \underline{00}$
$\quad\quad\quad\quad\quad\quad\ \ \ \ \ \ \ \ 0$

35. $\quad\quad\quad 1200$
$\quad\ \ \underline{\times\ 0.0825}$
$\quad\quad\quad 6000$
$\quad\quad\quad 2400$
$\quad\quad\ \ \underline{9600}$
$\quad\quad 99.0000$

37. $0.12\overline{)102}$ becomes

$\quad\quad\quad\quad\quad\quad\quad\quad\ \ 850$
$\quad\quad\quad\quad\quad\quad 12\overline{)10200}$
$\quad\quad\quad\quad\quad\quad\ \ \underline{96}$
$\quad\quad\quad\quad\quad\quad\ \ \ \ 60$
$\quad\quad\quad\quad\quad\quad\ \ \ \ \underline{60}$
$\quad\quad\quad\quad\quad\quad\ \ \ \ \ \ 00$
$\quad\quad\quad\quad\quad\quad\ \ \ \ \ \ \underline{00}$
$\quad\quad\quad\quad\quad\quad\ \ \ \ \ \ \ \ 0$

The customer used 850 kilowatt hours.

39. $12.8\overline{)275}$ becomes

$\quad\quad\quad\quad\quad\quad\quad\quad\ \ 21.4$
$\quad\quad\quad\quad\quad\quad 128\overline{)2750.0}$
$\quad\quad\quad\quad\quad\quad\ \ \underline{256}$
$\quad\quad\quad\quad\quad\quad\ \ \ \ 190$
$\quad\quad\quad\quad\quad\quad\ \ \ \ \underline{128}$
$\quad\quad\quad\quad\quad\quad\ \ \ \ 620$
$\quad\quad\quad\quad\quad\quad\ \ \ \ \underline{512}$
$\quad\quad\quad\quad\quad\quad\ \ \ \ 108$

Since $21.4 < 24.2$, this mileage per gallon is not as good as they claimed.

41. $\quad\quad \$18.65$
$\quad\ \ \underline{\times\quad 35.4}$
$\quad\quad\quad 7460$
$\quad\quad\quad 9325$
$\quad\quad\ \underline{5\,595}$
$\quad\ \ \$660.210$

The mixture used for this process costs $660.21.

Exercises A.5

1. $25\% = 0.25$
3. $0.78 = 78\%$
5. $0.05 = 5\%$
7. $9\% = 0.09$
9. $150\% = 1.5$
11. $28\% = 0.28$
13. $0.67 = 67\%$
15. $2 = 200\%$
17. $137\% = 1.37$
19. $0.007 = 0.7\%$
21. $62.4\% = 0.624$
23. $8.6\% = 0.086$
25. 30% of $70 = (0.30)(70) = 21$
27. 7.2% of $35 = (0.072)(35) = 2.52$

29. 0.8% of 5 = (0.008)(5) = 0.04

31. $\dfrac{1}{4} = 0.25 = 25\%$

33. $\dfrac{5}{8} = 0.625 = 62.5\%$

35. $\dfrac{8}{20} = \dfrac{2}{5} = 0.4 = 40\%$

37. $\dfrac{9}{0.06}$ becomes $0.06\overline{)9}$, which becomes

$$\begin{array}{r} 150 \\ 6\overline{)900} \\ 6 \\ \hline 30 \\ 30 \\ \hline 00 \\ 00 \\ \hline 0 \end{array}$$

39. $\dfrac{24.6}{0.30}$ becomes $0.30\overline{)24.6}$, which becomes

$$\begin{array}{r} 82 \\ 3\overline{)246} \\ 24 \\ \hline 6 \\ 6 \\ \hline 0 \end{array}$$

41. 20% of 1.50 = (0.20)(1.50)
    = 0.3
So the price after the reduction will be $1.50 − $0.30 = $1.20.

43. 6% of 12,500 = (0.06)(12,500)
    = 750
So in 5 years, the population of the town will be 12,500 + 750 = 13,250.

45. 140,000 − 133,000 = 7000
7000 is what percent of 140,000
$\dfrac{7000}{140,000} = 0.05 = 5\%$
There was a 5% decrease in the population.

47. 52% of 8,600 = (0.52)(8,600)
    = 4,472
48% of 9,250 = (0.48)(9,250)
    = 4,440
Therefore, more votes were cast in this town in 1988.

Appendix B
Using a Scientific Calculator

Appendix B Exercises

1. 861.55
3. 68.43
5. 65.5
7. 308.56
9. 4847.04
11. 42.01
13. 30.45
15. 27.1
17. 27.62
19. 0.15
21. 0.06

23. $89.95 + 0.0825(89.95) = \$97.37$.

25. $(186,000)(60) = 11,160,000$ miles.

27. 15% of $\$12 = (0.15)(\$12) = \$1.80$.

   The wholesaler marks up the price to $\$12 + \$1.80 = \$13.80$.

   12% of $\$13.80 = (0.12)(\$13.80) = \$1.66$.

   So the consumer pays $\$13.80 + \$1.66 = \$15.46$.

Appendix D
Introduction to Functions

Appendix D Exercises

1. Domain: $\{A, B, C, D\}$
   Range: $\{2, 3, 8, 11\}$
   Function

3. Domain: $\{F, G, 3\}$
   Range: $\{L, M, 2, 8\}$
   Not a function

5. Domain: $\{1, 3, 4, 9\}$
   Range: $\{5\}$
   Function

7. Function
   Domain: all real numbers

9. Function
   Domain: all real numbers

11. Not a function. For example if $x = 0$, then $y = 3$ or $y = -3$.

13. Function
    Domain: $\{x | x \neq 3\}$

15. Function
    Domain: all real numbers

17. Function
    Domain: all real numbers

19. $h = 1821 - 16t^2$
    For $t = 5$: $h = 1821 - 16(5)^2 = 1421$ feet
    For $t = 10$: $h = 1821 - 16(10)^2 = 221$ feet

21. $C = 125 + 0.0213n$
    For $n = 2500$: $C = 125 + 0.0213(2500) = 178.25$
    The cost is $178.25 thousand or $178,250.

23. $C = 2800 + 320m + 0.006m^2$
    For $m = 180$: $C = 2800 + 320(180) + 0.006(180)^2 = 60,594.4$
    The cost is $60,594.40 thousand or $60,594,400.

25. (a) $A = \text{width} \cdot \text{height}$
    $80 = w \cdot h$
    $\dfrac{80}{w} = h$
    (b) $h = \dfrac{80}{12} = \dfrac{20}{3}$ cm

27. (a) $C = 0.06p + 75$
    (b) $C = 0.06(875) + 75 = \$127.50$

29. (a) Since $w = $ width, then $2w - 3 = $ length.
    $P = 2w + 2l = 2w + 2(2w - 3) = 2w + 4w - 6 = 6w - 6$
    (b) $P = 6w - 6 = 6(9) - 6 = 48$ in.
    (c) $A = wl = w(2w - 3) = 2w^2 - 3w$
    (d) $A = 2w^2 - 3w = 2(9)^2 - 3(9) = 135$ sq. in.